Arc Volcanism : Physics and Tectonics

Advances in Earth and Planetary Sciences

Arc Volcanism: Physics and Tectonics

Proceedings of a 1981 IAVCEI Symposium
— Arc Volcanism —
August–September, 1981, Tokyo and Hakone

Edited by

D. Shimozuru

Earthquake Research Institute, University of Tokyo, Japan

and

I. Yokoyama

Faculty of Science, Hokkaido University, Sapporo, Japan

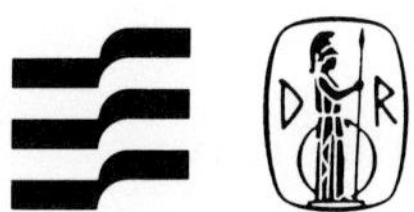

Terra Scientific Publishing Company/Tokyo
D. Reidel Publishing Company/Dordrecht, Boston, London

Library of Congress Cataloging in Publication Data

Symposium on Arc Volcanism (1981 : Tokyo, Japan and
    Hakone-machi, Japan)
    Arc volcanism

    (Advances in earth and planetary sciences)
    Includes index.
    1. Volcanism—Congresses.   2. Island arcs—Congresses.   I. Shimozuru, D.
(Daisuke), 1924- . II. Yokoyama, Izumi, 1924- . III. International Association
of Volcanology and Chemistry of the Earth's Interior. IV. Series.
QE521.5.S95   1981   551.2'1'09142   83-8654
ISBN 90-277-1612-9

Published by Terra Scientific Publishing Company (TERRAPUB),
307 Shibuyadai-haim, 4-17 Sakuragaoka-cho, Shibuya-ku, Tokyo 150, Japan,
in co-publication with D. Reidel Publishing Company, Dordrecht, Holland

Sold and distributed in the U.S.A. and Canada
by Kluwer Boston Inc.,
190 Old Derby Street, Hingham, MA 02043, U.S.A.,
in Japan by Terra Scientific Publishing Company (TERRAPUB),
307 Shibuyadai-haim, 4-17 Sakuragaoka-cho, Shibuya-ku, Tokyo 150, Japan

In all other countries, sold and distributed
by Kluwer Academic Publishers Group,
P. O. Box 322, 3300 AH Dordrecht, Holland

D. Reidel Publishing Company is a member of the Kluwer Academic Publishers Group

Preface

As it is called "Ring of Fire", more than 90 percent of the earth's volcanoes are located along the Circum Pacific region and other continental and island arcs. These volcanoes are characterized by a more or less linear alignment and by their explosive nature. During the past two decades, geological, geophysical and geochemical research on arc volcanism has progressed. Especially, the notable development of the concept of global tectonics stimulated the new interpretation on tectonic aspects of arc volcanism. On the other hand, to mitigate volcanic disasters, taking account of the considerable loss of life and welfare due to volcanic eruptions, becomes one of the urgent problems in volcanology.

Under these circumstances, the Symposium on Arc Volcanism was held during August 31-September 5,1981 at Tokyo and Hakone, and was sponsored by the Volcanological Society of Japan and the International Association of Volcanology and Chemistry of the Earth's Interior, I.U.G.G. Four subject topics, considering of 1) Physics and chemistry of arc magma, 2) Arc volcanism in time and space, 3) Geothermal and energetic aspects of arc volcanism, and 4) Prediction, hazards and environmental aspects of volcanic activities were proposed for this symposium. We received 193 scientists from abroad and the total number of participants was 470. The number of orally presented papers was 219. Basically, the organizing committee did not plan to publish all the papers as the proceedings of this symposium, however, it was felt that papers of the same discipline should be published in a combined form, and under this scheme, we have selected fifteen papers on physical and tectonical aspects which were orally presented during the symposium.

The editors are grately indebted to Mr. K. Oshida of Terra Scientific Publishing Company who arranged for the publication of this volume.

Editors
D. Shimozuru
I. Yokoyama

Contents

PHYSICS

Arc Volcanism: Physics and Tectonics, edited by D. Shimozuru and I. Yokoyama, 3–12.

A Model of Eruption Sequence and
Magma Supply Rate for Polygenetic Volcanoes

Naoyuki FUJII

Department of Earth Sciences, Kobe University,
Nada, Kobe 657, Japan

During a long-lasting activity of polygenetic central volcano, the magma intermittently rises through the same conduit from the upper mantle to the magma chamber. Although the magma supply rate is considered to be nearly constant in average, sequences of repose periods and erupted masses change randomly. It is assumed that the eruption is due to the thermal feedback instability of ascending magma driven by the pressure difference between the pressure in the magma chamber (Pc) and that in the magma source region. A proposed model assumes that (1) the variation of Pc is due to the accumulation of magma into the magma chamber during the repose period and (2) the change of Pc at the eruption is determined by the magma supply rate and Pc just before the eruption. The model generates the instability sequence which resembles the erupted mass sequence of some polygenetic volcanoes. It may suggest one possibility to predict the subsequent eruptive activities of polygenetic volcanoes deterministically.

1.   Introduction

Polygenetic central volcanoes are characterized by a long-lasting activity and large amount of magma supplied through the same conduit connecting the magma chamber to the magma source region (NAKAMURA, 1975). For each polygenetic volcano, the magma chamber is likely to exist in the upper crust or in the volcanic edifice itself, so that the magma chamber pressure is closely related with the tectonic stress and the magma effusion rate (WADGE, 1977, 1979; TANGUY, 1979; WILSON and HEAD, 1981). The supply rate of magma to the magma chamber is considered to be nearly constant in average for thousands of years and even longer (NAKAMURA, 1964; SWANSON, 1972). It is, however, apparent that the successive eruption intervals or the amount of the magma discharge irregularly fluctuate from event to event (SCHEIDEGGER, 1975; WICKMAN, 1976).

Knowledge of mechanisms and the actual magma supply rate to the magma chamber is rudimentary and allows for many different interpretations. There are two extreme cases for the time variation of the magma supply rate, i.e. the continuous and the episodic cases. When the magma chamber is open and the inflow rate of magma to the chamber from below is constant in the continuous case, the magma chamber pressure gradually increases and finally exceeds some critical pressure which is the condition of occurrence of the eruption (BLAKE, 1981). On the other hand, the episodic magma supply model assumes that the variation of magma outflow during the eruption period corresponds to the arrival of a batch of magma at an already inflated magma chamber (WADGE, 1981). The episodic batches of magma ascent are thought to be individual partial melting event and to have been transported rather rapidly through the mantle from the magma generating region (MARSH, 1978; WRIGHT and TILLING, 1980).

In the repose period, the former interpretation assumes continuous but un-observably small rate of magma supply so that the magma chamber would inflate very gradually. Whereas the latter assumes essentially no flow of ascending magma through the conduit which is reopened at each eruption period. As the repose period is usually longer than the eruption period, the eruption sequence has been treated as a stochastic process (i.e. the Markov model of WICKMAN, 1976) and statistical analysis was made for the 1971 eruptive activity of Stromboli (SETTLE and McGETCHIN, 1980). However, an eruptive activity of a polygenetic volcano depends on a triggering mechanism and the availability of magma that could be deterministic phenomena. Apparent random-ness does not always indicate the process is stochastic. For example, the sequence of great earthquake occurrence which has been generally treated as a stochastic process, can be simulated by a deterministic model with nonlinearly coupled physical variables (ITO, 1980).

The mechanisms that control dulation of repose periods of less than ten to hundreds of years and trigger the eruption are regarded as due to some instabilities of the magma flow rate through a conduit or the upward accummulation rate of melt phase from the magma source region. One of the possible mechanisms of such an instability which is related with the eruption, is a thermal feedback due to viscous heating for a flow under a constant stress (SHAW, 1969; SHAW and JACKSON, 1973; FUJII and UYEDA, 1974; HARDEE and LARSON, 1977; NELSON, 1981). Accumulation of magma into the magma chamber causes increase of the chamber pressure and depresses the magma ascent rate, so that the flow instability would be nonlinear coupled with the pressure difference between the magma chamber and the magma source region.

A preliminary analysis is made to demonstrate the sequence of repose periods and erupted masses of magma by using a simple nonlinear model in which the chamber pressure change is coupled with the magma ascent rate due to the thermal feedback instability.

2.   Outline of the Model Calculation

As the magma supply occurs using the same conduit through the active period, the pressure in the magma chamber (Pc) is in quasiequilibrium between the buoyancy force of the magma column and tectonic stress (or overburden load) of the volcanic edifice in the repose period. If the conduit connecting the magma chamber to the magma source region is kept open, the ascending flow of magma is controlled by the pressure difference between them. Near the magma source region in the upper mantle, it is likely that some extent of magma may be accumulated and formed a pool or a chamber presumably persisting for a long period (FEDOTOV, 1981) which is hereafter called a magma pool (Fig. 1). The volume of the magma pool would relate the availability of magma to the magma chamber.

It is assumed that the intermittent ascent of magma is caused by a thermal feedback instability due to a temperature dependent viscosity of magma. For a flow of magma with viscous heating under constant pressure difference along a cylindrical conduit, the macroscopic energy balance equation is expressed as:

$$C(\partial U/\partial t) = K\nabla_r^2 U + (r \times Gr/2)^2/\eta \tag{1}$$

where C is the heat capacity per unit volume, U is the local temperature, t is the time, K is the thermal conductivity, r is the distance away from the center of the conduit, and η is the viscosity. The pressure gradient (Gr) along the conduit is expressed as:

$$Gr = (Pm - Pc)/L \tag{2}$$

where Pm is the magma pool pressure and L is the length of the conduit.

By assuming the temperature dependence of viscosity (GRUNTFEST et al., 1964) as $\eta = \eta_0 \exp[-a(U - U_0)]$, temperature instability could occur for the nondimensional value of G' ($= aR^4Gr^2/4K\eta_0$) greater than 8.0 in which R is the radius of the conduit. In the case of a supercritical value of $G' \simeq 10$, more than one order of magnitude increases of the flow rate (Qc) are observed within the time intervals as short as unit reduced time, Kt/CR^2 (GRUNTFEST et al., 1964). The actual time required for such an instability for basaltic and andesitic magmas would range from several weeks to a few years (FUJII and UYEDA, 1974; NELSON, 1981).

It is noticed that the above results may change if the value of G' changes with time. The most probable change will come from the decrease of Gr and the increase of R with time. Because the heat is only lost by the conduction from the wall rock around the conduit, the change of R may not couple with the change of the initiation of the instability. When the accumulation of magma into the magma chamber increases Pc, the time variation of Pc can be expressed as:

$$Pc(t) = A \times \int_0^t Qc(y)\,dy \tag{3}$$

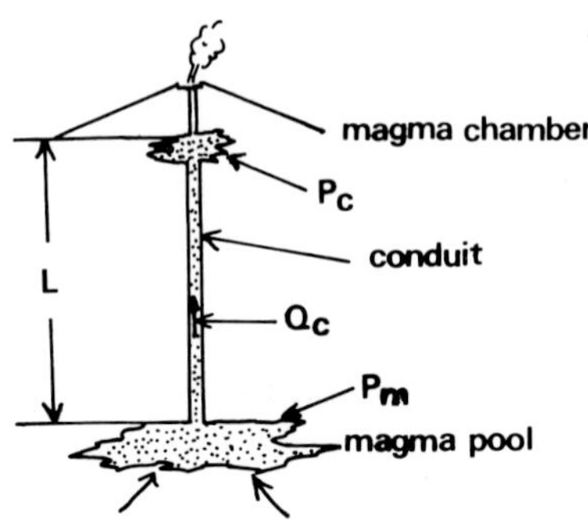

Fig. 1.   Schematic diagram of the magma chamber-conduit-magma pool system for polygenetic volcanoes.

where t is the time after the end of eruption period and A is a constant related with the mechanical property of the magma chamber during the repose period which is similar to the models of the Pc variation at the eruption period (Machado, 1974; Tryggvason, 1978; Wadge, 1981). The increase of Pc causes the decrease of Gr and depress the occurrence of the thermal instability so that the repose period will increase by this effect. It is likely that Gr just after the eruption is nonlinearly coupled with the growth rate of instability and Pc before the eruption. The occurrence of the instabilities in local temperatures and magma ascent rate are regarded as the onset of eruption. The change of Gr just before and after the eruption period can be expressed by:

$$Gr(+) = B \times Gr(-) \times (Gr(0) - Gr(-)) \tag{4}$$

where B is a constant and minus and plus signs respectively indicate the values just before and after the eruption. By assuming no change of Pm at the eruption, the change of Pc is obtained from Eqs. (3) and (4). The assumption made in Eq. (4) is that the change of Pc at the eruption can be proportional to the multiplication of the magma ascent rate $Qc(-)$ and $Pc(-)$ just before the eruption. $Qc(-)$ can be calculated from the ascent velocity distribution at the onset of the eruption. $Gr(0)$ is a constant pressure gradient (the maximum of Gr) corresponding to the minimum of Pc for no tectonic stress on the magma chamber. The erupted mass V is determined by the change of Pc just before and after the eruption as:

$$V = D \times (Pc(-) - Pc(+)) \tag{5}$$

where D is a constant related to the elastic property of the magma chamber at the eruption period (Mogi, 1958; Yokoyama, 1971; Swanson et al., 1976).

Following Gruntfest et al. (1964), the calculation was made by introducing nondimensional temperature, $a(U - U_0)$, distance, r/R, and time, Kt/CR^2, and the finite difference explicit method was applied for the numerical integration of non-dimensional form of Eq. (1). The amount of magma ascent rate is also numerically calculated from the ascent velocity distribution by assuming that the Prantl number is sufficiently large as made by Fujii and Uyeda (1974). The occurrence of instability in the local temperature is regarded as the eruption. Figure 2 schematically shows the variations of the magma chamber pressure Pc (top), the pressure gradient along the direction of flow Gr (middle), and the erupted mass sequence (bottom), respectively.

3.   Results and Examples of Erupted Mass Sequence

An example of the calculations is shown in Fig. 3, for a constant value of $B \times Gr(0) = 3.6$ and initial values of $G' = 10.0$ and $Gr/Gr(0) = 0.3$. The value of the constant, $B \times Gr(0)$, corresponds to a region of "chaotic" solutions in Eq. (4) as noted by May (1976). If this constant is less than 3.5700, the value of the parameter G' becomes periodic or stable and the erupted mass sequence becomes periodic. For the value of this constant greater than 4.0, G' grows infinitely and no repeated eruption sequence is observed. The variations of magma ascent rate (top) and erupted mass sequence (middle) with relative linear scales indicate that the pattern is not regular nor completely random. The division of time axis is five units of reduced time. In the case of

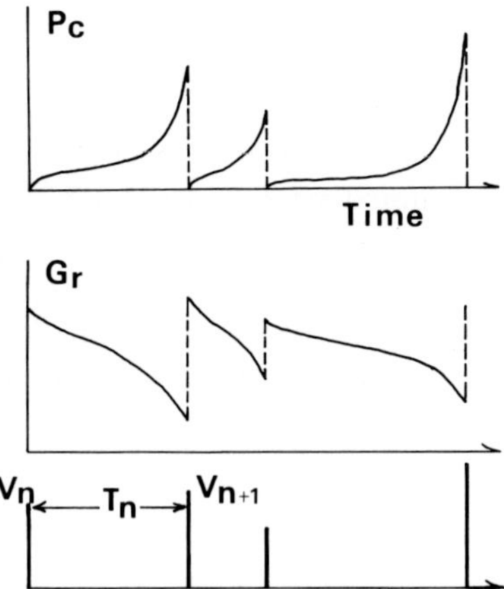

Fig. 2.   Schematic diagrams of the time variations of the magma chamber pressure (Pc), the pressure gradient along the direction of flow (Gr), and erupted mass sequence, qualitatively deduced from the present model.

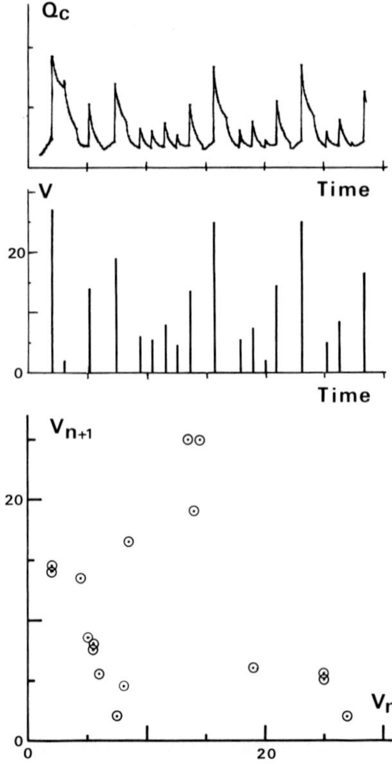

Fig. 3.   An example of the results showing the time variations of the magma ascent rate (top) and erupted mass sequence (middle). Successive erupted mass is shown in the bottom. The values of parameters used in this example are in the text.

a constant G' of about 10.0 (Gruntfest *et al.*, 1964), a longer repose period than one unit reduced time occurs intermittently, which could be the consequence of the variable Gr in Eqs. (1)–(3). As the unit of the erupted mass is arbitrary, but linear scale, about one order of magnitude variation of the erupted mass is observed. The bottom diagram in Fig. 3 shows the successive erupted mass sequence, which indicates no simple pattern due to the nonlinearity of this system. It is, however, noticed that the cumulative erupted mass represents a pseudo-steady pattern, as implied from Fig. 3 (middle).

For examples of eruption sequence which could indicate the time variation of the magma supply rate into the magma chamber, Izu-Oshima (Nakamura, 1964) and Vesuvio (Imbo, 1949; Yokoyama, private communication, 1981) are shown in Figs. 4.

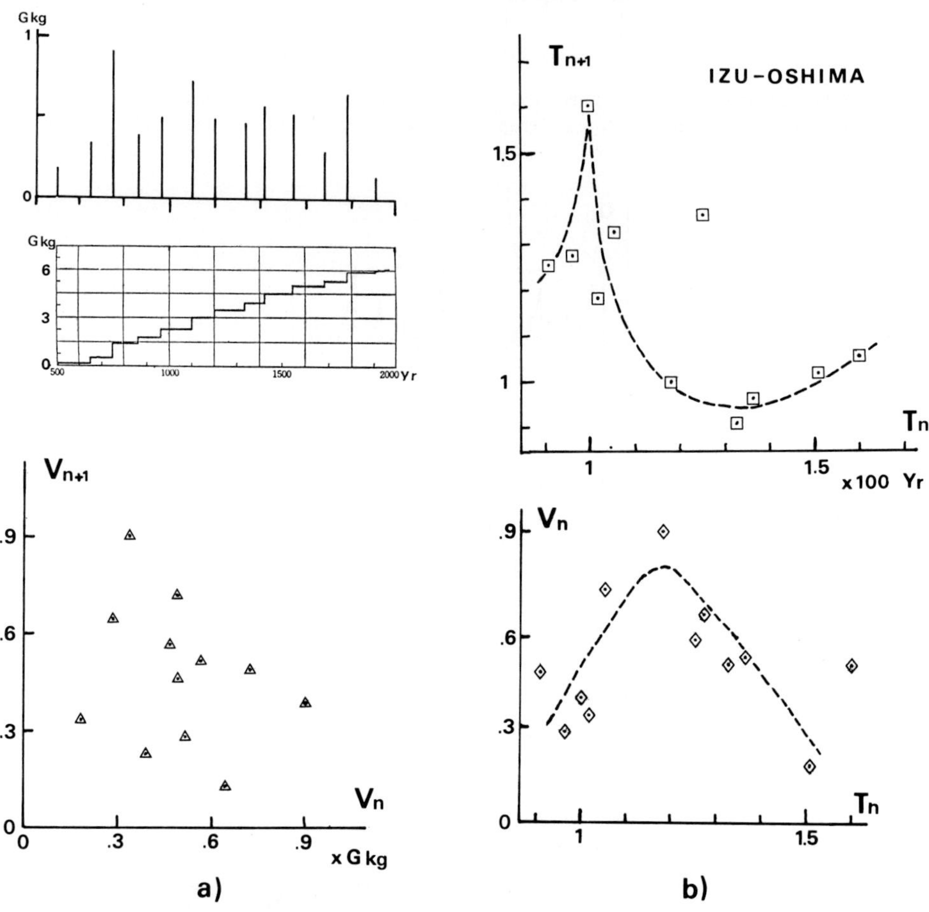

Fig. 4. Erupted mass sequence of Izu-Oshima (Nakamura, 1964). (a) Erupted mass of each event since 500 A.D. (top). Cummulative erupted mass since 500 A.D. (middle). Units of erupted mass are 10^9 kg (G kg). The time axis ranges about 1,500 years from 500 A.D. to present. The successive erupted mass is plotted in the bottom. (b) The diagram illustrating the successive erupted masses (top) and the successive repose period (bottom).

and 5., respectively. In Fig. 4a, cumulative and individual erupted mass of Izu-Oshima are shown in the middle and the top, respectively. From the cumulative erupted mass diagram, a constant rate of about $0.3\,km^3/100\,yrs$ is obtained since 500 A.D. (NAKAMURA, 1964). It is, however, apparent that the erupted mass varies from event to event. This nature becomes more clear when the successive erupted mass is plotted as shown in the bottom of Fig. 4a. The successive erupted mass plot for Izu-Oshima is more or less similar to that of the calculated example of Fig. 3 (bottom), though the pattern is not strictly similar each other. It would be noteworthy that the diagrams of the successive repose period (top of Fig. 4b) and the erupted mass versus subsequent repose period (bottom of Fig. 4b) seem to be a two-valued function. It might suggest that the mechanisms controlling the magma supply rate would be governed by

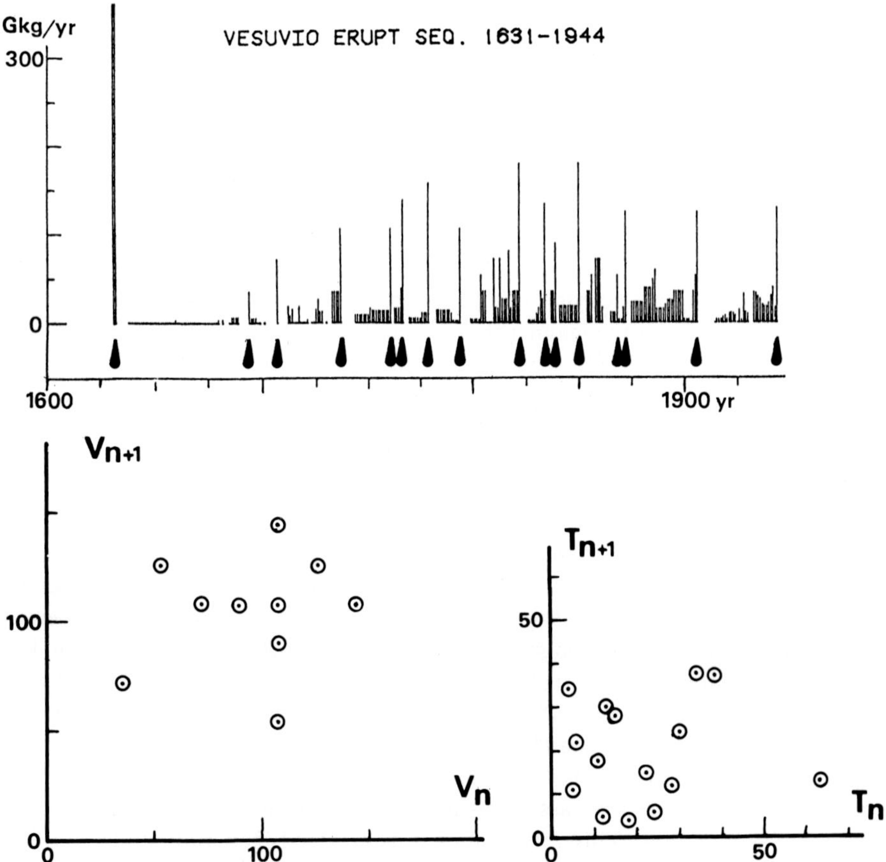

Fig. 5. Erupted mass sequence of Vesuvio (IMBO, 1949; retabulated by Yokoyama, private communication, 1981). Erupted mass for each event from 1630 A.D. (top). The bottom diagrams are the successive erupted mass (bottom left) and the successive repose period (bottom right). Black markers indicate that the event is taken for the analyses of diagrams in the bottom.

nonlinearly coupled variables which would lead to a chaotic behavior (K. Ito, private communication, 1981).

A different pattern of the erupted mass sequence is shown for the case of Vesuvio from 1630 A.D. (Imbo, 1949), which is retabulated by Yokoyama (private communication, 1981). As the persistent activity of Vesuvio can be considered as a repose period, eruptive events marked by black symbols are taken to construct the successive erupted mass sequence diagram (bottom left). Unlike the example of Izu-Oshima (Fig. 4.), the erupted mass sequence of Vesuvio seems to be largely affected by the 1633–34 eruption, which is far greater than the subsequent events and omitted from the bottom diagrams.

4. Discussions

The present model generates sequences of the erupted mass and the duration of repose period deterministically. The assumption of the nonlinear interaction of the magma chamber pressure in Eq. (4) is essential in this model. Calculated sequences shown in Fig. 3, seem to be nonperiodic for limited values of the constant $B \times Gr(0)$. The obtained sequence appears to resemble to actual ones of some polygenetic volcanoes, though the detailed structure could not coinside with a paticular data. It is, however, noticed that the model does not include any randomness a priori, in contrast with the stochastic model (Scheidegger, 1975; Wickman, 1976).

Although no direct evidence of the effect of thermal feedback instability is observed in the eruptive activity, it is likely that a flow instability should initiate the eruptive activity as far as the magma chamber and the conduit below were kept open to the magma source region. On the contrary, the episodic uprise of magmatic batch is likely to correspond with the eruption of monogenetic volcanoes. Otherwise another triggering mechanism in the magma source region should respond for each eruptive activity. The accumulation process of magma into the magma pool from the magma source region is beyond the scope of this paper, but implicitly we assumed that magma has been separated from the magma source region where the degree of partial melting is order of several to a few tens of percent.

Among the various types of volcanic activity of polygenetic volcanoes, a persistent activity and lava lake activity can be regarded as a repose period. Such a persistent activity may occur when the magma is not too viscous and the conduit is open, i.e. the open magma chamber (Wickman, 1976; Blake, 1981). The surface level of magma may indicate the existence of some quasiequilibrium level just above or within the magma chamber in such volcanoes. In the model considered here, we only note the eruption which largely affects the dynamical state of the magma chamber. Those eruption is characterized by a sudden growth of waxing flow followed by gradual decrease of waning flow. This variation of effusion rate is generally understood as an elastic response of the magma chamber and/or the upper crust surrounding it and the viscous flow through a feeding dyke or a conduit between the magma chamber and the surface (Machado, 1974; Tryggvason, 1978; Bjornsson et al., 1979; Wadge, 1981; Wilson and Head, 1981). The change of the magma chamber pressure deduced from geodetic observations of the deformation of volcanic edifice has been extensively described for some volcanoes (MOGI, 1958; FISKE and KINOSHITA, 1969; YOKOYAMA, 1971; SWANSON *et al.*, 1976). Combining these observations, the present approach could give a clue to explain the problem what physical factors actually control the amount of the erupted mass for cach event and the variation of the reposre period duration.

As it is difficult in this preliminary analysis to determine the actual values of parameters that control the variation of the erupted mass sequence for the individual volcano, further investigations are obviously needed to make constraints for the mechanism of magma supply to the magma chamber.

I. would like to thank Drs. K. Ito, K. Nakamura, and I. Yokoyama for their helpful discussions. I am indebted to Dr. I. Yokoyama who kindly made his retabulated unpublished data available for this study. This work was supported in part by the Grant in Aid for Scientific Research from Ministry of Education, Science and Culture, Japan.

REFERENCES

BJORNSSON, A., G. JOHNSEN, S. SIGURDSSON, and G. THORBERGSSON, Rifting of the plate boundary in north Iceland, 1975–1978, *J. Geophys. Res.*, **79**, 3367–3369, 1979.

BLAKE, S., Volcanism and the dynamics of open magma chambers, *Nature*, **289**, 783–785, 1981.

FEDOTOV, S. A., Magma rates in feeding conduits of different volcanic centers, *J. Volcanol. Geotherm. Res.*, **9**, 379–394, 1981.

FISKE, R. S. and W. T. KINOSHITA, Inflation of Kilauea Volcano prior to its 1967–1968 eruption, *Science*, **165**, 341–349, 1969.

FUJII, N. and S. UYEDA, Thermal instabilities during flow of magma in volcanic conduits, *J. Geophys. Res.*, **79**, 3367–3369, 1974.

GRUNTFEST, I. J., J. P. YOUNG, and N. L. JOHNSON, Temperatures generated by the flow of liquids in pipes, *J. Appl. Phys.*, **35**, 18–22, 1964.

HARDEE, H. C. and D. W. LARSON, Viscous dissipation effects in magma conduits, *J. Volcanol. Geotherm. Res.*, **2**, 299–308, 1977.

IMBO, G., L'attivita eruttiva vesuviana a relative osservazioni nel corse dell'intervallo intereruttivo 1906–44 ed in particolare del marzo 1944, *Annali dell'O.V.*, *Quinta Serie*, 185–379, 1949.

ITO, K., Periodicity and chaos in great earthquake occurrence, *J. Geophys. Res.*, **85**, 1399–1408, 1980.

MACHADO, F., The search for magmatic reservoirs, in *Physical Volcanology*, edited by L. Civetta, P. Gasparini, G. Luongo, and A. Rapolla, pp. 255–273, Elsevier, Amsterdam, 1974.

MARSH, B. D., On the cooling of ascending andesitic magma, *Philos. Trans. R. Soc. London, Ser. A*, **288**, 611–625, 1978.

MAY, R. M., Symple mathematical models with very complicated dynamics, *Nature*, **261**, 457–467, 1976.

MOGI, K., Relations between the eruptions of volcanoes and the deformations of the ground surface around them, *Bull. Earthq. Res. Inst.*, **36**, 99–134, 1958.

NAKAMURA, K., Volcano-stratigraphic study of Oshima Volcano, Izu., Bull. Earthq. Res. Inst., **42**, 649–728, 1964.

NAKAMURA, K., Volcano structure and possible mechanical correlation between volcanic eruptions and earthquakes, *Bull. Volcanol. Soc. Japan*, **20**, 229–240, 1975. (in Japanese with English abstract and figure captions).

NELSON, S. A., The possible role of thermal feedback in the eruption of siliceous magmas, *J. Volcanol. Geotherm. Res.*, **11**, 127–137, 1981.

SCHEIDEGGER, A. E., *Physical Aspects of Natural Catastrophes*, Chap. 3, p. 289, Elsevier, Amsterdam, 1975.

SETTLE, M. and T. R. McGETCHIN, Statistical analysis of persistent explosive activity at Stromboli, 1971: Implications for eruption prediction, *J. Volcanol. Geotherm. Res.*, **8**, 45–58, 1980.

SHAW, H. R., Rheology of basalt in the melting range, *J. Petrol.*, **10**, 510–535, 1969.

Shaw, H. R. and E. D. Jackson, Linear island chains in the Pacific: Result of thermal plumes or gravitational anchors?, *J. Geophys. Res.*, **78**, 8634–8652, 1973.

Swanson, D. A., Magma supply rate at Kilauea volcano, 1952–1971, *Science*, **175**, 169–170, 1972.

Swanson, D. A., D. B. Jackskon, R. T. Koyanagi, and T. L. Wright, The February 1969 east rift eruption of Kilauea Volcano, Hawaii, in *U. S. Geol. Surv. Prof. Paper*, 891, pp. 1–24, 1976.

Tanguy, J. C., The storage and release of magma on Mount Etna: a discussion, *J. Volcanol. Geotherm. Res.*, **6**, 179–188, 1979.

Tryggvason, E., Subsidence events in the Krafla area, *Rep. 78 02*, 1–65, Nord. Volcanol. Inst., Raykjavik, 1978.

Wadge, G., Storage and release of magma on Mount Etna, *J. Volcanol. Geotherm. Res.*, **2**, 361–384, 1977.

Wadge, G., Storage and release of magma on Mount Etna: reply to a discussion by J. C. Tanguy, *J. Volcanol. Geotherm. Res.*, **6**, 189–195, 1979.

Wadge, G., The variation of magma discharge during basaltic eruptions, *J. Volcanol. Geotherm. Res.*, **11**, 139–168, 1981.

Wickman, F. E., Markov models of repose-period patterns of volcanoes, in *Random Process in Geology*, edited by D. F. Merriam, pp. 135–161, Springer-Verlag, New York, 1976.

Wilson, L. and J. W. Head, III, Ascent and eruption of basaltic magma on the Earth and Moon, *J. Geophys. Res.*, **86**, 2971–3001, 1981.

Wright, T. L. and R. L. Tilling, Chemical variation in Kilauea eruptions 1971–1974, *Am. J. Sci.*, 777–793, 1980.

Yokoyama, I., A model for the crustal deformation around volcanoes, *J. Phys. Earth*, **19**, 199–207, 1971.

Arc Volcanism: Physics and Tectonics, edited by D. Shimozuru and I. Yokoyama, 13–27.
Copyright © 1983 by Terra Scientific Publishing Company (TERRAPUB), Tokyo.

Thermal Activities of Volcanoes in the Japan Arc
—A Nature and Geological Meanings

Tsuneomi KAGIYAMA

Earthquake Research Institute, University of Tokyo,
Yayoi 1-1-1, Bunkyo-ku, Tokyo 113, Japan

Non-eruptive thermal activities associated with Quaternary volcanoes in Japan are examined through the comparison with the volume of volcanoes, which is expected to reflect extrusive magmatism. The distributions of non-eruptive thermal activities were found to be similar to that of extrusive magmatism, which has already been established as the framework of arc volcanism; highest at just behind the volcanic front, rapidly decreases toward the inner side and essentially zero at the outer side of the volcanic front. It also appears that the ratio of intrusive and extrusive magmatism is almost constant within one arc, except the Izu-Bonin Arc, which is intercepted by the plate boundary. Volcanoes are classified into four types on the volume-thermal activity diagram: Type A, normal relation of volume and thermal activities; Type Ba, large volume, associated with normal hot springs but inactive at the crater zones; Type Bb, large volume but thermally inactive; Type C, small volume but thermally active. Geological meanings are presented for each type. The volcanoes in Hokkaido and the Northeast Japan Arc have similar features of thermal activities to each other, while those of Kyushu are clearly more active. The features of thermal activities of Izu-Bonin Arc are evidently different between both sides of the plate boundary. These regional differences and other geophysical and geological evidences suggest that the ratio of intrusive and extrusive magmatism is affected by regional stress.

1. Introduction

Besides the heat discharge associated with the eruptive activities, volcanoes keep discharging thermal energy throughout the non-eruptive stage. In Japan, hot springs, which are closely related with the Quaternary volcanism, also discharge significant amount of thermal energy. The rates of heat discharge by hot springs were compiled by SUMI (1977), and the total amount was obtained as 5.9 GW in Japan. For the crater zones of active volcanoes, the rates of heat discharge along the Northeast Japan Arc and the Southwest Japan Arc were estimated as 53 MW/100 km and 120 MW/100 km, respectively (KAGIYAMA, 1981). It appears that these values are of the same order as the thermal energy release rate at the eruptive stage estimated from the volume production rate of volcanoes by NAKAMURA (1974). SUGIMURA *et al.* (1963) presented the spatial distribution of solid volcanic products and proposed the Volcanic Front. For non-eruptive thermal activities, a close correlation of the spatial distributions between hot springs and volcanoes has been indicated only qualitatively. Since the amount of data about eruptive and non-eruptive activities increased in recent years, it is interesting to examine these spatial distributions quantitatively.

Intensity of thermal activity of volcanoes is different for individual volcanoes. For instance, the volcano Asama, which shows magmatic eruptions frequently, is discharging thermal energy with the rate of 100 MW at the non-eruptive stage (KAGIYAMA, 1981). On the other hand, the volcano Fuji which had magmatic eruption 275 years ago is dischaging thermal energy with the order of 1 MW. The difference of these thermal activities may be explained by the cooling of the heat source at depth. However, the volcano Hakone, which has intense fumarolic zones, has no eruptions in historic time at all. Our interests are also forcused on these differences and their geological backgrounds.

In the present study, characteristic features of non-eruptive thermal activities and their geological settings are argued on the aspect of arc volcanism through the comparison with the volume of volcanoes. These considerations may clarify the relation between intrusive and extrusive magmatism in the arc system. Though various kinds of heat source of non-eruptive thermal activities will be presented, these activities are reflected by intrusive magmatism, and the volume of volcanoes is the cumulative amounts of the eruptive process over a long time.

Non-eruptive heat discharge rates from crater zones and from hot springs are mainly from KAGIYAMA (1981) and SUMI (1977), respectively. The volume of volcanoes and other parameters are from the *List of Geodynamic Parameters of Quaternary Volcanoes of Japan, Mariana, Kulile, and Kamchatka* (1978). *Volcanoes of Japan* (1968, 1981) were used for the later discussions. The heat discharge rate from crater zones and that of hot springs are treated separately, because hot springs are more or less affected by human developments in contrast to the natural discharge process of crater zones. In the following discussions, volcanic arcs are separated into regions, as shown by Fig. 1.

2.   Spatial Distribution of Non-Eruptive Thermal Activities

It is well known that the total volume of volcanic products is highest at just behind the volcanic front, rapidly decreases toward the inner side and essentially zero at the outer side of the volcanic front (SUGIMURA *et al.*, 1963). We shall examine the spatial distribution of heat discharge rate of the non-eruptive stage.

Figure 2 shows the volume of volcanoes (upper figure), heat discharge rate of hot springs (middle figure), and heat discharge rate from crater zones (lower figure) related with the distance from the volcanic front. Hot springs on the outer side of the volcanic front are known as non-volcanic hot springs, and except for these activities, the non-eruptive thermal activities show a smoothed distribution of that of volume. Loose linear relation may be seen between the volume of volcanoes and the hot spring heat discharge rate in Fig. 3; the ratio of these amounts is almost constant with the distance from the volcanic front. The temperature of hot springs have similar frequency distributions within 100 km from the volcanic front (see Fig. 4). These facts suggest that the ratio of the eruptive and non-eruptive thermal energy release rate (i.e. intrusive and extrusive magmatism) is almost constant within 100 km from the volcanic front.

Figure 3 shows another interesting feature; the volume of volcanoes has bi-modal

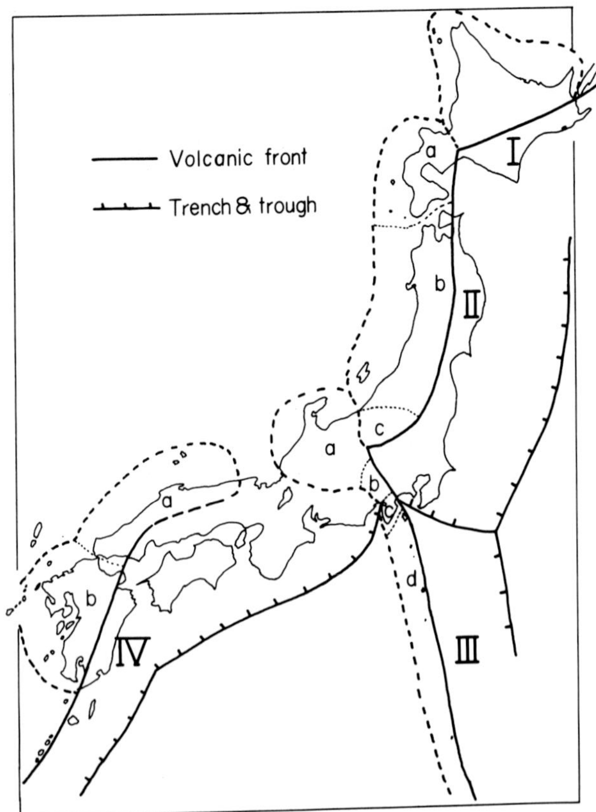

Fig. 1.   Tectonic settings around Japan. I: Hokkaido-Kulile Arc, II: Northeast Japan Arc, III: Izu-Bonin Arc, IV: Southwest Japan Arc; a, b, c and d indicate the sub-regions.

distribution, while SUGIMURA *et al.* (1963) presented the single mode. It may be caused by the difference in procedure of calculations; volume is accounted at 20 km intervals from the volcanic front for the present study in contrast with 50 km for the former. Further examinations are desired to clarify either the second peak represents the secondary volcanism associated with the subduction system, since the second peak appears only in the Northeast Japan Arc (Region II) and most of that peak consists of only two large volcanoes, Chokai and Iwaki.

Our second interest is to clarify the relationship between the spatial distribution of the volume of volcanoes and non-eruptive heat discharge rates along the volcanic front. Cumulative heat discharge rate from crater zones along the volcanic front around Japan were presented by KAGIYAMA (1981). From the average inclinations of cumulative curves, 53 MW/100 km and 120 MW/100 km were obtained for around Northeast Japan Arc and for Southwest Japan Arc, respectively (see Figs. 5 and 6). KAGIYAMA (1981) also indicated that the heat discharge rate was lower in Northeast

T. KAGIYAMA

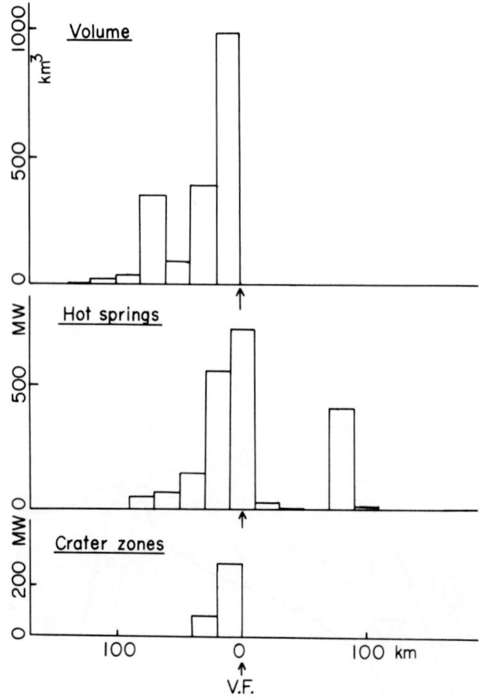

Fig. 2. Quantitative features of volume of volcanoes and thermal activities perpendicular to the arc trend in NE Japan Arc.

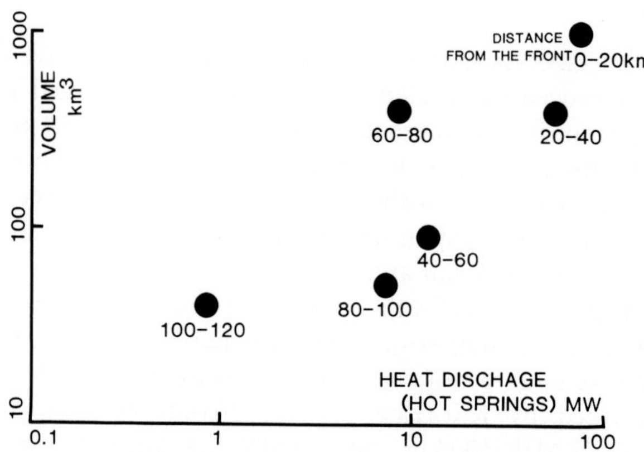

Fig. 3. Relation between the volume of volcanoes and heat discharge rate of hot springs presented in the Fig. 2.

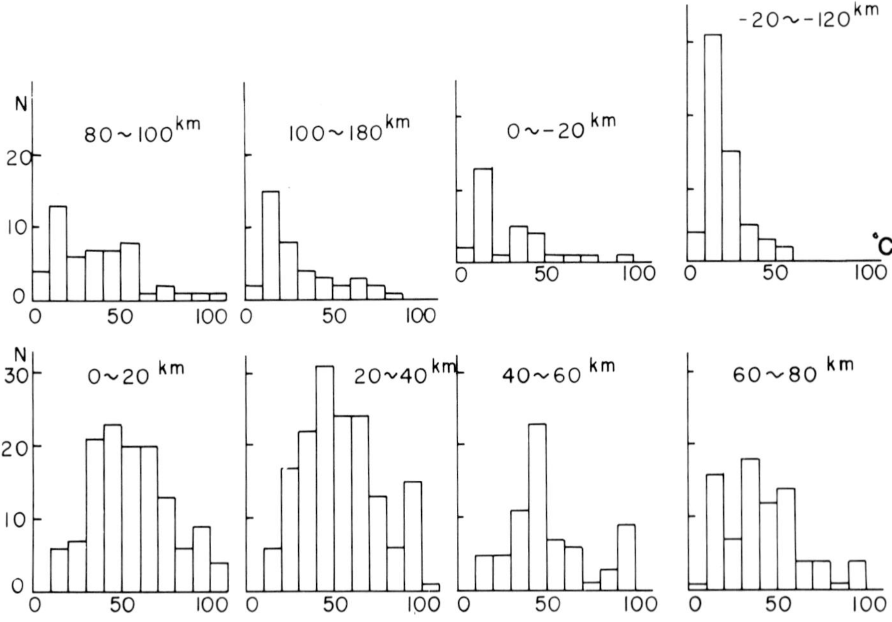

Fig. 4.   Frequencies of the temperature of hot springs in each segment.

Japan Arc (Region II) than in the junction regions between Izu-Bonin Arc and Hokkaido-Kulile Arc.

Cumulative volume around Northeast Japan Arc represented by the solid line in Fig. 5 is increasing linearly from Region IIb to Region I, which suggests that the activity of extrusive magmatism has no difference between these arcs. Smaller inclination of Region IIId may be caused by the underestimation of the volume of marine volcanoes in the area. Concerning the junction areas, extrusive magmatism is clearly more active from Region IIIc to IIc, though that of Region IIa is only slightly more active. The cumulative curve of hot springs shows the following features. Hot springs in Hokkaido (Region I) are less active compared with Northeast Japan (Region II). Since the volcanoes in the Region IIId are marine volcanoes, the amount of hot spring heat discharge may be underestimated. Concerning the junction point, hot springs from Region IIIc to IIc are more active similar to the volume and those around Region IIa are only slightly more active.

Cumulative volume and heat discharge rates of Southwest Japan Arc are presented in Fig. 6. Cumulative volume is divided into three regions; IVa, IVb, and southern volcanic islands. Smaller inclination of southern volcanic islands may be caused by underestimation for the same reason. Another smaller inclination of the Region IVa is consistent with the fact of lack of historic eruptions and Wadachi-Benioff zone beneath the volcanic belt. Similar tendencies are obtained for the cumulative heat discharge rate from crater zones and hot springs, respectively.

T. Kagiyama

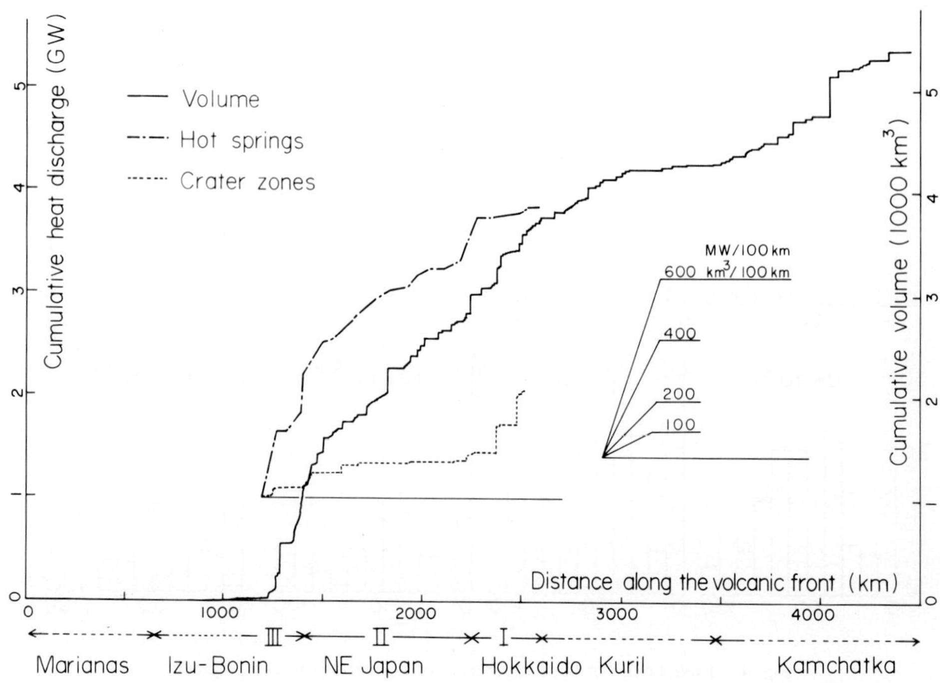

Fig. 5.　Cumulative volume and heat discharge rates along the volcanic front around NE Japan Arc.

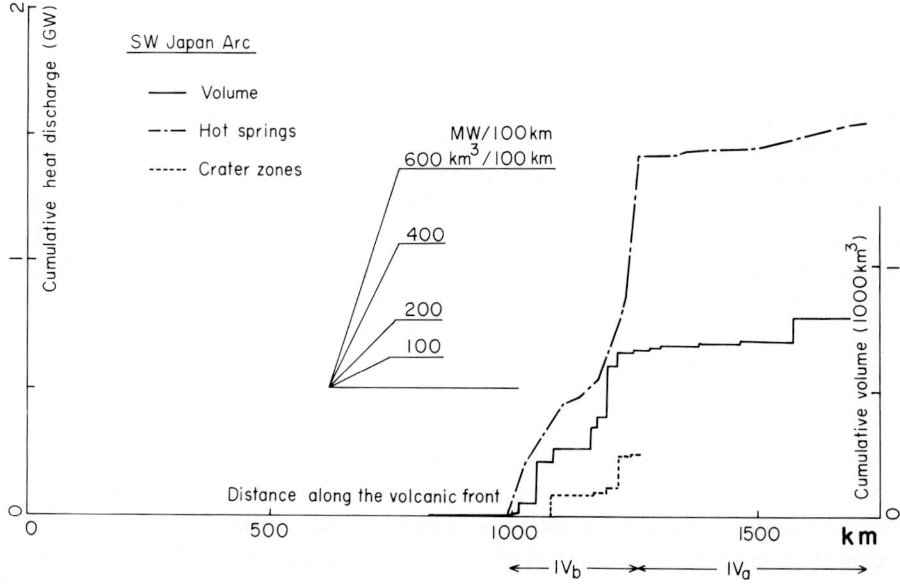

Fig. 6.　Cumulative volume and heat discharge rates along the volcanic front around SW Japan Arc.

From the inclinations of cumulative curves for each region, the linear relation is obtained as shown by Fig. 7 between eruptive and non-eruptive heat discharge rate. Close aspects suggest that Region IVb, Kyushu, is shifting in a thermally more active direction and Region I, Hokkaido, in a thermally less active direction. These facts will be examined later.

3.   Classification of Volcanoes Based on Energy Discharge

Characteristics of energy discharge will be examined for each volcano through the comparison between the volume and the rate of heat discharge. Hot spring heat discharge rate from one volcano is calculated with the following rule; a) hot springs on the volcanic body, b) hot springs at the sedimentary basin which is expected the hydrothermal system related with the volcano.

Relationship between volume of volcanoes and heat discharge rate of hot springs is plotted in Fig. 8a. It seems that volcanoes are divided into the following three groups; Group A, volcanoes which have normal or general volume-thermal activity relationship; Group B, large volume but thermally inactive volcanoes, and Group C, small volume but thermally active volcanoes. Similar relationship between volume of volcanoes and heat discharge rate from crater zones is seen in Fig. 9. The volcanoes which began their volcanic activities before 100,000 years tend to locate at the upper part of the diagram. Enough chronological data are needed to clarify the reason of this tendency.

Heat discharge rate from the volcano could also be assumed empirically from the photographs or other informations about the crater zones; a) no thermal activity at all or little, order of 1 MW or less, b) intense fumarolic activities enough to be observed as plume shape at the foot of the volcano, order of 100 MW or less, c) intense fumarolic activities but have no plume shape, order of 1 or 10 MW. This empirical relationship is applied to the volcanoes which have no estimates on heat discharge rates.

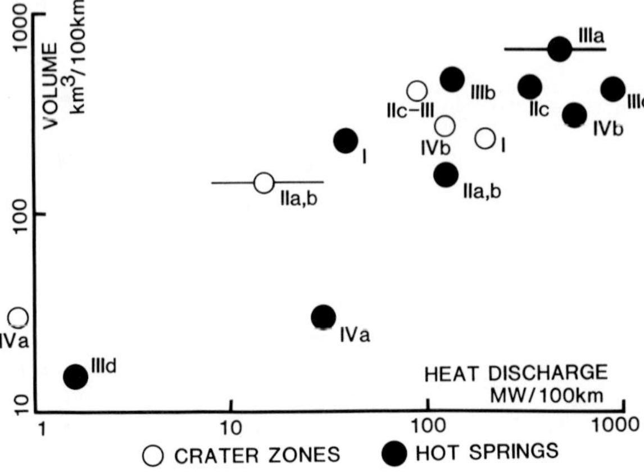

Fig. 7.   Relation between the volume of volcanoes and the heat discharge rate for each region.

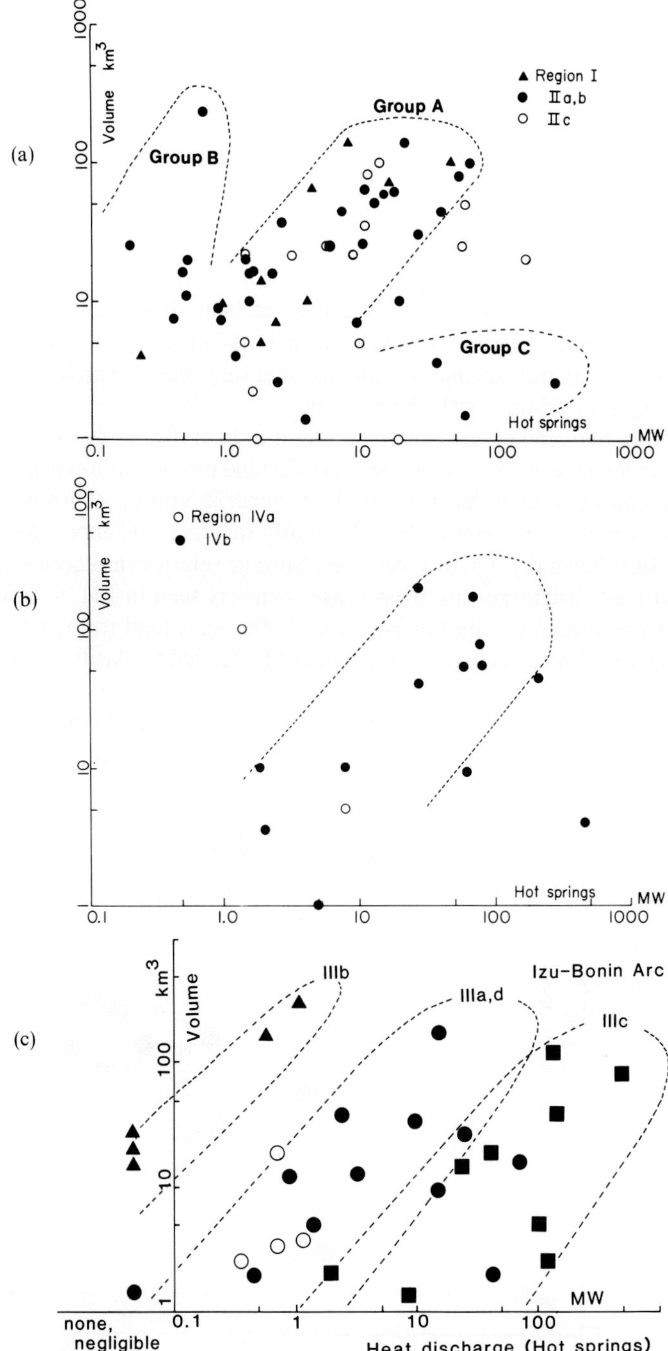

Fig. 8. Volume-thermal activity diagram for hot springs. Volcanoes are divided into three groups; a: Hokkaido and NE Japan Arc, b: SW Japan Arc, c: Izu-Bonin Arc (Izu-Ogasawara Arc).

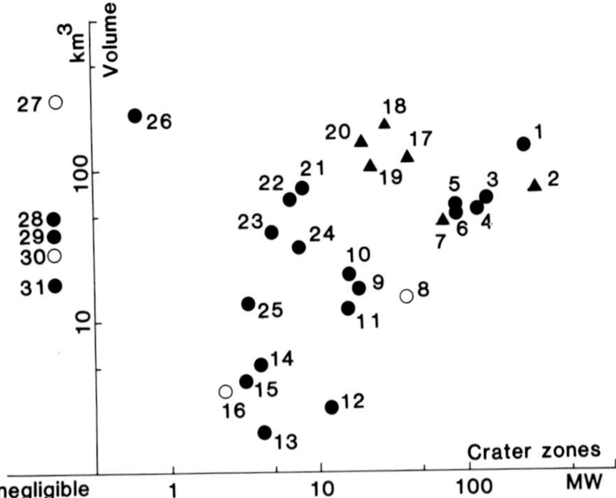

Fig. 9.   Volume-thermal activity diagram for crater zones. Prameters for each volcano are presented in Table 1. Open circles represent basaltic volcanoes, solid triangles, include the volume of pyroclastic flows, and solid circles, other andestic volcanoes.

Through those diagrams, the volcanoes are classified as four types: Type A, the volcano which is classified into Group A both for the crater zones and hot springs; Type Ba, the volcano which is thermally inactive at the crater zone but has hot springs; Type Bb, the volcano which has no thermal activity at all; Type C, the volcano which is included into the Group C for hot springs.

Type A is expected to represent the regional average ratio of intrusive and extrusive magmatism for the concerned area. The volcanoes Asama, Tokachi and Iwate are examples for Type A.

The volcanoes belonging to Type Ba are the volcanoes Haruna, Iwaki and Tateshina, for example, and have no magmatic eruptions in historic time. Only the volcanoes Akagi and Iwaki have steam explosions in historic time. These facts suggest that Type Ba volcanoes were chilled so that the fumarolic activities markedly decline at the crater zone, but have heat source to maintain hot springs.

Examples of Type Bb are the volcanoes Fuji, Chokai, Shokanbetsu, Rishiri, and Shiretoko. Four reasons, as described below, may be the possible characteristics for Type Bb.

1)   Though the volcanoes have in fact hot springs, those hot springs are not developed. The volcano Shiretoko in eastern Hokkaido or other volcanoes in solitary regions will be explained by the undevelopment of hot springs. These volcanoes should be classified into Type Ba in the geological meanings.

2)   The volcanoes are old or extinct, because most of Type Bb volcanoes have had no eruptions in historic time. Chronological data of the volcanic activities are poor, however, the volcanoes Shokanbetsu, Shari or Ashitaka are expected to be old or extinct, and to be explained by cooling. Most volcanoes of Type Bb will be explained by those two reasons. However, the volcanoes Fuji and Chokai need another reason,

T. KAGIYAMA

Table 1.   Heat discharge rates from crater zones and other parameters.

	Volcano	Heat discharge (MW)	Volume (km³)	Erosion[7] ratio (%)	Historic activity[8] in A.D.
1	Tokachi	250	140		M1962
2	Akan	290[2]	72		E1966
3	Asama	140	62.7	2.5	M1973
4	Kuju	120	53.9		E1738
5	Kirishima	85	53.8		E1959
6	Nasu	84	50.0		E1963
7	Toya (Usu)	70	44.0		M1978
8	Izu-o-shima	39	19.0	2.2	M1974
9	Iwate	19	16.0		M1719, E1939
10	Kusatsu-shirane	13	20.0		E1976
11	Tarumai[1]	16	11.6		M1909, E1978
12	Kuttara (Noboribetsu)	12	2.6		—
13	E-san	4.0	1.8		E1846
14	Eniwa[1]	4.1	5.0	4.9	—
15	Tsurumi	3.2[3]	3.9		E867
16	Miyake	2.2	3.4	2.3	M1962
17	Hakone	42[4]	118		—
18	Aso	29	197		M1979
19	Kutcharo	23[5]	101		—
20	Shikotsu	20	142		
21	Unzen	8.0[6]	77.2		M1792
22	Daisetsu	6.6	64.7		—
23	Hokkaido-komaga-take	4.9	37.3		M1929, E1942
24	Azuma	7.2	30.0		E1977
25	Akita-komaga-take	3.3	13		M1970
26	Chokai	0.59	232	5.7	M1801, E1974
27	Fuji	—	300	0.9	M1707
28	Iwaki	—	52	12.3	E1863
29	Shari	—	40	12.6	—
30	Ashitaka	—	28	28.7	—
31	Yotei	—	16.1	6.5	—

1) Post caldera cones of Shikotsu. 2) EHARA and YOKOYAMA (1974). 3) EHARA and YUHARA (1981). 4) OKI and HIRANO (1974). 5) KAWAMURA *et al.* (1978). 6) YUHARA *et al.* (1981). 7) SUZUKI (1969). 8) *Volcanoes of Japan* (1981).
   M: Latest eruption with essential products.
   E: Latest activity without essential products or details unknown.

Table 2.   Classification of volcanoes based on the thermal activity.

Type	Thermal activity of crater zone	Hot spring activity
A	Group A	Group A
Ba	Group B	Group A
Bb	Group B	Group B
C	Group A, B, C	Group C

because these volcanoes have magmatic eruptions in historic time and cannot be considered to be extinct volcanoes.

3) The volcanoes have heat source, but have no hot springs for the lack of available water to maintain them. It is frequently indicated that volcanoes which have new and uneroded volcanic body or which are completely covered with unpermeable layers, such as lava flows, have weak hot spring activity. (e.g. YUHARA and EHARA; 1982). At the volcano Fuji or Yotei, which has small value of R (0.9 and 6.5), erosion ratio by SUZUKI (1969), most of the precipitation wells up as normal water and the volcanoes which have a large value of R have active hot springs. However, some volcanoes, such as Chokai, have weak hot springs or evidently no hot springs, even though they have a large value of R and magmatic eruptions in historic time.

4) Volcanoes retain no heat source after the magmatic eruptions at depth. Some volcanoes such as Nishino-shima cool rapidly compared with other volcanoes after their magmatic eruptions. This fact will be explained by the difference of the volume or nature of heat source which remains at shallow depth to maintain the non-eruptive heat discharge. Two possible reasons are presented as follows; viscosity of magma and regional stress field.

For viscous magma, it will be hard to extrude or to drain back, while in contrast it will be easy for less viscous magma. The volcanoes Fuji and Ashitaka which extrude basaltic magma have low activities of non-eruptive heat discharge in fact. However, the volcanoes Izu-oshima and Miyake which also extrude basaltic magma have normal thermal activities. This fact will be explained by the height of magma head. Though the number of basaltic volcanoes is limited, viscosity of magma may control the non-eruptive thermal activities.

At compressive stress field, magma will mostly extrude or drain back without storage at shallow depth for the difficulties to insure the space for storage. It is impressive that the volcanoes of Type Bb are large stratovolcanoes or tend to be isolated. NAKAMURA (1977) proposed that the volcano at tensile stress field tend to be monogenetic volcano and at compressive stress field, to be polygenetic volcano, because at compressive field, magma will rise through the path or fissure system which was used in past events, while at tensile stress field, magma will rise freely. By the same considerations, it is expected that the volcano at more compressive stress field tend to be a large stratovolcano and at less compressive stress field, to be a group of smaller volcanoes. These features and considerations imply that some volcanoes classified into Type Bb are controlled by regional stress.

Most of type C volcanoes have lava domes. The following explanations are plausible for their active nature. 1) Viscous magma which makes lava domes may be more intrusive. 2) Fissure system to supply the available water for hot springs may be created during the dome formations. 3) Underestimation of the volume of the volcano. We can find several mountains which look like lava domes but are not regarded as Quaternary volcanoes around Type C volcano frequently. In these cases, the volume of the volcano which is related with the hot spring activities may be underestimated.

4. Regional Differences of Thermal Activities

Obviously, four types of non-eruptive heat discharge, as mentioned above, reflect the difference of energy discharge process of each volcano. Let us argue regional differences of these processes and their geological meanings. Three features are clarified through the comparisons of volume-hot spring activity diagrams for each region in Figs. 8.

1) Volcanoes hold one portion for each arc except Izu-Bonin Arc. This fact suggests that volcanoes have a similar ratio of eruptive and non-eruptive energy discharge within one arc, and therefore, the energy discharge processes are dominantly controlled with the factor which is common within the arc. It is also impressive that Izu-Bonin Arc, which is intercepted by the convergent boundary of Philippine Sea Plate and Eurasian Plate, cannot be defined as one portion in the figure.

2) Hokkaido (Region I) and Northeast Japan Arc (Region II) hold the same portion in the figure, while that of Kyushu (Region IV b) is clearly separated from those regions to thermally more active direction. Because the density of human activity is exceedingly high in Japan, it cannot be expected that this fact reflects the regional difference of the development of hot springs, though it is impossible to deny. According to Fig. 3, hot springs in Hokkaido seem to be less active than any other region. However, this discrepancy can be explained by the fact that the percentage of Group B volcanoes in Hokkaido is evidently higher. Therefore, these arcs have similar features of energy discharge processes with each other, which is consistent with the result of cumulative volume as shown by Fig. 6. Geodetic and seismological investigations suggest that the stress field of Northeast Japan is compressive and thrust earthquakes occur at the coast of the Japan Sea, while Kyushu is rather at tensile stress field and, especially in central Kyushu, normal fault earthquakes occur frequently. Therefore, in Kyushu, it will be easy to store the heat source at shallow depth or to create a fissure system to supply available water for hot springs.

3) The portion of Izu-Bonin Arc (Region III) is divided into four sub-regions, as shown by Fig. 8c; to the north of the volcano Tateshina (Region IIIa), from the volcano Yatsuga-take to the volcano Ashitaka (Region IIIb), Hakone and Izu Peninsular volcanoes (Region IIIc) and the volcanic islands to the south of Izu-oshima (Region IIId). The volcanoes of Regions IIIa and IIId hold the same portion as those of Hokkaido and Northeast Japan Arc. Region IIIc is clearly separated to a thermally more active direction, while Region IIIb, to a thermally less active. It is impressive that the boundary between Regions IIIb and IIIc is known to be the convergent boundary of two plates. These facts suggest that the collision of the two plates affect the energy discharge processes of these regions and does not affect Regions IIIa and IIId, which stand apart from the boundary. Similar differences are also obtained for these regions. MATSUDA (1962) presented that the crustal deformation since middle Miocene is evidently different between the northern side and southern side of the boundary; extremely deformed at northern side and relatively stable as a block on the southern side (Fig. 10). The crowdedness of volcanoes also has difference between both sides of the boundary as shown in Fig. 11; volcanoes are isolated from each other in Region IIIb, while in Region IIIc, volcanoes are crowded. These facts seem to support the

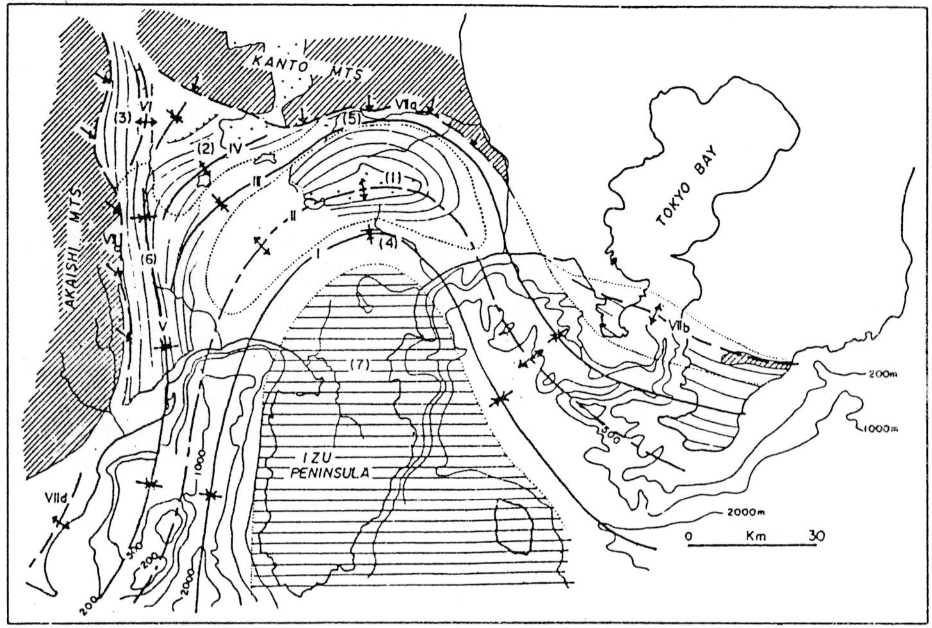

Fig. 10.   Tectonic map of the South Fossa Magna; I-VI: belts of uplift and subsidence since middle Miocene time, (1) Tanzawa Mountains, (2) Misaka Mountains, (4) Ashigara area, (5) Katsura River Valley, (6) Fuji River Valley, (7) Izu platform; VIIa-d: marginal thrusts, and axes of uplift since early Miocene time. (after Matsuda (1962)).

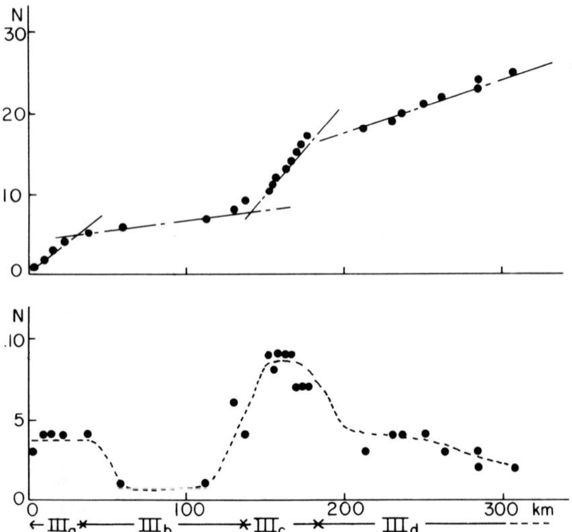

Fig. 11.   Crowdedness of the volcanoes of Izu-Bonin Arc. Upper figure indicates the number of volcanoes from the junction point between NE Japan Arc. Lower figure indicates the number of volcanoes within 30 km from one volcano. Both figures clearly show that the volcanoes are separated each other in the region IIIb and are crowded in the region IIIc, respectively.

following idea; in Region IIIb, it will be hard to store the heat source at shallow depth or to create a fissure system for hot springs and, therefore, the non-eruptive thermal activities will decrease more rapidly than any other region.

5. Conclusions

Non-eruptive thermal activity, which is expected to represent the intrusive magmatism, shows a similar spatial distribution to that of the volume of volcanoes; highest at just behind the volcanic front and rapidly decreases toward the inner side. The ratio of thermal activity with respect to the volume is almost constant within one arc.

The volcanoes in Hokkaido and Northeast Japan Arc have similar features in thermal activities, while those of Kyushu are clearly thermally more active.

The features of thermal activities of Izu-Bonin Arc are evidently different on both sides of the plate boundary; thermally active at southern side and thermally inactive at northern side.

Volcanoes are classified into four types with respect to the relation of the volume and the rate of heat discharge: Type A, normal volcanoes; Type Ba, normal but seem to be cooled volcanoes; Type Bb, cooled or cold volcanoes; and Type C, hot volcanoes. Difference between these types may be explained by individual factors of volcanoes such as erosion ratio or morphological characteristics. However, these differences occur on a regional scale. Therefore, regional factors will be dominant for the nature of thermal activities. Regional difference of thermal activities, crowdedness of volcanoes, and other geodetic or geophysical implications suggest that the ratio, (the rate of heat discharge)/(the volume) (i.e. the ratio of intrusive and extrusive magmatism) reflects the regional stress. The viscosity of magma is also expected to be a dominant factor, though positive evidences are not presented in the present paper.

Our knowledge about thermal structure of heat source of non-eruptive thermal activities or chronology of volcanic activities is insufficient to a further understanding of the energy discharge process, and these investigations will clarify the relations of intrusive and extrusive magmatism in the arc region and their geological meanings quantitatively.

I am grateful to Dr. Kazuaki Nakamura for helpful suggestions. I also thank Prof. D. Shimozuru, Dr. T. Watanabe and Dr. T. Kato for reviewing the manuscript and encouragements.

REFERENCES

EHARA, S. and I. YOKOYAMA, Thermal field on and around active volcanoes in Hokkaido, Japan, in *Proc. U.S.-Japan Coop. Sci. Seminar, The Utilization of Volcano Energy*, pp. 307–328, 1974.

EHARA, S. and K. YUHARA, Two volcanic belts in Kyushu and the descending of the Philippine Sea plate, *Bull. Volcanol. Soc. Jpn.*, **26**, 138, 1981 (in Japanese).

KAGIYAMA, T., Evaluation methods of heat discharge and their applications to the major active volcanoes in Japan, *J. Volcanol. Geotherm. Res.*, **9**, 87–97, 1981.

KAWAMURA, M., M. SEKIOKA, and M. AOKI, Digital analysis of MSS thermal image, *Bull. Volcanol. Soc. Jpn.*, **23**, 280–281, 1978 (in Japanese).

List of Geodynamic Parameters of Quaternary Volcanoes of Japan, Mariana, Kulile, and Kamchatka, Contribution from Geodynamics Project of Japan 78–2, edited by S. Aramaki and T. Ui, 1978.

MATSUDA, T., Crustal deformation and igneous activity in the South Fossa Magna, Japan, *AGU Geophys. Monogr.*, **6**, 140–150, 1962.

NAKAMURA, K., Preliminary estimate of glocal volcanic production rate, in *Proc. U.S.-Japan Coop. Sci. Seminar, The Utilization of Volcano Energy*, pp. 273–285, 1974.

NAKAMURA, K., Volcanoes as possible indicators of tectonic stress orientation-Principle and proposal, *J. Volcanol. Geotherm. Res.*, **2**, 1–16, 1977.

OKI, Y. and T. HIRANO, Hydrothermal system and seismic activity of Hakone Volcano, in *Proc. U.S.-Japan Coop. Sci. Seminar, The Utilization of Volcano Energy*, pp. 13–40, 1974.

SUGIMURA, A., T. MATSUDA, K. CHINZEI, and K. NAKAMURA, Quantitative distribution of late Cenozoic volcanic materials in Japan, *Bull. Volcanol.*, **26**, 125–140, 1963.

SUMI, K., The relationship of the rates of convective heat discharge to the tectonic provinces in Japan, *Bull. Geol. Surv. Jpn.*, **28**, 277–325, 1977 (in Japanese).

SUZUKI, T., Rate of erosion in strato-volcanoes in Japan, *Bull. Volcanol. Soc. Jpn.*, **14**, 133–147, 1969 (in Japanese).

Volcanoes of Japan (First Edition), edited by Isshiki, N., T. Matsui, and K. Ono, Geol. Surv. Jpn., 1968.

Volcanoes of Japan (Second Edition), edited by K. Ono, T. Soya, and K. Mimura, Geol. Surv. Jpn., 1981.

YUHARA, K., S. EHARA, and K. TAGAMORI, Estimation of heat discharge rates using infrared measurements by a helicopter-borne thermocamera over the geothermal areas of Unzen Volcano, Japan, *J. Volcanol. Geotherm. Res.*, **9**, 99–109, 1981.

YUHARA, K. and S. EHARA, Geothermal fields of arc volcanism in Japan, *Bull. Volcanol. Soc. Jpn.*, **26**, 185–203, 1982 (in Japanese).

Arc Volcanism: Physics and Tectonics, edited by D. Shimozuru and I. Yokoyama, 29–41.
Copyright © 1983 by Terra Scientific Publishing Company (TERRAPUB), Tokyo.

Gravimetric Studies and Drilling Results at the Four Calderas in Japan

Izumi YOKOYAMA

Usu Volcano Observatory, Hokkaido University
Sapporo 060, Japan

Almost all the calderas in Japan, about 15 in number, have been surveyed gravimetrically. At four of them, drilling results are available for cross-checking the gravimetric discussions. The four calderas are of low gravity anomaly type. The followings are confirmed:

1)   Within the calderas, coarse material is usually accumulated a few kilometers in depth. The caldera deposits consist of pumice, tuff and rocks of pre-caldera volcanoes (all are named "fall-back"). At some calderas, there remain only sparse caldera deposits because of erosion by rivers.

2)   The caldera boundaries are not always faults, and dip inward at low angles with a few exceptions of rather steep angles: the configuration of the basements beneath calderas is funnel-shape.

3)   It is concluded that pre-caldera volcanoes could not collapse into magma reservoir through narrow vents at the centers of the calderas.

1.   Introduction

Violent volcanic eruptions which were accompanied by ejection of voluminous pumice and ash, and led to formation of calderas are one of the characteristics of arc volcanisms. Discussions on the mechanisms of such gigantic eruptions may contribute to achievement of prediction of volcanic eruptions in our days. As the basis for discussions of the eruption mechanisms, we should know quantitative subsurface structure beneath calderas and volcanoes in general. Gravimetric surveys on calderas have proved to be useful for the discussions of their subsurface structure and further, of the mechanism of their formation. Gravity anomalies provide us with information about mass anomalies of caldera deposits within the calderas and about possible configurations of the basements beneath the calderas. However, the material of the deposits and the exact configurations still remain undeterminative. In these respects, test drillings afford us useful clues to the problems.

The calderas in Japan may be classified into two types from the standpoint of gravity anomaly: the calderas of high gravity anomaly type such as Ooshima (YOKOYAMA and TAJIMA, 1957), Batur in Bali (YOKOYAMA and SUPARTO, 1970), and Hawaiian calderas, associated with the effusive eruption of basaltic magmas, and the calderas of low gravity anomaly type such as Kuttyaro (YOKOYAMA, 1958), Nigorikawa (Ando, 1982), Hakone (Yokoyama and Saito, 1965), Aso (Kubotera et al., 1969) and many other calderas, formed by explosive eruptions of siliceous magmas. As for the high gravity anomaly type, Yokoyama (1969a) discussed the subsurface structure of Ooshima caldera. In this paper, the discussions will be confined to the four calderas of low gravity anomaly type, i.e. Kuttyaro, Nigorikawa, Hakone and Aso calderas, and be cross-checked with the results of the drillings there.

The present author (Yokoyama, 1969b) discussed three of the four calderas from the same standpoint. This paper is a revised one.

2. Kuttyaro Caldera

Kuttyaro caldera in the eastern part of Hokkaido, measuring about 20 km in diameter, is one of the largest calderas in Japan; and its western half is a lake of which the mean depth is about 40 m. Pumice which was ejected at the time of the catastrophe of the caldera formation (29,000 ~ 30,000 years B.P.) is found widely spread around the caldera. A gravity survey was first carried out in March 1958 on the surface of the frozen lake, and thereafter it was supplemented repeatedly. The distribution of Bouguer anomalies on this caldera is shown in Fig. 2 where the dotted square contains about 100 gravity points occupied for prospection in 1963. The gravity anomalies were

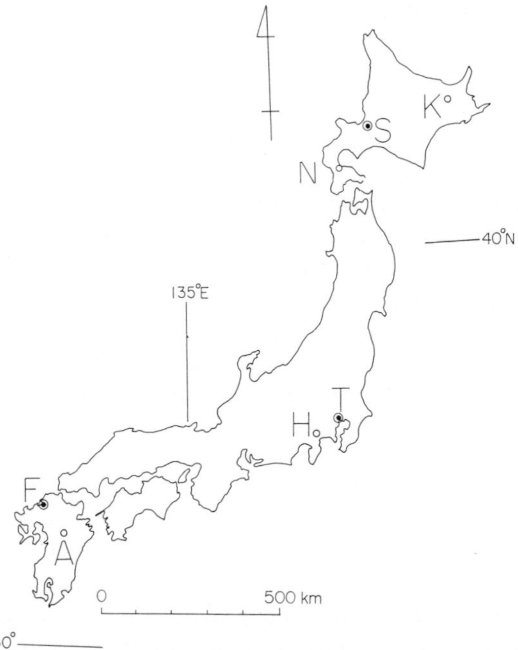

Fig. 1. The four calderas discussed in this paper. K: Kuttyaro caldera, N: Nigorikawa caldera, H: Hakone caldera, A: Aso caldera, S: Sapporo, T: Tokyo, F: Fukuoka.

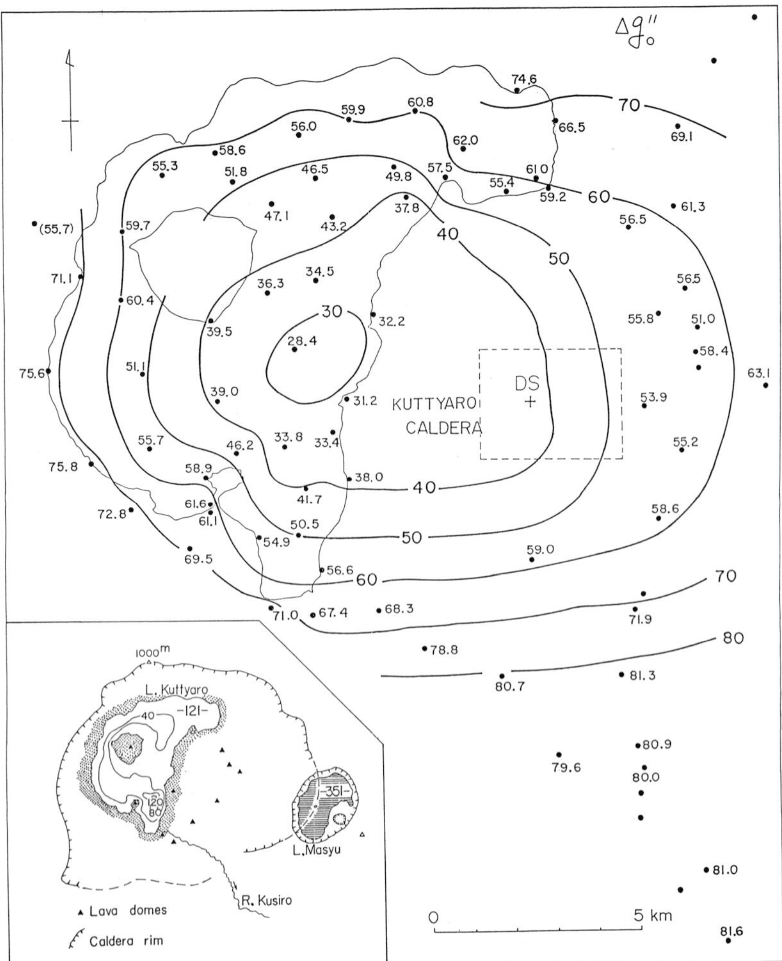

Fig. 2. Bouguer gravity anomalies on Kuttyaro caldera (unit: mgal). The dotted square contains about 100 gravity points and a drilling site (DS).

already analyzed and a schematic profile of the subsurface structure of the caldera was presented by YOKOYAMA (1959) as shown in Fig. 3, where the density contrast $(\rho_0 - \rho_1)$ between the country rocks and the deposits was assumed to be 0.3 g/cc.

In 1963, a borehole of 1,000 m deep was drilled to exploit steampower inside the caldera (at the center of the dotted square in Fig. 2). NISHIDA and YOKOYAMA (1965) studied several physical properties of the bored cores. One of them, the density distribution along the borehole is shown in Fig. 4. At the drilling site, the depth of coarse material which is the origin of the observed gravity anomaly is expected to be less than 2 km as shown in Fig. 3. Therefore, almost all material to the depth of 1,000 m must be coarse caldera deposits. In fact, there are many parts consisting of tuffaceous

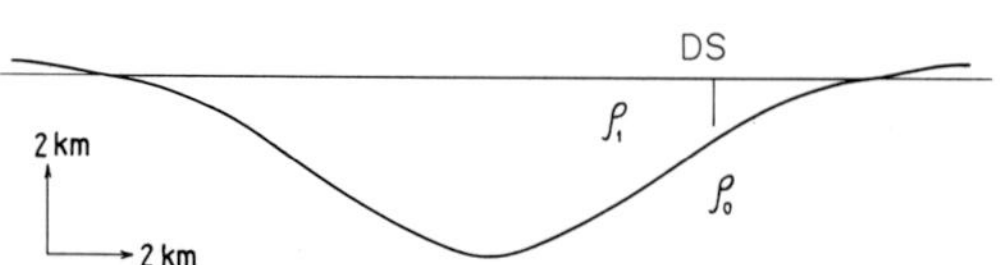

Fig. 3.   A schematic cross section of Kuttyaro caldera. $\rho_0 - \rho_1 = 0.3$ g/cc.

ejecta as shown in Fig. 4. Density increases from 1.7 to 2.2 g/cc with depth inside the caldera, while it may usually increase from 2.0 to 2.5 g/cc outside the caldera. Therefore, the before-mentioned density contrast 0.3 g/cc may be reasonable.

In Fig. 4 it is noticeable that many lava fragments are deposited at depths deeper than 500 m and the total core length of the lava fragments may reach a few tens of meters along the lower 500 m. This suggests that the deeper beneath the caldera, the more lithic fragments one may find, and this does not support the theory of the later collapse of old volcanoes than pumice eruptions in caldera formation though the

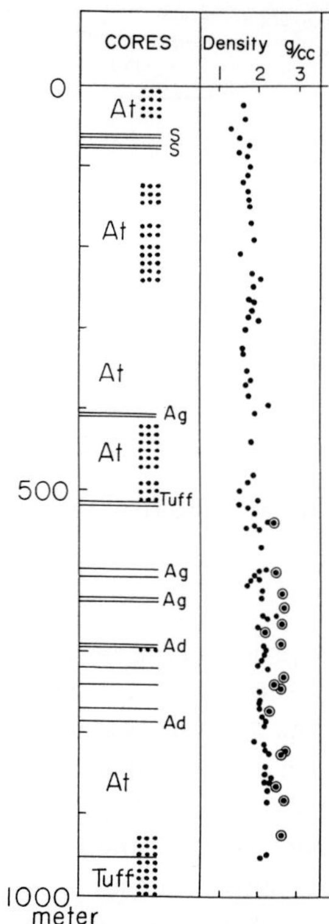

Fig. 4.   The bored cores and their density inside Kuttyaro caldera. At: agglomerate tuff, S: sand stone, Ag: agglomerate, Ad: andesite. Double circles denote lava fragments.

mechanism of the earlier deposition of the fragments is not yet clear. The composition of the caldera deposits along a borehole shown in Fig. 4 shows that we can not deduce any definitive conclusion about composition and distribution of volcanic ejecta on and around calderas only from surface observations.

3.   Nigorikawa Caldera

Nigorikawa caldera is situated in the southern part of Hokkaido, measuring about 2.5 km in diameter and was formed about 12,000 years B.P. Various prospectings have been carried out and nine exploratory wells were drilled in the caldera to get information about the geological structure and occurrence of the geothermal resources. The deepest well reaches a depth of 2,230 m.

The distribution of Bouguer gravity anomalies on this caldera is shown in Fig. 5 where the crust density is assumed as 2.2 g/cc. A low gravity anomaly amounting to 6 mgal is clear. A detailed discussion about geology of this caldera revealed by the drillings was reported by ANDO (1981). The followings are quoted from his report.

The caldera deposits at Nigorikawa are divided into five parts: alluvium deposits (ca. 100 m thick), lake deposits (ca. 350 m thick), landslide deposits (max. 150 m thick), post-caldera central dome lava, and fall-back deposits (max. 1,300 m thick). The fall-back deposits are composed of lapilli tuff, tuffbreccia or volcanic breccia, and contain many accidental lapillies or blocks from the Miocene formations and the pre-Tertiary

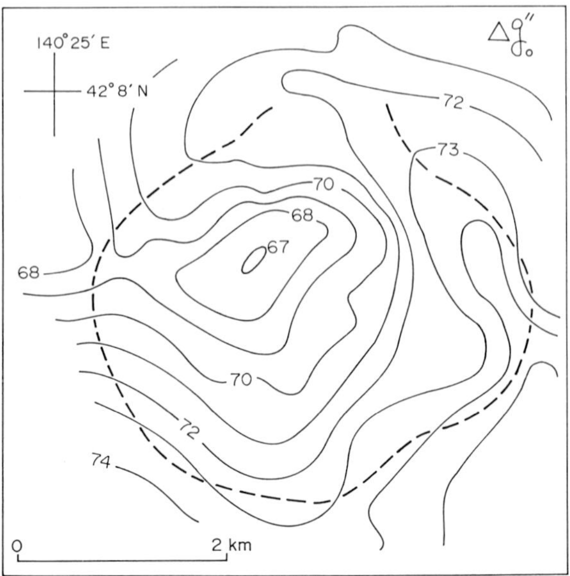

Fig. 5.   Bouguer gravity anomalies on Nigorikawa caldera (Unit: mgal) after Japan Metals and Chemicals Co. Ltd.

formations, both being the country rocks. The size of the rock fragments ranges from 1 m or more to 1 cm or less and concentrates around several centimeters.

The caldera has a funnel-shaped structure bounded by a steep wall with an angle of 60° to 70°, and the depth of the bottom is estimated at 1,700 ~ 1,800 m. ANDO (1981) considers that the volcanic eruptions forming the caldera was not followed by the collapse of the volcanic edifice to the magma reservoir.

4. Hakone Caldera

Hakone caldera is situated at a distance of about 80 km from Tokyo, and its diameter measures about 9 km. The geology of Hakone volcano was fully studied by KUNO (1952), and his geologic sketch map is shown in Fig. 6. It is peculiar to this caldera that a large area of the caldera is occupied by many post-caldera cones, and a part of the caldera has been eroded to the basement by the rivers, as shown in Fig. 6. According to Kuno, Hakone volcano consists of two concentric calderas, old and young, and central post-caldera cones, and the old caldera is of "Glen Coe type" and

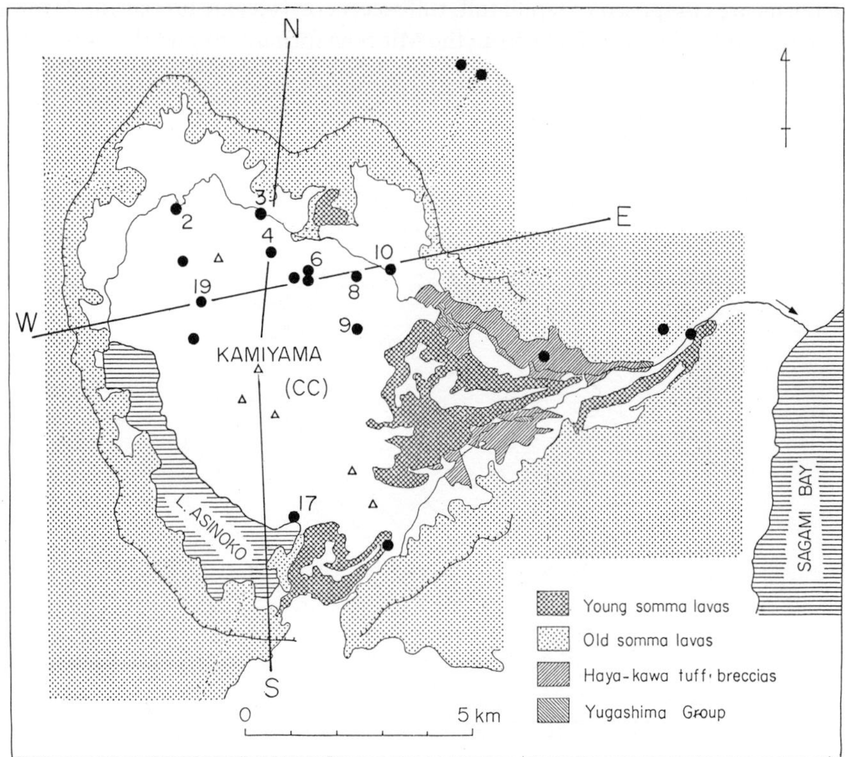

Fig. 6.　A geologic sketch map of Hakone caldera after KUNO (1952). Solid circles denote drilling sites.

the young caldera of "Krakatau type." In the author's classification, calderas of the former type belong to high gravity anomaly type and those of the latter type to low gravity anomaly type. The author doubts that a caldera of low anomaly type could be formed at a caldera of high anomaly type because the latter eruption should eviscerate the pre-existing massive accumulations within the caldera. The reverse sequence may have been possible. The present discussion is concerned only with the Kuno's young caldera which was formed in Pleistocene.

Gravity surveys on Hakone caldera were first carried out by YOKOYAMA and SAITO (1968), and thereafter were supplemented by YOKOYAMA (unpublished) and by HIRAGA *et al.* (1970) repeatedly. The distribution of Bouguer anomalies is shown in Fig. 7. In this district, the regional anomaly is relatively high; this means that the basements are upheaved in this region. And the residual anomalies due to the caldera structure is not so conspicuous amounting to about -10 mgal as shown by a gravity profile in Fig. 8 perhaps because the caldera deposits had been eroded away by the rivers and a characteristic low anomaly may be largely canceled by existence of the post-caldera cones.

On the other hand, about 20 boreholes have been drilled mainly inside the caldera. Their sites are shown in Fig. 6. According to the descriptions of the bored cores reported by ŌKI *et al.* (1968), one can draw many geological cross sections across the caldera. One of them, an east-west cross section passing Kamiyama, a central cone, is shown in Fig. 9, where the surface of the Tertiary basements are rather shallow within the caldera, as expected from the regional gravity anomaly. Also at many sites, the

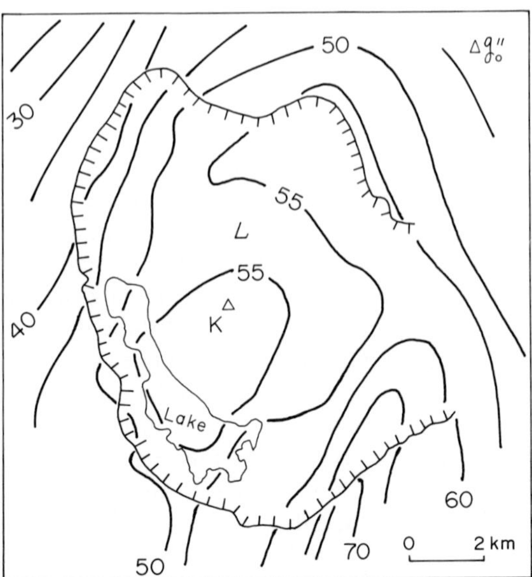

Fig. 7.   Bouguer gravity anomalies on Hakone caldera, compiled from YOKOYAMA and SAITO (1965), and HIRAGA *et al.* (1970). Unit is mgal and crust density is assumed as 2.4 g/cc.

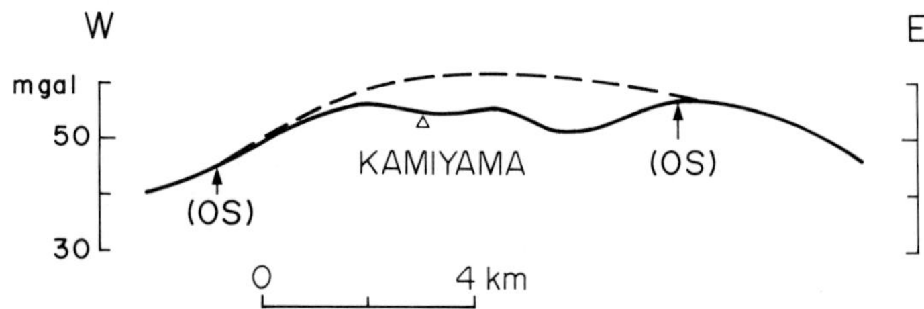

Fig. 8.   A gravity profile across Hakone caldera passing Kamiyama (a central cone) in the east-west direction. OS denotes the Old Somma.

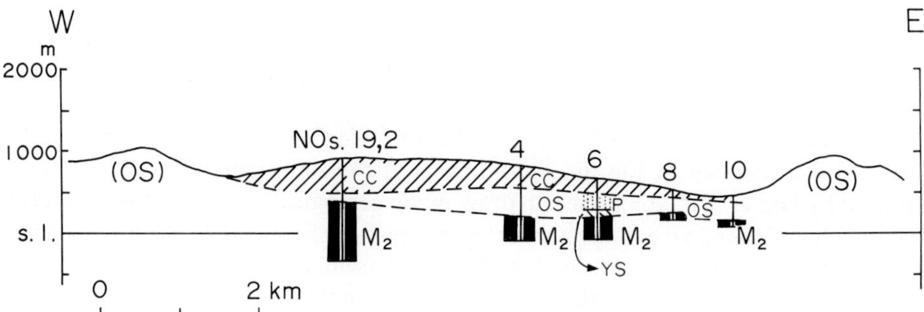

Fig. 9.   A schematic east-west cross section obtained from drilling results. M_2 denotes Yugashima group (older Miocene), YS Young Somma lavas, CC central cone lavas, and P pumice flow deposits.

Tertiary basements were covered directly by the central cone lavas, and among the caldera deposits, there remain a little quantity of pumice and no lake deposits which are common to many calderas. Before the post-caldera cones were formed, there might have been caldera deposits which consisted of the ejecta of the pumice eruption and the rocks of the old volcano, and are called the fall-back by the author. The author agrees with Kuno et al. (1970) in interpreting that much of the material within the caldera could have been removed by River Haya-kawa traversing the margin of the caldera.

Kuno et al. (1970) also discussed the above drilling data: the author wishes to comment on their interpretation. One of their text figures (Kuno et al., 1970, Fig. 3) is reproduced in Fig. 10 which represents a north-south cross section across the caldera. In this cross section, the Tertiary basements dip towards the center of the caldera: the problem is what is the cause of the depression. Kuno et al. (1970) suppose the down collapse of the basements without evidence. An alternative cause is the upward evisceration of the basements. Further, they assume that the original topography of the precaldera volcano is a profile obtained by extending the outward slope of the Old Somma toward the center of the caldera as shown by broken lines in Fig. 10. However, this is meaningless because we have no evidence for this assumption. Therefore, the amounts of subsidence of the basements estimated by referring to the above

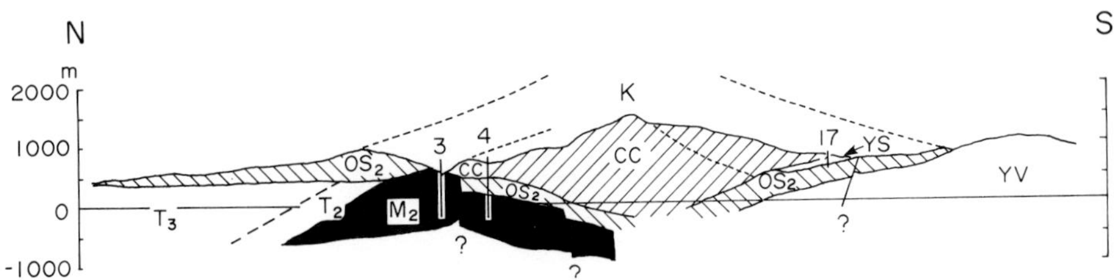

Fig. 10.　A north-south cross section surmised from drilling results by KUNO *et al.* (1970). T₂ and T₃ denote Tertiary rocks, and YV Pleistocene rocks. The other symbols are the same as in Fig. 9.

extrapolated profiles (KUNO *et al.*, 1970, Fig. 4 and Table 2) are not definitive.

In Fig. 10, KUNO *et al.* (1970) assume questionable faults at the Tertiary basements and thereafter, they proceed to assume the stepwise collapse towards the center of the caldera (KUNO *et al.*, 1970, Fig. 5). They presume a narrow central vent of about 2 km in diameter at the middle of the caldera, which has not been confirmed by the drillings. According to them, the pre-caldera volcano should have collapsed into the magma reservoir to compensate the void caused by ejection of voluminous pumice and ash. If they adopt their structure model, such collapses can scarcely have occurred.

5.　Aso Caldera

Aso caldera in Kyushu is comparable to Kuttyaro caldera in diameter. This caldera was formed 20,000 ~ 36,000 years B.P. There are several post-caldera cones within the caldera, and one of them, Nakadake has active craters.

The author (YOKOYAMA, 1963) already published a note on the gravity anomalies at this caldera observed by TSUBOI *et al.* (1956), mentioning the following prominent features:

1)　Low gravity anomalies amounting to about 30 mgal are concentric with the center of the caldera and indicate the existence of coarse material to a depth of about 3 km beneath the caldera.

2)　From the distribution of the gravity anomaly shown in Fig. 11, it is deduced that the caldera boundaries at the rim are not vertical faults but incline aslant towards the center of the caldera.

3)　In order to interpret the transportation of the pyroclastic flows as far as 60 km from the caldera, one must suppose the existence of some extraordinary and, in some cases, lateral explosive energy.

These have been proved to hold good for the before-mentioned three calderas.

After TSUBOI *et al.* (1956), KUBOTERA *et al.* (1969) carried out more detailed gravity surveys on the Aso volcanic region and obtained the results shown in Fig. 12.

To date, two boreholes have been drilled just inside the northern caldera wall. The drilling sites are indicated by letters A and B in Fig. 12. A is 170 m deep with a granite basement below a depth of 154 m while B is 600 m deep with a granite basement below

 I. YOKOYAMA

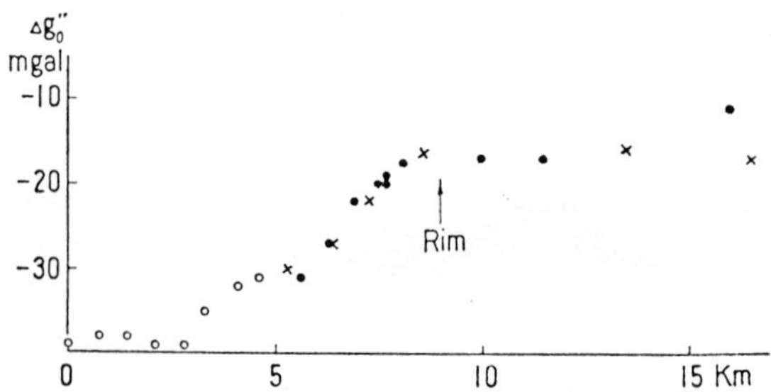

Fig. 11. Gravity profiles on Aso caldera. The abscissa is radial distance from the center of the caldera. Hollow circles: central route to the craters, Solid circles: western route, Crosses: eastern route.

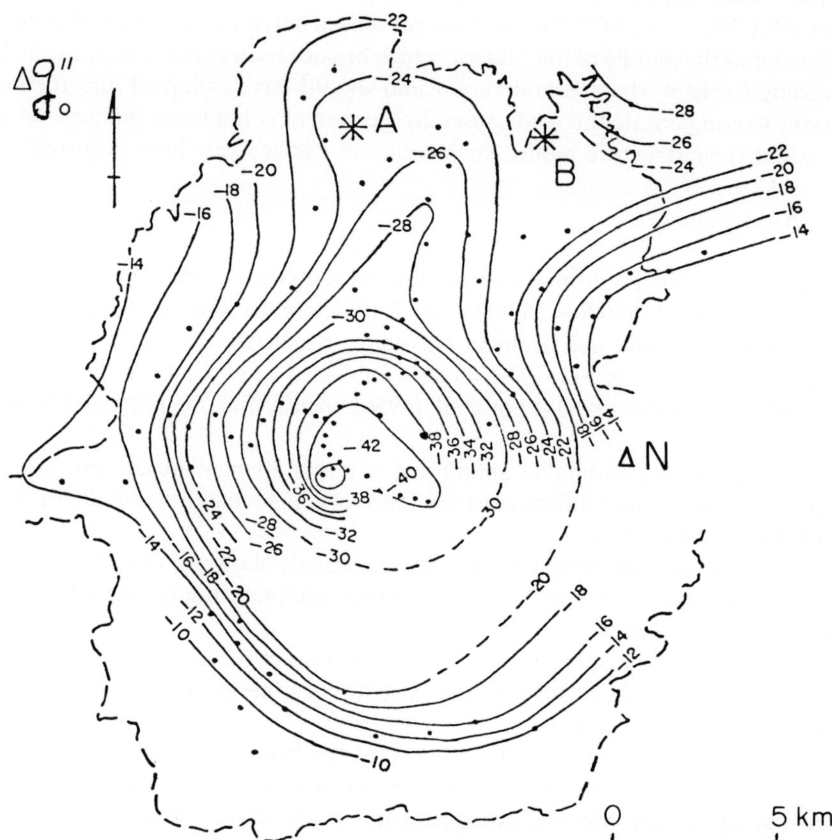

Fig. 12. Bouguer gravity anomalies on Aso caldera (unit: mgal) after KUBOTERA *et al.* (1969). A and B denote the drilling sites, and N Neko-dake (a pre-caldera cone).

a depth of 480 m; the upper surface connecting both the granite is almost horizontal. These show that the basements just inside the caldera are rather shallow and agree with the result deduced from the gravity anomalies that the caldera rim is not a steep fault, and also supports that the basements beneath the caldera are of funnel-shape.

As for topographies of caldera rims, the present author has supposed that these are not so significant tectonic structure and some of them have been expanded by erosions. This is verified at this caldera by the following fact: Neko-dake at the east rim, shown in Fig. 12, was thought to be a post-caldera cone until recently because it gives the impression that it is situated inside the caldera. In fact, it is a pre-caldera volcano and is a result of selective erosion worked at the foot of the volcano. Judging from the shape of the rim, the original Aso caldera may have been formed by plural eruption centers.

6.   Discussion and Conclusion

In the foregoing, the gravimetric estimation of the caldera-filling material and the configurations of the basements beneath the four calderas in Japan are verified by the drilling results though the drilled holes are limited in number and depth.

On the other hand, many geophysical, geological and geochemical data on caldera structures in California and New Mexico, U.S.A. have been accumulated. Apart from the conventional classification of calderas, the investigations on Long Valley caldera, California, will be referred to, which is geologically classified as a resurgent caldera. To the present author, such genetic classification of calderas seems too early before their subsurface structures are clarified quantitatively and to sufficient depths.

Long Valley caldera is an elliptical depression, measuring about 32 km east-west and about 17 km north-south, and the major caldera-forming eruption occurred 0.7 Myr B.P. On the caldera, the advanced reconnaissances have been carried out by various methods including deep drillings. KANE *et al.* (1976) made a gravity study on the caldera and found low residual Bouguer gravity anomalies amounting to 40 mgal at the maximum, and their contours being approximately parallel to the caldera rim. They interpret the anomalies as originating from one or two circular discs with density contrast 0.45 g/cc. HILL (1976) obtained P wave velocity structure under the caldera by a seismic refraction experiment. The basement depth beneath the caldera floor was estimated at $2 \sim 4$ km. He tentatively identified later P wave arrivals as a reflection from a low-velocity horizon at a depth of $7 \sim 8$ km beneath the western section of the caldera, and related this horizon with the roof of the magma chamber in the form of a trapped melt or a crystallized lens of low-velocity residual material.

Inside Long Valley caldera, two deep wells were drilled by Union Oil Co. Both the wells penetrated Bishop tuff of about 1,000 m thick, and one deep well penetrated metasediments from 1,378 to 1,605 m while the other apparently penetrated a shallow silicic intrusive body from 1,420 to 1,846 m (after GAMBILL, 1981). From the results of the above-mentioned investigations on Long Valley caldera, it may be said that this caldera has similar subsurface structure to the four Japanese calderas so far as the upper structures are concerned. The differences, if any, among these calderas in the deeper structure should contribute to the discussion on the origin of various calderas.

Since Yokoyama (1966) proposed an explosion hypothesis for caldera formation on the basis of subsurface structure of several calderas of low gravity anomaly type, a few geologists have contended the hypothesis.

Kuno *et al.* (1970) denied the author's supposition that caldera deposits consist of fall-backs of pumice ejecta, referring to the cores drilled at Hakone caldera. The author answers that much of the original caldera deposits of low density within the caldera have been removed by River Haya-kawa as mentioned by Kuno *et al.* (1970, p. 722) and replaced by the central cones. Each caldera has its own tectonic condition: at Aso caldera, still abundant caldera deposits remain in spite of river erosions and formation of many central cones.

Williams and McBirney (1979, p. 222) contended that the author's conclusions were inconsistent with the relative scarcity of pre-volcanic and other lithic ejecta observed around calderas and ring complex. As for the former objection, the author proposes that some parts of lithic fragments are to be found among caldera deposits within calderas as seen at Kuttyaro and Nigorikawa calderas. As for the latter objection, the author has found that caldera walls are sometimes false and resulted from erosion as seen at Aso caldera, and supposes that the dip-angles of caldera walls would depend on explosivity, depth of explosion centers and tectonic conditions.

Williams (1941) proposed that calderas of Krakatau type were formed by collapses of pre-caldera volcanoes into magma reservoirs. Does he think of a nest of ant lions as a possible subsurface structure of calderas of the type? The author doubts that the fall-backs can drop down to underlying magma reservoirs through narrow vents.

Recently, Yokoyama (1981) discussed the subsurface structure of Krakatau islands and also interpreted its 1883 eruption. Self and Rampino (1982) contended some points in this paper, and Yokoyama (1982) replied to them: Self and *Rampino* lay stress on scarcity of lithic fragments among the 1883 eruption ejecta on Krakatau islands; Yokoyama doubts the surface observations of the ejecta at calderas on the evidence of the drilling results at Kuttyaro caldera. Self and Rampino support the idea of Williams (1941) that the collapse at Krakatau was along two intersecting submarine grabens; Yokoyama supposes that the grabens are not clear in bathymetric charts and that the main motivity for the caldera formation is violent explosions which produced a gravity low.

In conclusion, it may be said that the results of the drillings carried out at the four calderas do not contradict the geophysical interpretations of the gravity anomalies observed there, but may refute the conventional idea of caldera formation:

1) The caldera walls are not steep faults. Some of them have been expanded outward by erosion over long ages.

2) Beneath calderas, there are deposits of coarse material which are composed of fall-backs of both juvenile ejecta and old fragments. The depths of these deposits usually range from 1 to 4 km.

3) The basements beneath calderas are funnel-shaped, and not piecemeal nor chaotic. The lithic fragments could scarcely collapse down to the deeper magma reservoir through narrow vents at the middle of calderas. We are inevitably led to suppose that the lithic fragments should be found partly among the caldera deposits and partly among the widely spread ejecta outside the calderas.

REFERENCES

ANDO, S., An example of the structure of Crater Lake-type caldera—Nigorikawa cladera, southwest Hokkaido, Japan, in *Abstracts, 1981 IAVCEI Symposium—Arc Volcanism—*, pp. 9–10, 1981.

GAMBILL, D. T., Preliminary HOT DRY ROCK geothermal evaluation of Long Valley Caldera, California, in *Hot Dry Rock Site Selection Report*, No. 1, pp. 1–22, Los Alamos National Laboratory, 1981.

HILL, D. P., Structure of Long Valley caldera, California, from a seismic refraction experiment, *J. Geophys. Res.*, **81**, 745–753, 1976.

HIRAGA, S., H. TAJIMA, S. HIRATA, and M. KASAI, Gravity survey in Hakone Volcano, Part I, *Bull. Hot Spring Res. Inst. Kanagawa Pref.*, 11, 33–38, 1970 (in Japanese).

KANE, M. F., D. R. MABEY, and R. L. BRACE, A gravity and magnetic investigation of the Long Valley caldera, Mono County, California, *J. Geophys. Res.*, **81**, 754–762, 1976.

KUBOTERA, A., H. TAJIMA, N. SUMITOMO, H. DOI and S. IZUTUYA, Gravity survey on Aso and Kuju volcanic region, Kyushu District, Japan, *Bull. Earthq. Res. Inst.*, **47**, 215–255, 1969.

KUNO. H., *Explanatory Textbook for Geologic Map of Atami*, Geological Survey of Japan, 1952.

KUNO, H., Y. OKI, K. OGINO, and S. HIROTA, Structure of Hakone caldera as revealed by drilling, *Bull. Volcanol.*, **34**, 713–725, 1970.

NISHIDA, Y. and I. YOKOYAMA, A note on physical properties of boring cores dug at Kuttyaro Caldera, Hokkaido, *Geophys. Bull. Hokkaido Univ.*, **14**, 53–58, 1965 (in Japanese).

ŌKI, Y., K. OGINO, T. OHGUCHI, S. HIROTA, T. HIRANO, and M. MORIYA, Geological investigation of a drill hole at Moto-Hakone and its relationship to the hydrothermal system of Hakone Volcano, *Bull. Hot Spring Res. Inst. Kanagawa Pref.*, **6**, 21–34, 1968 (in Japanese).

SELF, S. and M. R. RAMPINO, Comments on "A geophysical interpretation of the 1883 Krakatau eruption" by I. Yokoyama, *J. Volcanol. Geotherm. Res.*, **13**, 379–383, 1982.

TSUBOI, C., A. JITSUKAWA and H. TAJIMA, Gravity survey along the lines of precise levels throughout Japan by means of a Worden gravimeter, Part IX, Kyushu District, *Bull. Earthq. Res. Inst.*, Suppl. Vol. IV, Part VIII, 1956.

WILLIAMS, H., Calderas and their origin, *Bull. Dept. Geol. Sci., Univ. Calif.*, **25**, 238–346, 1941.

WILLIAMS, H. and A. R. MCBIRNEY, *Volcanology*, Freeman, Cooper & Co., San Francisco, 1979.

YOKOYAMA, I., Gravity survey on Kuttyaro Caldera Lake, *J. Phys. Earth.*, **6**, 75–79, 1958.

YOKOYAMA, I., Gravity anomaly on the Aso Caldera, in *Geophys. Paper Dedicated to Prof. K. Sassa*, pp. 687–692, 1963.

YOKOYAMA, I., Crustal structures that produce eruptions of welded tuff and formation of calderas, *Bull. Volcanol.*, **29**, 51–60, 1966.

YOKOYAMA, I., The subsurface structure of Ooshima Volcano, Izu, *J. Phys. Earth*, **17**, 55–68, 1960a.

YOKOYAMA, I., Gravimetric studies and test drillings at three calderas in Japan, Atti della Associazione Geofisica Italiana, XVIII Convegno, 1–13, 1960b.

YOKOYAMA, I. and T. SAITO, Preliminary report on a gravimetric survey on Volcano Hakone, Japan, *J. Fac. Sci., Hokkaido Univ., Ser. VII*, **2**, 239–245, 1965.

YOKOYAMA, I. and S. SUPARTO, Volcanological survey of Indonesian volcanoes, Part 5, A gravity survey on and around Batur caldera, Bali, *Bull. Earthq. Res. Inst.*, **48**, 317–329, 1970.

YOKOYAMA, I. and H. TAJIMA, A gravity survey on Volcano Mihara, Ooshima Island by means of a Worden gravimeter, *Bull. Earthq. Res. Inst.*, **35**, 23–33, 1957.

YOKOYAMA, I., A geophysical interpretation of the 1883 Krakatau eruption, *J. Volcanol. Geotherm. Res.*, **9**, 359–378, 1981.

YOKOYAMA, I., Author's reply to the comments by S. Self and M. R. Rampino, *J. Volcanol. Geotherm. Res.*, **13**, 384–386, 1982.

Arc Volcanism: Physics and Tectonics, edited by D. Shimozuru and I. Yokoyama, 43–61.
Copyright © 1983 by Terra Scientific Publishing Company (TERRAPUB), Tokyo.

Comparative Study of Earthquake Swarms Associated with Major Volcanic Activities

Hm. Okada

*Usu Volcano Observatory, Hokkaido University, Sohbetsu,
Hokkaido 052–01, Japan*

Comparative study of earthquake swarms associated with major volcanic activities allows us to summarize the dynamical processes ranging a co-seismic volumetrical deformation to a co-seismic sector collapse of a volcano. The magnitude scale is the most basic parameter to describe the time-space development of a volcanic earthquake swarm. Once this scale is settled, various swarms can be compared with each other on the homogeneous basis, in order to derive, if possible, the empirical generalization of the subsurface processes beneath the volcano.

A magnitude 5 event triggered a cataclysmic sector collapse of St. Helens on May 18, 1980. Same at Bandai-san in 1888. The time sequence from earthquake to collapse is well witnessed in both cases. Similar sector collapse accompanied with M 5 earthquake occurred also at Bezymianny in 1956, at Sheveluch in 1964, and at Unzen in 1792. Thus, it may be concluded that the well established sector collapses are co-seismic and its magnitude is around 5.

The similarity between the premonitory seismic activities of St. Helens and Bezymianny is particularly noteworthy. The magnitude-frequency distribution in these cases are non-loglinear and their temporal variations are evident. Such a non-loglinear M-n relation has been accumulated in the past volcanic earthquake swarms and following conclusions are obtained.

First, there tend to exist a maximum limit in magnitude at around 5, except those in which regional tectonics is required for their direct causes. This can be interpreted as a consequence of a limited size of a volcanic edifice or a limited localized volcanism, and is consistent with the general relation between earthquake magnitude and fault length (linear dimension).

An earthquake family, earthquakes having similar wave-form, is another important feature for the non-loglinear M-n relation. The large earthquakes which form a peak in M-n distribution are observed in the caldera collapse at Fernandina in 1968 and in the recent doming of Usu. Detailed studies of Usu revealed that this peak corresponds to one or two major earthquake families, originating from repeated stick-slip motions. The similarity of seismograms of magnitude 5 events at Fernadina had been pointed out and is interpreted by the episodic repeated subsidence along the caldera wall.

The co-seismic deformation directly observed at Usu (and probably at St. Helens) are the recent topics in modern instrumental observation. A large amount of episodic deformation is always co-seismic. The accumlation of these co-seismic deformation resulted in a bulging, a doming or a caldera collapse, indicating the long-term stability of a system under the magma pressure.

The scaling of magnitude has some problems. The anomalous M_s-m_b relationship is found for some of the volcanic M 5 earthquakes. For example, the May 18 event at St. Helens has $M_s = 5.2$ and $m_b = 4.7$, suggesting considerable excess energy in low-frequency compared with ordinal earthquakes. This is just the opposite case to the low-frequency deficiency phenomena for nuclear explosions.

1. Introduction

Detailed seismometrical and geodetical data accumulation of the recent major volcanic activities brought a new scope toward physical understandings of the dynamical processes of certain volcanic activities. Sufficiently similar earthquake swarm activity prior to the cataclysmic eruptions of St. Helens in 1980 and Bezymianny in 1956 strongly suggests the possibility that the physical process involved is a predictable one (Geophysical Program Univ. Wash., 1980).

Fortunately for the people, but unfortunately for volcanologists' experiences, major volcanic activities occurred less frequently and often they take place in the remote unaccessible region mostly due to severe natural environment. However, we have now nearly 100 years of seismometrical observations since the world first *in situ* observation in 1888 on the summit area of Bandai, Japan by Sekiya and Kikuchi (1890). The 100 years experiences may not be sufficient to extract the decisive conclusion on physical processes of the major activities. Those observations are often incomplete or not accurate in early days. Recent modernized instrumentation allowed us to obtain detailed and precise data, which often emphasize the peculiarity of their specific phenomena rather than to correlate them to some generalized physical processes.

In this paper, we would like to summarize and to discuss certain regularities in the time-space development of a volcanic earthquake swarm associated with major volcanic activities. The magnitude scale is the most basic parameter for comparing the various swarms on the homogeneous basis. We do not attempt to review the complete activities. Special emphasis will be made on the developement of an earthquake family with regard to the deformation dynamics.

2. Earthquake Family

In a text book titled "*Manual of Seismology*," Davison (1921) had proposed a classification of volcanic earthquake origin; (i) new fractures, (ii) explosions, (iii) sudden injection of lava, and (iv) relative displacement of rock masses. This classification was derived from the relatively short-term observational history in seismology, but it accounts, as he mentioned, for all the known phenomena of volcanic earthquakes; the shallowness and small size of the foci, frequent repetition in the same region, the intensity and brevity of the shock, the swarm type occurrence, and limited deformation zone. Recent development in researches on volcanic tremor (Aki *et al.*, 1977; Chouet, 1981; Aki and Koyanagi, 1981) may indicate the theoretical and quantitative approach to Davison's third class events. Davison thought the fourth class is more important for explaining the frequent repetition of large volcanic earthquakes in the same region. Such repeated displacements can be highly possible, if the rock masses were not destroyed completely but were able to withstand the repeated slips for a considerable length of time responding to the shallow underground pressure system. This situation is what OKADA *et al.* (1981) favoured for interpreting the intensive earthquake swarm associated with the doming deformation at Usu volcano in terms of stick-slip producing earthquake families.

An earthquake family is a group of earthquakes having a similar wave form at fixed seismic stations. It has been occasionally experienced in the past observations, and in most cases, a set of a few earthquakes were reported accidentally by the seismogram examiners. HAMAGUCHI and HASEGAWA (1975) examined systematically more than 50,000 aftershocks of the great Tokachi-oki earthquake of 1968 and found that 72 earthquakes over 80 days have a similar wave-form which were located in a limited zone with radius of less than 5 km. This kind of earthquake group was named by them "earthquake family" considering the common spacial and dynamical nature of their sources.

One of the first significant examples of an earthquake family associated with volcanic activity can be traced back to the report of one previous dome formation activity of Usu volcano (MINAKAMI *et al.*, 1951). A group of earthquakes which were named by Minakami "C-type earthquakes," are resembling one another in full detail as if they were copies of the same earthquake. They had occurred only during the stage of lava dome extrusion (Octover 1944–September 1945). They suggested such earthquakes are located about 500 m beneath the extruding lava dome and have a special dynamical source mechanism which can be easily repeated. We have no evidence of earthquake family associated with the 1910 activity (Meiji-shinzan formation) of Usu volcano.

Much knowledge on an earthquake family was obtained from the recent observation of Usu volcano (WANO and OKADA, 1980; MIZUKOSHI and MORIYA, 1980; OKADA *et al.*, 1981; HARADA, 1981; UMEHARA and OKADA, 1981). The followings are their major conclusions:

(1)   Limited number of earthquake families. About 80% of the 2,300 large volcanic earthquakes with magnitude greater than 3.5 can be classified into only four families. The largest earthquakes form one particular family. Considerable portion of smaller events can be still classified into a limited number of families. Each family shows significant rise and fall in their activity which is corresponding to that of the doming deformation and the eruptions.

(2)   Source identity for a family. The hypocentral identity for each earthquake family is well established within the accuracy of locating resolution (less than 100 m). Hence, it is concluded that each family is originated from the episodic stick-slip at the same fault. An identical source mechanism for a family is also consistent with this conclusion.

(3)   Optimum magnitude for a family. Earthquakes of an any particular family tend to exist over a narrow range of their size distribution. The range of optimum magnitude for a family may vary considerably with time.

(4)   Scaling law for a family. The wave form for a family earthquakes varies only

in the amplitude and not in the spectral shape. This scaling can be interpreted by the variation of the amount of displacement at the identical stick-slip plane.

3.   Intrinsic Non-Loglinear M-n Distribution

One of the most fundamental law in seismology is a statistical log-linear relationship between the earthquake frequency $n(M)$ and magnitude M

$$\log n(M) = a - bM \tag{1}$$

where a and b are constants (GUTENBERG and RICHTER, 1944). Suppose any seismic activity which satisfies the relation (1), then the temporal seismicity could be completely described by an earthquake frequency curve (such as hourly frequency) and the constant b, and no more specific phenomena could be expected other than those caused by the statistical fluctuation. This is not attractive statistics. Seismologists are often making an effort evaluating the variation of b-value in time and space. Some physical basis explaining the difference in b-value are proposed by MOGI (1967) and SCHOLZ (1968). Diagram of the cummulative frequency $N(M)$ (number of earthquake with M or larger) is more commonly used. The relation (1) is equivalent to formula

$$\log N(M) = a' - bM \tag{2}$$

where

$$a' = a + \log (b \ln 10) \tag{3}$$

So, we can evaluate same b-value from both diagrams. However, it must be emphasized that it is valid only under the condition, if formula (1) hold completely.

Less frequent attention has been focused on the observational evidences of the intrinsic non-loglinear relationship between $n(M)$ and M. Okada et al. (1981) pointed out that there exist some cases which are characterized by the intrinsic lack of earthquakes over a particular magnitude range. Such non-loglinearity can be hardly distingushed from the $N(M)$ diagram, because $N(M)$ curve shows never a peak or a trough but forms only a small bend. For example, in the case of Fernadina volcano (Galapagos) in 1968, FILSON et al. (1973) evaluated 'b-value' for two magnitude ranges: $b = 0.68$ for $M_s = 3.3-4.5$, $b = 1.91$ for $M_s = 4.6-5.1$. Similar values were obtained by FRANCIS (1974) for the same swarm: $b = 0.67$ for $M_s = 3.5-4.4$, $b = 1.78$ for $M_s = 4.5-5.2$. They plotted the cummulative frequency $N(M)$ instead of $n(M)$, and $N(M)$ distribution apparently shows two log-linear lines. So, they got two 'b-values.' But, as shown in Fig. 1, the $n(M)$ distribution is characterized by the intrinsic non-loglinear peak-trough structure (OKADA et al., 1981). In both papers, they actually realized that the large earthquakes formed one particular group and were very active only during the period when the subsidence of caldera floor had been extensively in progress. Similar peak-trough $n(M)$ structure can be found also during the recent doming activity of Usu volcano (Fig. 1, OKADA et al., 1981). In the above two cases, Fernandina and Usu, the peak-forming large earthquakes are characterized by the similar wave form, the major earthquake family, and are in close connection with a large amount of surface deformation, though the sense of deformation is opposite.

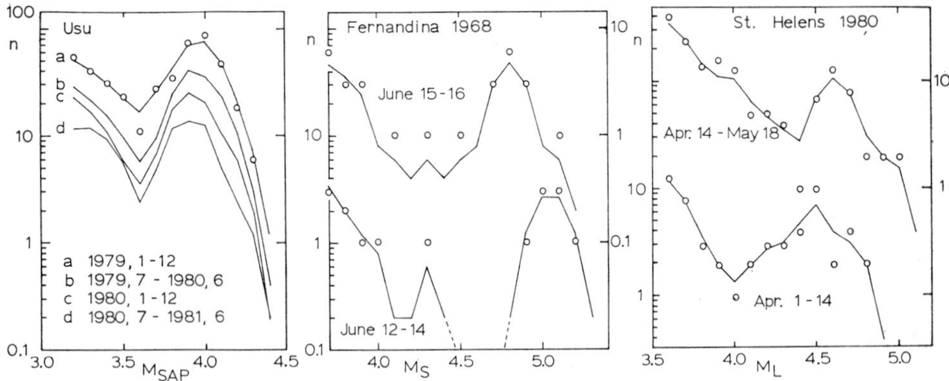

Fig. 1.   Typical examples of non-loglinear magnitude-frequency relation. Persistent occurrence of large volcanic earthquakes forms a distinct peak and may result a large amount of surface deformation in a form of doming (Usu), cardela collapse (Fernandina), and bulging (St. Helens). Open circles indicate the raw data and solid lines indicate the smoothed distribution by using the formula $n(M) = 0.2*(n(M - 0.1) + 3*n(M) + n(M + 0.1))$.

More recently, we have experienced another example of the intrinsic non-loglinear $n(M)$ relation. During the bulge forming stage before the cataclysmic eruption of St. Helens on May 18, 1980, large earthquakes with $M = 4$–5 continuously occurred for two months, while the activity of smaller earthquakes decreased drastically (Geophysical Program Univ. Wash., 1980). Similar situation had occurred also during the six months bulge forming stage before the similar cataclysmic eruption of Bezymianny volcano (Kamchatka) on March 30, 1956 (Gorshkov, 1957, 1959). Gorshkov showed the famous figure (probably very famous after the experience of St. Helens) illustrating the nearly constant seismic energy discharge inspite of the significant decrease of small earthquake frequency. These two examples should not be interpreted by the temporal variation of b-value, but by the development of peak-trough structure in $n(M)$ distribution. Those special large earthquake groups are probably forming the major earthquake families and are closely associated with the large amount of bulging deformation. This idea is strongly supported by the Gorshkov's observation that all large earthquakes occurred repeatedly with the similar wave form with the minutest details and were originated from the same source and cause (p. 102 of Gorshkov, 1959), which is exactly the same situation we found at Fernandina and Usu.

The magnitude versus cummlative frequency diagrams were constructed for the all past activities of Usu volcano (Fig. 2). Data of 1977–1980 activity are taken from the Seismological and Volcanological Bulletin of Hokkaido (Sapporo Meteorological Observatory, S.M.O., 1977–1980). Data for 1943–1945 activity are recently evaluated by Japan Meteorological Agency (1980) and kindly supplemented by Dr. Seino of S.M.O. Magnitude data for 1910 activity are newly evaluated by using the table presumably compiled by Omori (Earthquake Investigation Committee, 1913).

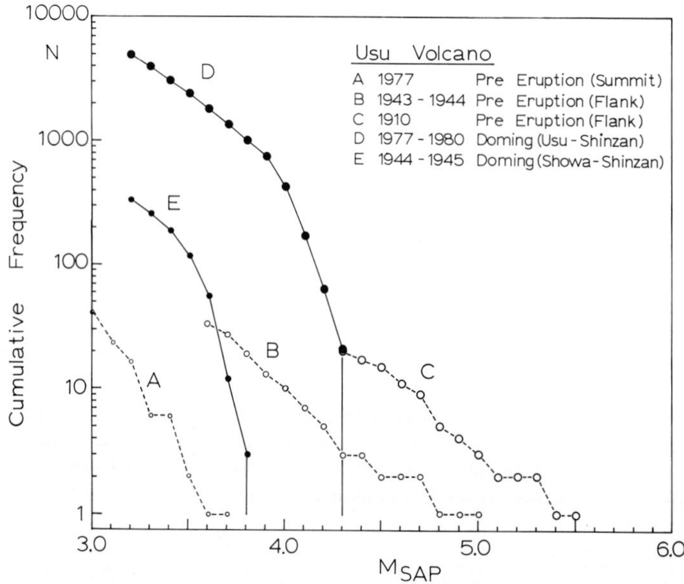

Fig. 2. Magnitude-cummulative frequency relation for the past three activities of Usu volcano. We were luckily able to utilize the extremely homogeneous magnitude scale M_{SAP} which is determined from the seismometrical observations at Sapporo about 70 km north of the volcano. Doming earthquakes (D to E) are characterized apparently by the steeply fall off type non-loglinearity, while pre-eruption earthquakes (A to C) indicate roughly log-linearity.

Very luckily, we can use the extremely homogeneous magnitude scale for these past activities, since all large earthquakes during the past three activities were registered by the seismometers at Sapporo (SAP) about 70 km from the volcano. The type of seismographs are Omori tromometer with magnification $V = 30$, Wiechert with $V = 60-80$ and JMA-59-type with $V = 100$, for 1910, 1943–1945 and 1977–1980, respectively. The magnitude M_{SAP} adopted here is determined from the single station data at SAP by using Tsuboi's formula (TSUBOI, 1954).

At a glance of this simple diagram, two different earthquake groups are evident; pre-eruption earthquakes (A, B, and C) and doming earthquakes (D and E). Distribution of pre-eruption earthquakes (A–C) shows roughly loglinear relationship, though the samples are not so numerous. For the case C, 647 pre-eruption earthquakes were felt at Date city (former Nishi-Monbetsu) about 8 km SSE of the volcano (OMORI, 1911). The magnitude of these felt earthquakes are roughly $M \geq 3$, so this assures the extension of loglinear relationship down to M 3. The slopes of B and C for pre-flank-eruption stage show the normal value, while for A, pre-summit-eruption stage b-value is a little higher. The distributions D and E show extremely steep bend at $M = 4.0-4.3$ and $M = 3.5-3.8$, respectively. For case D, about 1,000 events have their magnitude larger or equal to 3.8. In such case, Gutenberg-Richter's formula usually predicts an earthquake with M as large as 6.5 and several shocks with $M \geq 6.0$, which were strong enough to cause severe damages. But, it was not the case. The 21 largest events have only M 4.3. For the case E, it is 3.8 instead of 5 or more. Now, we have an additional example of clear non-loglinearity which is associated with the earthquake family (C-type of Minakami) and large deformation (Showa-Shinzan formation).

Another important feature found in Fig. 2 is that the differences of the largest magnitude among the activities A-E. M 5.5 for C and M 5.0 for B are the largest. Both are in the stage of pre-flank-eruption, while that of the summit eruption A is characterized by the smallest magnitude 3.7. This may be interpreted as an indication of easier passage through the central vent than fracturing the flanks of a strato volcano.

In short conclusion of this section, we would like to propose a simple illustration "MT-diagram (Magnitude-time diagram)" for general comparison of the swarm development. FILSON $et\ al.$ (1973) adopted the similar illustration. Sample MT-diagrams are displayed for Usu and St. Helens in Fig. 3 and Fig. 4. The large earthquakes show a distinctive development with time, that can not be interpreted by the temporal variation of b-value, but it must account for the principal physical process directly.

4.   Coseismic Deformation

From the previous discussions on the persistent large volcanic earthquakes, it is clear that the localized large amount of deformation is always evident in those examples in a form of doming, bulging or caldera subsidence. Next, we would like to study the direct relation between those large earthquakes and deformations.

The most well established evidences were obtained from the continuous in-strumental observations of the recent Usu volcano. Nearly exact parallelism was observed in the rise and fall patterns of upheaval rate of new dome and the discharge rate of seismic energy (YOKOYAMA and SEINO, 1979). The short term correspondence was first obtained by HARADA $et\ al.$ (1979) from the 44-hours semi-continuous geodetic measurement in December 1977. They found the episodic doming deformation of the north rim corresponding to the large earthquakes ($M = 3.5–4.0$). The amount of such episodic deformation is as large as 3–4 cm and flow type of deformation immediately followed it. OKADA $et\ al.$ (1981) supplemented the data for the later stage and confirmed that the doming deformation is episodic and always accompanied with large earthquakes. The amount of those deformation reached as large as 10–30 cm. The similar diagram for 1980 is shown in Fig. 5. Though the geodetical measurements are less frequent, it is still obvious that the large step-like deformations in HK-NR distances (mostly 10–30 cm) are associated always with the large earthquakes marked with solid triangles. According to UMEHARA and OKADA (1980), those events form one of the major earthquake families (KB-family), while other events with $M = 3.8–4.0$ form the other major family (OU-family). These two families are located at the corners of the U-shaped major faults indicating the major barriers which can be displaced repeatedly under the pressure of the intruding or expanding magma body (WANO and OKADA, 1980; OKADA $et\ al.$, 1981). Instrumental observation of 1944–1945 doming activity of Usu volcano also indicates the general correspondence between deformation and seismic energy (IDE and SEINO, 1980). Similar correspondence is also clear for

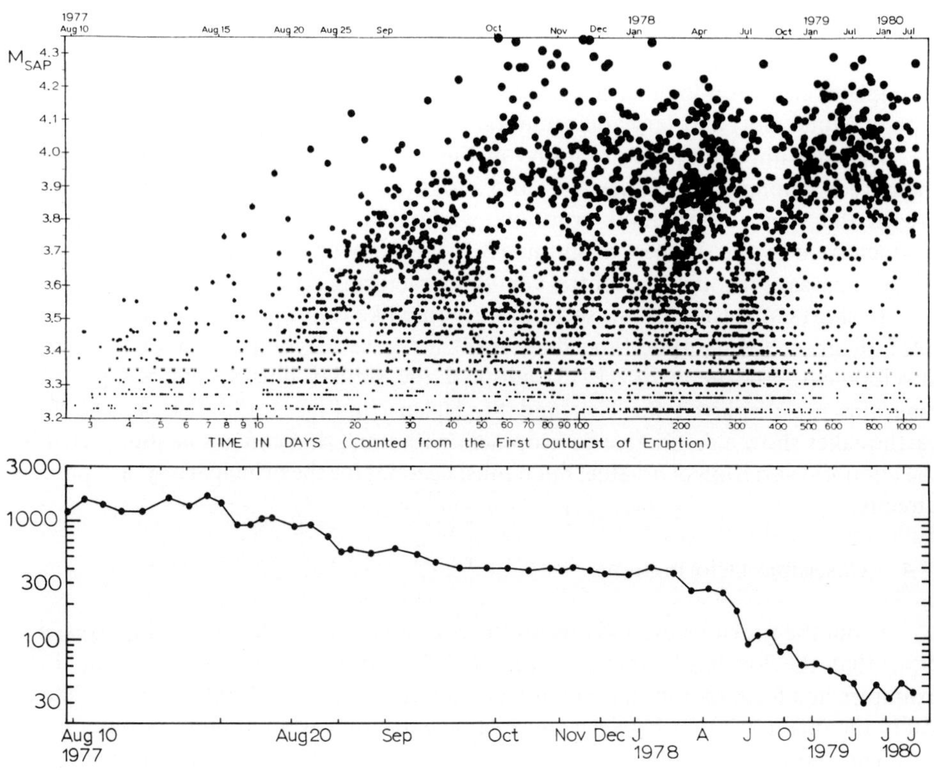

Fig. 3. MT-diagram (Magnitude-time diagram) of 4,700 large earthquakes ($M \geq$ 3.2) and daily frequency of earthquakes ($M \geq$ 0.5) at Usu volcano. The time scale in the middle abscissa indicates the logalithmic time interval in days counted from the first outburst of the eruption at 09:12, August 7, 1977. Note the significant structural change in MT-diagram corresponding to the frequency variation.

Fernandina (Filson *et al.*, 1973) and Bezymianny (Gorshkov, 1959).

In the case of St. Helens, Lipman *et al.* (1980a, b) made the precise geodetic measurements at 15 min intervals over 8 hour period in order to correlate the major earthquakes with bulging deformation. The data of those distance measurements between Coldwater 2 and Goat Saddle are illustrated in Fig. 6 together with the earthquake data (Endo *et al.*, 1980). One possible and our preferable interpretation is shown by the fine dotted line, indicating the episodic deformation ($\simeq$ 2 cm) corresponding to M $3\frac{1}{2}$ earthquakes. Continuous movement without any clear correlation with earthquakes may also possible (Lipman *et al.*, 1980a, b). In the latter case, the average deformation rate is 0.37 cm/day. The time interval of this observation does not cover the major earthquakes ($M = 4.5$–5) while the average deformation rate encompassing large events is as large as 1.5–2.5 cm/day (Lipman *et al.*, 1980b), so we can expect the large episodic deformation associated with major earthquakes which form a peak in the $n(M)$ distribution. Thus, the comparison between Usu and St. Helens strongly suggests a hypothesis that the localized large amount of deformation of

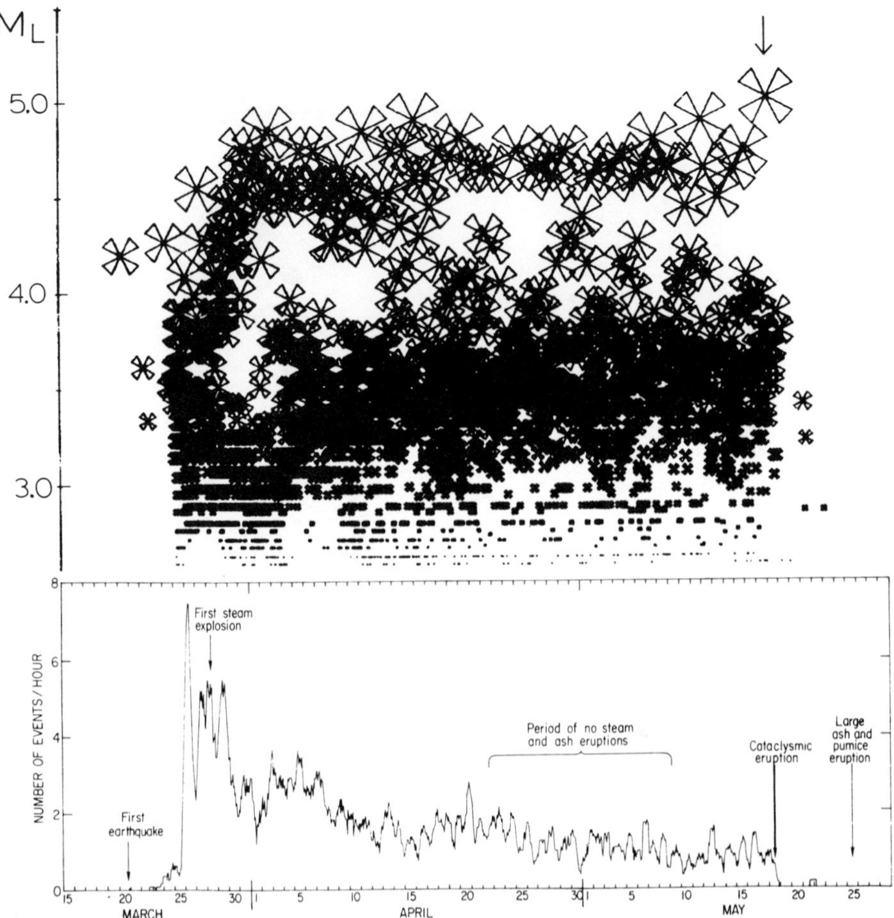

Fig. 4.   *MT*-diagram of 2,374 earthquakes (USGS Local Magnitude File, ENDO *et al.*, 1980) and hourly frequency of earthquakes (GEOPHYSICAL PROGRAM, UNIV. WASH., 1980). Note the significant structural development of the non-loglinear $n(M)$ distribution.

a volcano accompanied with the intensive earthquake swarm is generally coseismic and the major earthquake family may contribute to a large amount of deformation. Such earthquakes indicate the nearly identical stick-slip motion along the major barriers which resist the deformation until a certain level of stress build up. This system can be recycled as long as the mechanical and the energy system both remain functioning. For St. Helens and Bezymianny, the mechanical system gradually approached to its critical point that the gravitational unstability could not withstand the repeated slips and finally failed. In the case of recent Usu, the deformed region is still gravitationally stable, except for the tip of the northern summit crater rim.

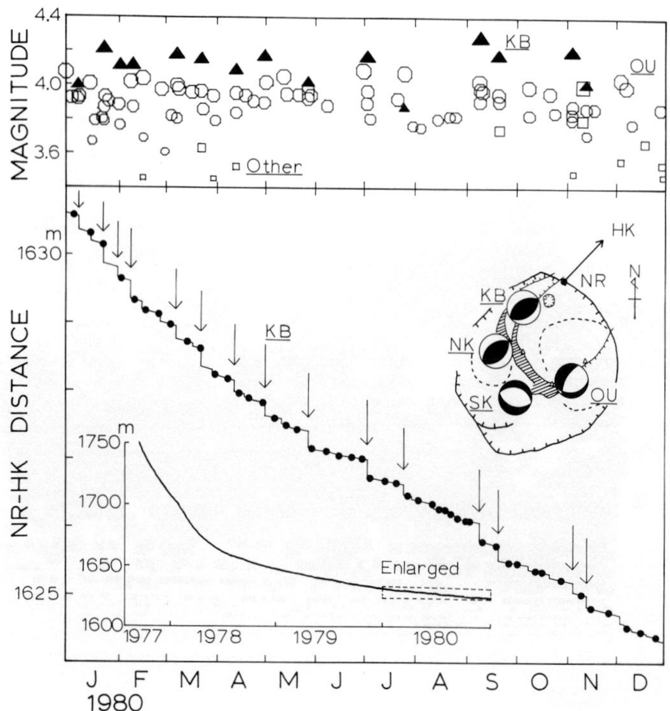

Fig. 5. Correspondence between doming deformation (data after Usu Volcano Observatory) and earthquakes at Usu volcano in 1980. Two major earthquake families are specified either by solid triangles (KB-family) or by open octagons (OU-family), and open squares indicate other earthquakes (Umehara and Okada, 1981). Though the distance measurements are less frequent, they show large amount of episodic deformation which is always corresponding to the KB-family as indicated by arrows. Inserted map shows the location and the source mechanism (upper hemisphere, solid part compression) for the major four families (Umehara and Okada, 1981).

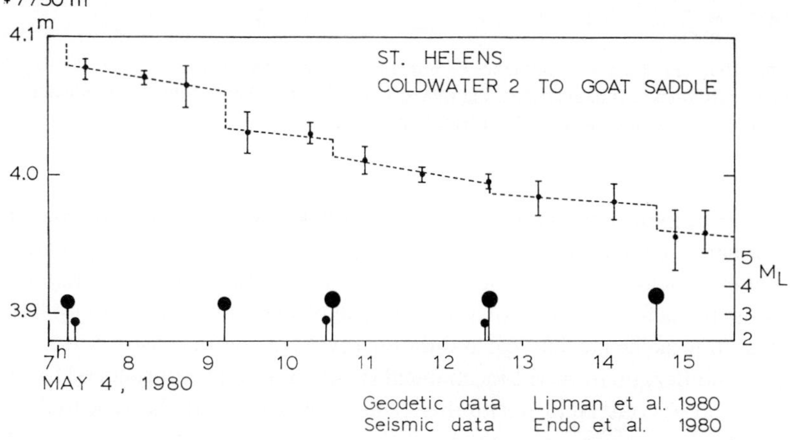

Fig. 6. Correspondence between bulging deformation and earthquakes at St. Helens.

5.   Coseismic Sector Collapse of a Volcano

A magnitude 5 earthquake triggered a cataclysmic sector collapse of St. Helens. The precise time schedule from the earthquake to the collapse is reconstructed from the various sources of information (CHRISTIANSEN, 1980; CHRISTIANSEN and PETERSON, 1980; GLICKEN *et al.*, 1981; MOOR, 1981).

A similar volcanic sector collapse immediately following the large volcanic earthquake occurred at Bandai, Japan on July 15, 1888. The time sequence of this case was well documented by SEKIYA and KIKUCHI (1980). A Buddhist priest, Mr. Tsurumaki, who was staying at the Nakanoyu Spa only about 100 m away from the collapsed amphitheater rim near the summit, escaped luckily from death and wrote to Sekiya his valuable but terrifying experiences. Mr. Tsurumaki and all the people rushed out from the hutte immediately after the severe earthquake and were wandering, looking around at what had happened. Then, in minutes (or seconds) a terrible explosion occurred. During the first 200–300 m of his emergency descent, Mr. Tsurumaki experienced another two explosions. From eye-witnesses in the neighbouring villages, 15 to 20 explosions, each lasting for a minute or more, were known to have occurred successively and the last explosion was directed almost horizontally. NAKAMURA (1978) interpreted that this last explosion marked the outbreak of the volcanic dry avalanche. From the above accounts, it is concluded that a large earthquake had occurred earlier than the collapse and can be regarded as a trigger event as in the more well established case of St. Helens.

SEKIYA and KIKUCHI (1890) sent inquiry letters to the neighbouring area and found that this trigger earthquake was felt in a limited area with a radius of about 48 km. This gives us a rough estimate of the magnitude of the earthquake at around 5, assuming the event is a typical low-frequency volcanic earthquake. This evaluation can be confirmed independently from the seismometrical measurements in Tokyo, about 220 km south of the volcano. Detailed seismological reports in this era were published by OMORI (1902a, b). The minimum reported amplitude was 0.2 mm for the observed peak-to-peak maximum ground motion on the Gray-Milne seismograph. This threshold level corresponds to about M 5.3, if we assume the events originated at Bandai. Much smaller events were also listed as small or very small events, that marks the threshold level at about 5.1. Hence, we may conclude that the magnitude of the major Bandai earthquake which triggered the eruption and the collapse is well below 5.5, or can be regarded as around 5.

In the above two cases at St. Helens and Bandai, a magnitude 5 earthquake triggered a cataclysmic sector collapse. Here, we face the question whether this correspondence is merely accidental or it is due to some inevitable physical consequence of the underground fracturing process. The final answer may not be given here, since no clear observation of the immediate distances are known. However, before experiencing another occurrence within the next few decades, we had better summarize whether the sector collapse is always or mostly accompanied with a magnitude 5 earthquake or not. In this respect, two volcanoes in Kamchatka, Bezymianny and Sheveluch, are important because in both cases a large volcanic earthquake occurred at about the time of the collapse.

CORSHKOV (1959) noted that at the time of the paroxysmal explosion of March 30, 1956 a rather strong earthquake took place. We could not find the magnitude of this earthquake in the literatures, but were able to find the amount of seismic energy discharge due to this event as 2×10^{19} erg (GORSHKOV, 1961), which is equivalent to M 5.0 using the general relation between seismic energy and magnitude. This shock was not felt at Kliuchi, about 43 km NEN of the volcano, suggesting its magnitude was not significantly greater than 5. Another independent estimation can be possible using teleseismic observations. In the seismological bulletin of Uppsala (Sweden) (BATH, 1959), readings of the 20 sec period surface waves are found for this earthquake; 0.9 μ for NS and 1.7 μ for Z-component. The M_s estimation is 5.0.

For the case of Sheveluch in 1964, the strongest earthquake took place at 07:07 on November 12, which was considered to be closely connected with the gigantic explosion and the dome destruction (GORSHKOV and DUBIK, 1969). TOKAREV (1967) suggested that this large event may only marked the initiation of weak explosions and partial destruction of lava plug (the first stage), and some time interval is necessary for getting much larger eruption and the major destruction of dome complex (the second stage). This seems to be in good agreement with regard to the time sequence of St. Helens and Bandai. The size of this earthquake is estimated as $M = 5.5$ (energy class $K = 12.5$) (FEDOTOV et al., 1967) or $m_b = 5.5$ (ISC). The shock was felt at Kliuchi ($\simeq 50$ km) and Kozyrevsk ($\simeq 80$ km) at intensity 3–4 (equivalent Japanese intensity $I_J = 1–2$) (GORSHKOV and DUBIK, 1969).

Three historical collapses among the Japanese volcanic activities were examined. No detailed information on earthquakes associated with the collapse was available for Oshima-Komagatake in 1640 and Oshima-Oshima in 1741. Oshima-Oshima, an uninhabited small island is situated about 55 km off southwestern Hokkaido, so it may be reasonable to understand that no reports on earthquakes are available in the case of M 5 or much smaller earthquakes. Many old documents exist for the case of Unzen, Kyushu in 1792 (KATAYAMA, 1974). Two strong earthquakes were felt at Shimabara in the evening (about 20 o'clock) of May 21. This did not give the local inhabitants specific alarm because of the frequent experiences of such large earthquakes in the preceeding months. It was followed, however, immediately by a great roar, like a thousand claps of thunder. Everything happened in complete darkness and they had to wait till morning to realize what they had actually gone through. Therefore, a precise time table of the collapse can not be constructed. The death toll counted more than 14,500, mostly drowned by the tsunami in Shimabara Bay (Ariake Sea). It is the most catastrophic disaster of all Japanese volcanic activities. The magnitude of the earthquake accompanied with the collapse may be roughly 5–5.5.

Tsunami is one of the main factors of disaster in the case of volcanic sector collapse. It is obvious that a "large" volcanic M 5 earthquake is not enough to produce tsunami. It is generally considered that such tsunami is caused by the sudden impact of a large amount of collapsed material (debris) rushing into the sea. The tsunami caused by the 1741 Oshima-Oshima eruption is the largest known tsunami in the Japan Sea. The tsunami energy is estimated by Hatori and KATAYAMA (1977) at roughly as 10^{23} erg. Such huge energy can not be generated by a sector collapse, because the potential energy of debris is too small (0.1 km^3 in volume, KATSUI et al., 1977). Hence,

Hatori and Katayama proposed the coincidental occurrence of a M 7.5 earthquake for the cause of the tsunami. For explaining the lack of earthquake reports in the old documents, they suggested that the event is a low frequency tsunamigenetic one.

A small tsunami but widely reported in the northern half of the Pacific was generated at the paroxismal explosion of Bezymianny on March 30, 1956 (IIDA *et al.*, 1967; SOLOVIEV, 1978). Probably this event is unfamiliar to most volcanologists (GORSHKOV, 1957, 1959; VLODAVETZ and PIP, 1959; SIMKIN *et al.*, 1981). If we regard this case as an earthquake generating tsunami, a great earthquake as large as M 8.0 would be expected from the tsunami amplitude at teleseismic distances (ABE, 1979), but as previously discussed, the largest earthquake at that time was certainly a M 5 event. The volcano is situated about 75 km from the Pacific coast, while even the blast area is limited to a distance less than 25–30 km. Therefore, the potential energy caused by the collapse is not responsible to the tsunami. But yet a possible explanation, we prefer, is the collapse judging the important role of the collapse dynamics including the directed blast. Both the collapse and the directed blast were directed toward the ESE direction, which is the direction of the Pacific Sea. Strong air waves were reported (GORSHKOV, 1959; MURAYAMA 1967). In the case of the 1883 Krakatau eruption, the tsunami at large distances is interpreted by the coupling of strong air waves (EWING and PRESS, 1953; HARKRIDER and PRESS, 1967).

The above case study of volcanic sector collapse, although the samples are still few, in number, suggests that the collapse is generally accompanied with a M 5–5.5 earthquake. No clear case exists without such an earthquake. The most well established cases, St. Helens and Bandai, suggest that the event is not a result of the collapse but a trigger of the collapse.

6.   Discussion: Volcanological Significance of Large Volcanic Earthquakes

In the previous sections, we have discussed volcanic earthquakes with special emphasis on magnitude, which we think is the most basic parameter to describe the time-space development of the volcanic earthquake swarm. A magnitude 5 earthquake seems to have special significance in some major volcanic activities. Some triggered a volcanic sector collapse. Some occurred at the time of the culminating stage of the swarm immediately before the major eruptions. Some continue to occur for a while resulting in a large amount of deformation.

The recent development of earthquake source studies indicates that the size of the fault from which seismic waves are radiated is the most important factor to the magnitude. In Fig. 7 linear dimension L (the fault length or the size of deformation area) is plotted against magnitude M. The data of shallow tectonic earthquakes with M_s 5.8 are taken from the table of earthquake fault parameters compiled by GELLER (1976). Large open circles and small plus marks indicate the events which occurred inland of Japan and other events, respectively. Two large squares indicate the rock-bursts (UTSU, 1969). Large solid circles indicate the largest magnitude for each volcanic earthquake swarm discussed here. The size of the bulging zone or that of the collapsed amphitheater are taken as a linear dimension for the cases of St. Helens, Sheveluch, Bezymianny, and Bandai. For Fernandina the size of the collapsed caldera, and for the

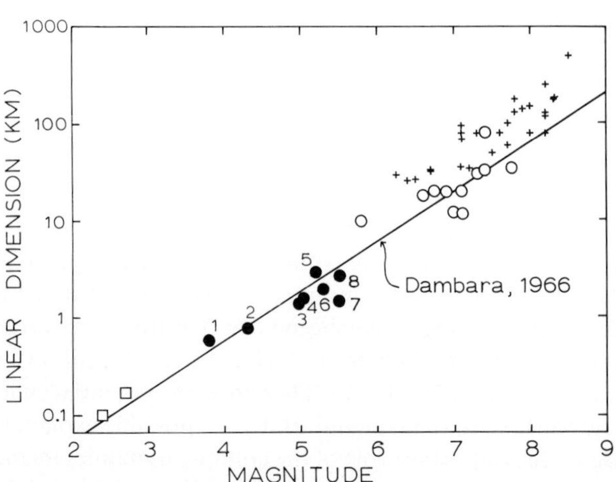

Fig. 7.   Linear dimension (L in km) versus magnitude M. Solid circles indicate volcanic swarms (No. 1 Usu 1977 Pre-eruption, No. 2 Usu 1977–1980 Doming, No. 3 Bandai 1888, No. 4 Bezymianny 1956, No. 5 Fernandina 1968, No. 6 St. Helens 1980, No. 7 Sheveluch 1964, No. 8 Usu 1910 Pre-eruption). open circles and plus marks indicate Japanese inland shallow tectonic events and other shallow tectonic events, respectively (Geller, 1976). Two open squares indicate rock burst (Utsu, 1969). The straight line indicates the relation log $L = 0.51$ $M - 2.27$ (Dambara, 1966). Most data except plus mark events can be well expressed by this formula.

past two domings of Usu, the dome size are taken as L. For the 1910 pre-eruption swarm at Usu, the horizontal extent of the craterlets on the northern flanks are taken. The linear dimension L ranges roughly 1.5–2.0 km for the volcanic M 5–5.5 events. Because there certainly exists the size limitation of the strato volcano, it is reasonable to expect such limitation for L around 2–3 km. Numerous empirical relations between M and L have been proposed. In Fig. 7, the relation log $L = 0.51$ $M - 2.27$ (Dambara, 1966) is shown by the straight line for comparison. General correspondence between M and L is obvious for the considerable wide range of magnitude ($M = 2$–8) for rock-burst, volcanic earthquakes and Japanese inland shallow earthquakes. All those data are consistent with the empirical relation. The data indicated by plus marks have systematically larger L.

The physical meaning of the fault length as an earthquake source dimension is somewhat different from the size of surface deformation zone. The estimation of L may be in the accuracy within factor 2. The magnitude scale used here is mostly M_s basis, but Japanese volcanic events and rock-bursts are based on the Japanese standard M_{JMA} or its equivalent estimates. Although those factors cause the scatter of the data, it is still clear to conclude the general correspondence between M and L including volcanic earthquakes.

The depth of large volcanic earthquakes has been accurately determined recently by the *in situ* close-spaced seismometer networks. At St. Helens major earthquakes during the bulging stage were located mostly in a volcanic structure at depth of around

3 km, and the May 18 event is as shallow as 1.3 km (MALONE *et al.*, 1980). Numerous M 4 events at Usu are located inside the summit crater basin at depths less than 2 km (OKADA *et al.*, 1981). These reliable estimations are very shallow. GORSHKOV (1959) suggested the depth of the Bezymianny event at about 50 km based on the two independent methods; the apparent incident angle at Kliuchi ($\Delta = 43$ km), and the relation between seismic intensity at the epicenter and the magnitude (GORSHKOV, 1961). Neither estimates seem to be sufficiently reliable. A set of seismograms at Kliuchi are shown in a report by TOKAREV (1961), which are characterized with intensive surface waves, probably Rayreigh waves, with period 1.4–3.0 sec suggesting its shallow origin. Hence, it is highly possible, though not verified completely, that the depth of earthquakes during the bulging stage of Bezymianny is shallow as similar to the case of St. Helens. The major large earthquakes at Sheveluch is located also within the volcanic structure at depths less than 5 km (TOKAREV, 1967). The depth estimation of the Fernandina earthquakes is not settled yet. SIMKIN and HOWARD (1970), FILSON *et al.* (1973), and FRANCIS (1974) favored the shallow origin around 1 km for major M 5 earthquakes, while KAUFMAN and BURDICK (1980) estimated the depth close to 14 km.

The shallowness and the relative large size of M and L indicate that those volcanic earthquakes are somewhat different dynamical nature compared with tectonic earthquakes. The surface wave magnitude M_s versus body wave magnitude m_b relation is examined in Fig. 8. M_s and m_b are calculated from 20-sec surface waves and short period (1 Hz) body waves, respectively. M_s-m_b relation had been extensively studied during 1960s for discrimination between underground nuclear explosion and earthquakes (BOLT, 1976). Small dots in Fig. 8 indicate the data from nuclear explosions (MARSHALL and BASHAM, 1972) and the horizontal bars indicate the average range of m_b against M_s for world-wide shallow earthquakes (PDE 1973–1975, depth ≤ 50 km) (NOGUCHI and ABE, 1977). The M_s–m_b data for volcanic earthquakes are mostly based on ISC or USGS bulletins and plotted by the solid symbols. The samples are still limited in number and their quality remains mostly low. However, it is found in Fig. 8 that some volcanic M 5 earthquakes show significantly large M_s compared those of shallow tectonic earthquakes and nuclear explosions. The May 18 event at St. Helens has $M_s = 5.2$ and $m_b = 4.7$, indicating the considerable excess energy in low frequency. Nuclear explosions are originated from the shallowest depths, but they are characterized with highly deficient generation of low frequency waves, even compared with the ordinal earthquakes.

Comparison between those anomalous volcanic earthquakes and nuclear explosions should indicate the clue toward the better understanding of the dynamical process beneath volcano. The pre-existing structure as a dynamical system responding to the magmatic pressure may be the main difference from the case of nuclear explosion, in which the medium sorrounding the nuclear bombs is free from any forthcoming detonation.

7.   Conclusions

Through the comparative study of earthquake swarms associated with major volcanic activities, we can draw the following conclusions.

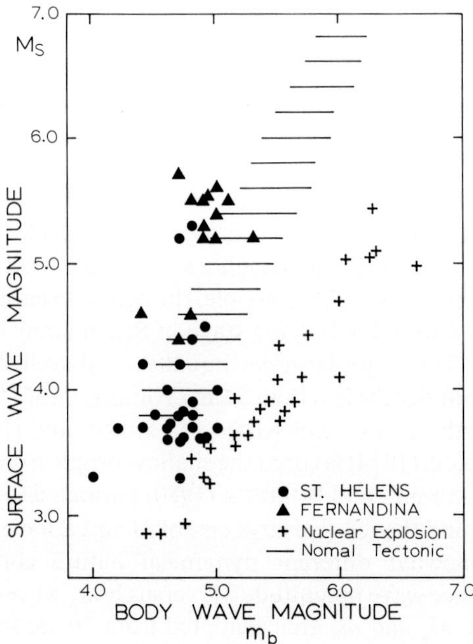

Fig. 8. Surface wave magnitude M_s versus body wave magnitude m_b. Solid symbols indicate volcanic earthquakes at Fernandina and St. Helens. Data for tectonic earthquakes (horizontal bars) and nuclear explosions (plus marks) are shown for comparison.

(1)   The persistent occurrence of large ($M = 4$–5) volcanic earthquakes is not exceptionally rare, but is commonly observed in certain volcanic activities associated with large amount of deformation of a volcano in a form of doming, bulging and caldera collapse.

(2)   Those large earthquakes form a major earthquake family having the similar wave form and the optimum magnitude, which causes a distinct peak in their magnitude-frequency distribution.

(3)   Large amount of episodic deformations are observed at the time of those large earthquakes. Some magnitude 5 earthquakes, hence, can trigger the cataclysmic sector collapse of a volcano, which is nearly ready to fail due to the accumulating gravitational unstability.

The author is greatly indebted to Prof. I. Yokoyama and Dr. K. Abe. Discussions with Drs. Y. Katsui, I. Moriya, K. Nakamura and S. Malone are extremely fruitfull. Dr. E. Endo provided us pre-prints and the data file of local magnitude of St. Helens.

REFERENCES

ABE, K., Size of great earthquakes of 1837–1974 inferred from tsunami data, J. Geophys. Res., **84**, 1561–1568, 1979.

AKI, K., M. FEHLER, and S. DAS, Source mechanism of volcanic tremor: Fluid-driven crack model and their application to the 1963 Kilauea eruption, *J. Volcanol. Geotherm. Res.*, **2**, 259–287, 1977.

AKI, K. and R. KOYANAGI, Deep volcanic tremor and magma ascent mechanism under Kilauea, Hawaii, *J. Geophys. Res.*, **86**, 7095–7109, 1981.

BATH, M., *Seismological Bulletin 1956*, Uppsala, Sweden, 1959.

BOLT, B. A., *Nuclear Explosions and Earthquakes*, Freeman Co., 1976.

CHRISTIANSEN, R. L., Eruption of Mt. St. Helens: Volcanology, *Nature*, **285**, 531–533, 1980.

CHRISTIANSEN, R. L. and D. W. PETERSON, Chronology of the 1980 activity of Mount St. Helens, Washington, *EOS*, **61**, 1113, 1980.

CHUET, B., Ground motion in the near field of a fluid-driven crack and its interpretation in the study of shallow volcanic tremor, *J. Geophys. Res.*, **86**, 5985–6016, 1981.

DAMBARA, T., Vertical movements of earth's crust in relation to the Matsushiro earthquake, *J. Geod. Soc. Japan*, **12**, 18–45, 1966 (in Japanese).

DAVISON, C., *A Manual of Seismology*, Cambridge Univ. Press, London, 1921.

EARTHQUAKE INVESTIGATION COMMITTEE, Precursory earthquakes to the eruption of Usu volcano in 1910, *Rep. Imp. Earthq. Invest. Comm.*, **68**, 128–132, 1913 (in Japanese).

ENDO, E. T., S. D. MALONE, L. L. NOSON, and C. S. WEAVER, Locations, magnitudes, and statistics of the March 20–May 18 earthquake sequence, (preprints), 1980.

EWING, M. and F. PRESS, Further study of atmospheric pressure fluctuations recorded on seismographs, *Trans. Am. Geophys. Union*, **34**, 95–100, 1953.

FEDOTOV, S. A., P. I. TOKAREV, M. F. BOBKOV, and I. P. KUZIN, Earthquakes in Kamchatka and Komandorskie Islands according to the data of detailed seismological observations, in *Earthquakes in USSR in 1964*, pp. 166–184, 1967 (in Russian).

FILSON, J., T. SIMKIN, and L. LEU, Seismicity of a caldera collapse: Galapagos Islands 1968, *J. Geophys. Res.*, **78**, 8591–8622, 1973.

FRANCIS, T. J., A new interpretation of the 1968 Fernandina caldera collapse and its implication for the mid-oceanic ridges, *Geophys. J. Roy. Astron. Soc.*, **39**, 301–318, 1974.

GELLER, R. J., Scaling relation for earthquake source parameters and magnitudes, *Bull. Seismol. Soc. Am.*, **66**, 1501–1523, 1976.

GEOPHYSICAL PROGRAM UNIV. WASH., Eruption of Mt. St. Helens: Seismology, *Nature*, **285**, 529–531, 1980.

GLICKEN, H., B. VOLGHT, and R. J. JANDA, Rockslide-debris avalanche of May 18, 1980, Mount St. Helens volcano, in *Abstr. 1981 IAVCEI Symposium—Arc Volcanism—*, p. 109, 1981.

GORSHKOV, G. S., Eruption of Bezymianny volcano, *Bull. Volcanol. Station*, **26**, 2–72, 1957 (in Russian).

GORSHKOV, G. S., Gigantic eruption of the volcano Bezymianny, *Bull. Volcanol.*, **20**, 77–109, 1959.

GORSHKOV, G. S., On the relation between the volcanological and seismological phenonmena at the eruptions of Bezmianny volcano in 1955–1956, *Bull. Volcanol. Station*, **31**, 32–37, 1961 (in Russian).

GORSHKOV, G. S. and Y. M. DUBIK, Gigantic directed blast at Sheveluch volcano (Kamchatka), *Bull. Volcanol.*, **34**, 261–288, 1969.

GUTENBERG, B. and C. F. RICHTER, Frequency of earthquakes in California, *Bull. Seismol. Soc. Am.*, **34**, 185–188, 1944.

HAMAGUCHI, H. and A. HASEGAWA, Reccurent occurrence of the earthquakes with similar wave forms and its related problems, *J. Seismol. Soc. Japan*, **28**, 153–169, 1975 (in Japanese).

HARADA, T., Stress field in Usu volcano deduced from focal mechanism solutions, *Bull. Volcanol. Soc. Japan*, **26**, 93–110, 1981 (in Japanese).

HARADA, T., H. YAMASHITA, and H. WATANABE, Quasi-continuous observation of changes in distance caused by the 1977–1978 eruption of Usu volcano, Hokkaido, *Geophys. Bull. Hokkaido Univ.*, **38**, 31–40, 1979 (in Japanese).

HARKRIDER, D. G. and F. PRESS, The Krakatoa air-sea waves: an example of pulse propagation in coupled systems, *Geophys. J. Roy. Astron. Soc.*, **13**, 149–159, 1967.

HATORI, T. and M. KATAYAMA, Tsunami behavior and source areas of historical tsunamis in the Japan Sea, *Bull. Earthq. Res. Inst.*, **52**, 49–70, 1977 (in Japanese).

IDE, S. and M. SEINO, Comparison between seismic activities of Usu volcano in 1943–1945 and 1977–1978, in *The eruption of Usu volcano, August 1977–December 1978, Tech. Rep. Japan Meteorol. Agency*, **99**, 75–78, 1980 (in Japanese).

IIDA, K., D. C. COX, and G. PARARAS-CARAYANNIS, Preliminary catalog of tsunamis occurring in the Pacific Ocean, in *Data Report No. 5*, pp. 1–131, Hawaii Inst. Geophys., Univ. Hawaii, 1967.

JAPAN METEOROLOGICAL AGENCY, The eruption of Usu volcano (August 1977–December 1978), *Tech. Rep. Japan Meteorol. Agency*, **99**, 1–204, 1980 (in Japanese).

KATAYAMA, N., Old records of natural phenomena concerning the 'Shimabara Catastrophe,' *Sci. Rep. Shimabara Volcano Obs., Fac. Sci. Kyushu Univ.*, **9**, 1–53, 1974 (in Japanese).

KATSUI, Y., I. YOKOYAMA, S. EHARA, H. YAMASHITA, K. NIIDA, and M. YAMAMOTO, Oshima-Oshima, Report of the volcanoes in Hokkaido, Part 6, Committee for Prevention of the Natural Disasters of Hokkaido, Sapporo, 1977 (in Japanese).

KAUFMAN, K. and L. J. BURDICK, The reproducing earthquakes of the Galapagos Islands, *Bull. Seismol. Soc. Am.*, **70** 1759–1770, 1980.

LIPMAN, P. W., J. G. MOOR, and D. A. SWANSON, Bulging of the north flank of Mount St. Helens volcano before the 5/18 eruption: Geodetic data, *EOS*, **61**, 1135, 1980a.

LIPMAN, P. W., J. G. MOOR, and D. A. SWANSON, Bulging of the north flank before the May 18 eruption: geodetic data, (preprints), 1980b.

MALONE, S. D., E. ENDO, C. S. WEAVER, and J. W. RAMEY, Seismic monitoring for eruption prediction, (preprints), 1980.

MARSHALL, P. and P. W. BASHAM, Discrimination between earthquakes and underground explosions using an improved M_s scale, *Geophys. J. Roy. Astron. Soc.*, **28**, 431, 1972.

MINAKAMI, T., T. ISHIKAWA, and K. YAGI, The eruption of volcano Usu in Hokkaido, Japan, *Bull. Volcanol.*, **11**, 45–157, 1951.

MIZUKOSHI, I. and T. MORIYA, Broad band and wide dynamic range observation of Usu volcano earthquake swarm: Occurrence of similar earthquakes and smoothing of fault motion, *J. Seismol. Soc. Japan*, **33**, 479–491, 1980 (in Japanese).

MOGI, K., Earthquakes and fractures, *Tectonophys.*, **5**, 35–55, 1967.

MOOR, J. G., The pyroclastic surge of May 18, 1980, Mount St. Helens, Washington, in *Abstr. 1981 IAVCEI Symposium—Arc Volcanism—*, p. 238, 1981.

MURAYAMA, N., Propagation of atmospheric pressure waves produced by the explosion of volcano Bezymianny of March 30, 1956 and transport of the volcanic ashes, *Q. J. Seismol.*, **33**, 1–11, 1967 (in Japanese).

NAKAMURA, Y., Geology and petrology of Bandai and Nekoma volcanoes, *Sci. Rep. Tohoku Univ.*, Ser. 3, **14**, 67–119, 1978.

NOGUCHI, S. and K. ABE, Earthquake source mechanism and $M_s - m_b$ relation, *J. Seismol. Soc. Japan*, **30**, 487–507, 1977 (in Japanese).

OKADA, Hm., H. WATANABE, H. YAMASHITA, and I. YOKOYAMA, Seismological significance of the 1977–1978 eruptions and the magma intrusion process of Usu volcano, Hokkaido, *J. Volcanol. Geotherm. Res.*, **9**, 311–334, 1981.

OMORI, F., Macro-seismic measurement in Tokyo (1), *Publ. Earthq. Invest. Comm.*, **10**, 1–102, 1902a.

OMORI, F., Macro-seismic measurement in Tokyo (2), *Publ. Earthq. Invest. Comm.*, **11**, 1–77, 1902b.

OMORI, F., The Usu-san eruption and earthquakes and elevation phenomena, *Bull. Imp. Earthq. Invest. Comm.*, **5**, 1–38, 1911.

SCHOLZ, C. H., The frequency-magnitude relation of microfracturing in rock and its relation to earthquakes, *Bull. Seismol. Soc. Am.*, **58**, 399–415, 1968.

SEKIYA, S. and Y. KIKUCHI, The eruption of Bandai-san, *Sci. Imp. Univ. Tokyo, Japan*, **3**, 91–171, 1890.

SIMKIN, T. and K. A. HOWARD, Caldera collapse in the Galapagos Islands, 1968, *Science*, **169**, 429–437, 1970.

SIMKIN, T., L. SIEBERT, L. MCCLELLAND, D. BRIDGE, C. NEWHALL, and J. H. LATTER, *Volcanoes of the World*, pp. 1–233, Smithonian Inst., Hutchinson Ross Publ. Co., 1981.

SOLOVIEV, S. L., Fundamental data of tsunamis in the Russian coast of the Pacific Ocean during 1737–1976, in *Study of Tsunami in the Open Sea*, Nauka, Moskva, 1978 (in Russian).

TOKAREV, P. I., Energy estimation of the strong earthquakes at Bezymianny volcano, *Bull. Volcanol. Station*, **31**, 38–45, 1961 (in Russian).

TOKAREV, P. I., The giant eruption of the Sheveluch volcano on November 12, 1964, and its forerunners, *Izvestiya, Phys. Solid Earth*, **9**, 572–579, 1967.

Tsuboi, C., Determination of the Gutenberg-Richter's magnitude of earthquakes occurring in and near Japan, *J. Seismol. Soc. Japan*, **7**, 185–193, 1954 (in Japanese).

Umehara, H. and H. Okada, Earthquake activity at Usu volcano as revealed from wave-form pattern analysis, in *Abstr. 1981 IAVCEI Symposium—Arc Volcanism—*, pp. 390–391, 1981.

Utsu, T., Seismic investigation of rock-burst in Bibai Coal Mine, Hokkaido, *J. Seismol. Soc. Japan*, **22**, 76–79, 1969 (in Japanese).

Vlodavetz, V. I. and B. I. Pip, Catalogue of the active volcanoes of the world including solfatara fields, Part 8, *Kamchatka and continental areas of Asia*, pp. 1–110, Int. Volcanol. Assoc., Italy, 1959.

Wano, K. and H. Okada, Peculiar occurence of Usu earthquake swarm associated with the recent doming activity, *J. Seismol. Soc. Japan*, **33**, 215–226, 1980 (in Japanese).

Yokoyama, I. and M. Seino, Prediction of developments in the 1977–1978 activities of Usu Volcano with consideration for energy discharge, *J. Fac. Hokkaido Univ.*, Ser. 7, **6**, 187–200, 1979.

Arc Volcanism: Physics and Tectonics, edited by D. Shimozuru and I. Yokoyama, 63–80.
Copyright © 1983 by Terra Scientific Publishing Company (TERRAPUB), Tokyo.

A Mechanism of Successive Eruption
as Inferred from Seismic Data Associated with
the 1973 Eruptive Stage of Asama Volcano

Hiroshi IMAI

Earthquake Research Institute, University of Tokyo,
Yayoi 1–1–1, Bunkyo-ku, Tokyo 113, Japan

During the activity of Asama Volcano in February-April 1973, a succession of small and gentle eruptions was observed on February 16 to 18. The succession of eruptions are here termed successive eruptions. Each eruption occurred regularly at a 10 to 60 sec time interval accompanied by no detonation sound. The visual and the seismic observational data suggest that the eruptions may take place by a similar mechanism. The earthquakes associated with successive eruptions may be termed implosion earthquakes, because the direction of initial motion indicates downward or pull in all bearings. Temporal variation in the ratio of the maximum trace amplitude in each horizontal component to the vertical one indicates the vertical migration of effective magma head. On the basis of the above results, an interpretation of the successive eruptions is conducted by a model system of two-phase flow which consists of liquid magma and gas bubbles. An alternative model is also presented on the basis of the syphonage. Although the proposed model may appear to be oversimplified and speculative, it explains the visual and the seismic observational data satisfactorily.

1. Introduction

Volcanic eruptions are regarded as resulting from mass and heat transportations from depth and their manner and cause are believed to be highly dependent on the concentration of volatiles in magma and chemical nature of magma, essentially water, which exsolves into bubbles in magma. The problem is one of the most basic and essential themes in volcanology. VERHOOGEN (1951) first investigated theoretically the role of bubble growth in magma see also, SCRIVEN, 1959, MCBIRNEY and MURASE, 1970, BENNETT, 1974, SHIMOZURU, 1978, SPARKS, 1978). However, it might be sure that the eruption mechanism can be changable according not only to each volcano but also to individual eruptions. Therefore, we are obliged to make a case study on the mechanism of eruption occurred at a volcano. In this paper we treat the 1973 successive eruptions of Asama Volcano, which is located in the central part of Honshu, Japan (Fig. 1).

2. The 1973 Eruptions of Asama Volcano

Asama Volcano is one of the most active volcanoes in Japan and is also an

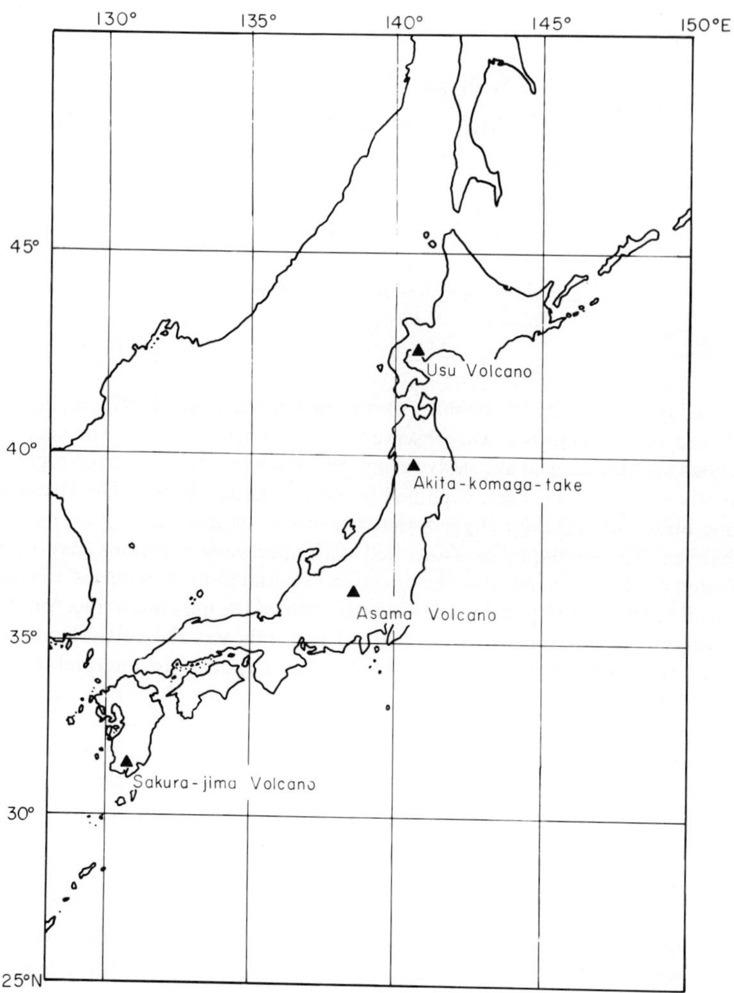

Fig. 1.   Locations of volcanoes appeared in this article.

andesitic volcano characteristic of island arc volcanism, as is Sakura-jima Volcano and many others in Japan. It commenced activities with an explosive eruption at the summit crater on February 1, 1973. A series of explosive eruptions continued until April 26, 1973. SHIMOZURU *et al.* (1975) briefly outlined the seismic events associated with the eruptive activities and also described manifestation of individual eruptions. The explosion earthquakes, which exceeded 10^7 J in seismic energy, associated with single eruptions have already been analyzed (IMAI *et al.*, 1979, IMAI, 1980) in terms of spectra and foci and the rise time and the maximum amount of possible pressure increase at the explosion site. During the course of explosive stage, successions of small eruptions, which are called successive eruptions (SHIMOZURU, 1979), took place in

16–18 February (Fig. 2). The successive eruptions are more or less regular repetition of gentle upflow of a series of volcanic clouds without any detonation sound, on the contrary, a single explosive eruption associates with a strong detonation sound. The occurrence time intervals of those eruptions mostly in the range of 10 to 60 sec and the magnitudes were fairly small in comparison with single explosive eruptions. Synchronously with each of the successive eruptions, earthquakes were recorded; the seismic magnitude was 1.6 at the largest. Obviously these earthquakes were directly related to the eruptions. The number of eruptions exceeded 3,900, judged from associated earthquakes (SHIMOZURU *et al.*, 1975). No bread-crust bombs were ejected but much ash fall was observed at the vicinity of the volcano during the successive eruptive stage. SHIMOZURU (1979) also pointed out that magnitude-frequency relation of the earthquakes associated with the successive eruptions obeyed a formula: $N = -mA$, where N, m, and A denote the frequency of the earthquakes, the coefficient, and the maximum trace amplitude, respectively. Thus, the relation did not obey the established formula for usual tectonic earthquakes. SHIMOZURU *et al.* (1975) also observed that long period waves are predominating during the initial part of successive eruption earthquakes. From the facts mentioned above, it seems that the earthquakes are not the result of destruction of solid such as a sudden break of the volcanic vent cap (SELF *et al.*, 1979) but the result of fracture of liquid such as a disruption of liquid magma. Therefore, the above two types of eruptions, explosive eruption and successive eruption, are considered to be discriminated essentially relative to their mechanisms.

3.   Seismic Data and Method of Analysis

The present arguments are made mostly on the basis of seismograms obtained by

Fig. 2.   A photograph of gentle successive eruption clouds of Asama volcano, taken from north of the volcano on Feb. 17, 1973 (Photo by D. Shimozuru).

three component, medium-period ($T_0 = 5$ sec), displacement (Magnification $= 500$) seismographs at NAK which is one of the seismic stations of Asama Volcano Observatory (A.V.O.) and is located ca. 4.2 km east of the summit crater as shown in Fig. 3. The E-W component represents the radial component, whereas the N-S component represents the transverse component. The seismic waves were registered on smoked paper drums at A.V.O. with a paper speed of 1 mm/sec. The frequency characteristics of the overall recording system have been shown previously in the article by IMAI *et al.* (1979, Fig. 3). The distortion of the recording signals were corrected for linear ramp, shift of the zero line and arcuation of the recorded line due to a finite arm length ($a = 120$ mm) of galvanometers (IMAI *et al.*, 1979). Then, power spectral analyses were carried out on the coda of the seismic signals by means of the Fast Fourier Transform method (F.F.T.) under the correction of the frequency response of the overall recording system. The data window used in the calculations was a box-car shaped window smoothed at both edges by a cosine-curve.

To determine the onset time and the direction of the initial motion of each earthquake, the seismograms of short-period ($T_0 = 1$ Hz), displacement, vertical component seismographs installed at the seismic stations GIP, SAN, HOT, and BUT are used, where BUT was a temporary station.

4. Similarity of the Seismic Waveforms

For the purpose of the present study, the seismograms of earthquakes associated with successive eruptions ought to be strictly discriminated from those of earthquakes

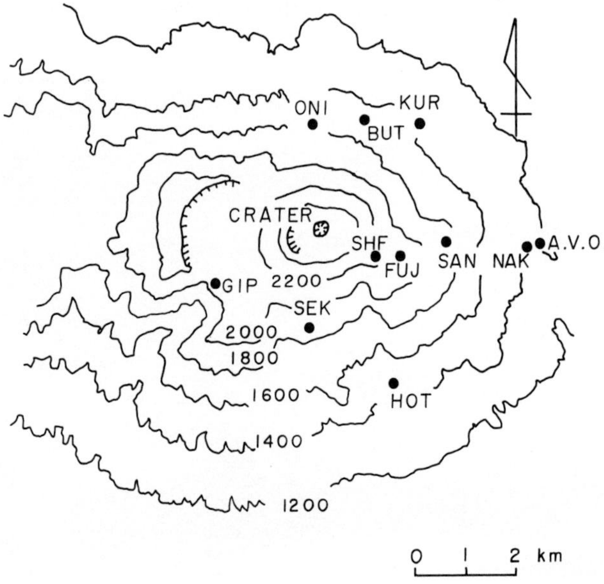

Fig. 3. Locations of seismic stations and Asama Volcano Observatory (A.V.O.). Seismic signals are telemetered by underground cables to A.V.O.

of other origins. In general, a waveform similarity is expected among the earthquakes in the same focal region and by the same source mechanism (TSUJIURA, 1979a, b; OKADA *et al.*, 1981). OKADA *et al.* (1981) referred to a group of earthquakes, which occurred at Usu Volcano (Fig. 1), having similar waveforms as an "earthquake family" after HAMAGUCHI and HASEGAWA (1975). The idea of the waveform similarity has also been reported on the basis of ordinary tectonic earthquakes. The seismograms are investigated under the assumption that the same analogy is applicable to volcanic earthquakes. The results show that the successive eruption earthquakes can be sorted into four types on the basis of the apparent period of the wavelet of the initial part as listed in Table 1. Figure 4 shows examples of the vertical component of Type-1 to

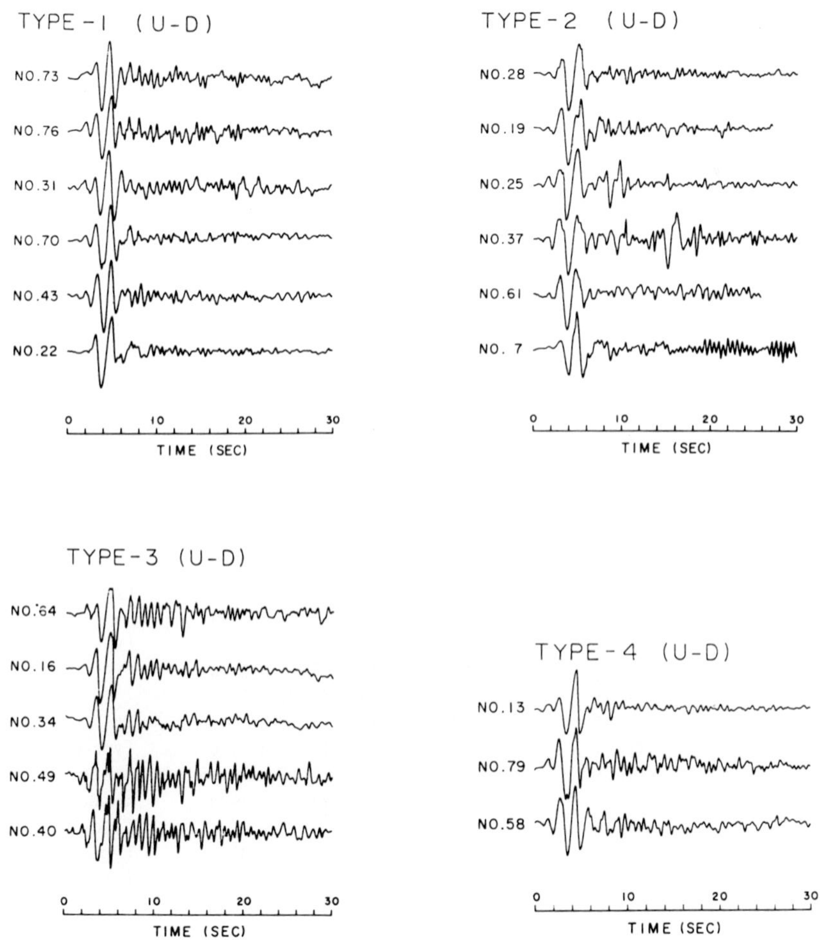

Fig. 4.   Normalized vertical component seismograms of Type-1 to Type-4.

Type-4 seismograms normalized to the maximum trace amplitude for convenience of comparison. The number at the head of the seismograms corresponds to the same number in Table 1. None of seismograms in one group were assigned to any other group. These morphological sorting enable us to expect the earthquakes to have occurred in the same focal region and by the same source mechanism.

Figure 5 shows the normalized F.F.T. power spectra of the coda of each seismogram excluded the first 8 sec on the time scale represented below the seismograms in Fig. 4. The values of characteristic frequencies of power spectra are listed in Table 2. At a glance, the analyzed coda spectra might not be classified into the same four types. However, the characteristic similarity of coda spectral patterns in each type is found, particularly in the E-W component as is pointed out by Imai *et al.* (1979) for the seismograms of explosion earthquakes recorded at the same station, NAK. Since the seismic station NAK is located on the ridge extending in the east-west direction, the topographic effect on the seismic waves may appear most in the N-S component and least in the E-W component. Therefore, the similarity of coda spectral

Table 1. Seismic data list of the successive eruption earthquakes of the 1973 Asama Volcano activity.

Component			Arrival time	Type
U-D	N-S	E-W		
NO. 28	NO. 29	NO. 30	Feb. 17, 06 h 55 m	Type-2
NO. 73	NO. 74	NO. 75	Feb. 17, 07 h 43 m	Type-1
NO. 76	NO. 77	NO. 78	Feb. 17, 07 h 52 m	Type-1
NO. 13	NO. 14	NO. 15	Feb. 17, 08 h 27 m	Type-4
NO. 79	NO. 80	NO. 81	Feb. 17, 08 h 42 m	Type-4
NO. 31	NO. 32	NO. 33	Feb. 17, 08 h 44 m	Type-1
NO. 70	NO. 71	NO. 72	Feb. 17, 09 h 07 m	Type-1
NO. 19	NO. 20	NO. 21	Feb. 17, 09 h 50 m	Type-2
NO. 64	NO. 65	NO. 66	Feb. 17, 09 h 58 m	Type-3
NO. 16	NO. 17	NO. 18	Feb. 17, 09 h 59 m	Type-3
NO. 34	NO. 35	NO. 36	Feb. 17, 10 h 02 m	Type-3
NO. 55	NO. 56	NO. 57	Feb. 17, 22 h 23 m	
NO. 43	NO. 44	NO. 45	Feb. 17, 22 h 41 m	Type-1
NO. 25	NO. 26	NO. 27	Feb. 17, 23 h 15 m	Type-2
NO. 67	NO. 68	NO. 69	Feb. 17, 23 h 22 m	
NO. 52	NO. 53	NO. 54	Feb. 17, 23 h 23 m	
NO. 22	NO. 23	NO. 24	Feb. 17, 23 h 24 m	Type-1
NO. 4	NO. 5	NO. 6	Feb. 17, 23 h 38 m	(tremor)
NO. 1	NO. 2	NO. 3	Feb. 17, 23 h 39 m	(tremor)
NO. 37	NO. 38	NO. 39	Feb. 17, 23 h 45 m	Type-2
NO. 61	NO. 62	NO. 63	Feb. 18, 00 h 02 m	Type-2
NO. 58	NO. 59	NO. 60	Feb. 18, 00 h 21 m	Type-4
NO. 49	NO. 50	NO. 51	Feb. 18, 00 h 31 m	Type-3
NO. 46	NO. 47	NO. 48	Feb. 18, 00 h 33 m	
NO. 10	NO. 11	NO. 12	Feb. 18, 00 h 57 m	
NO. 40	NO. 41	NO. 42	Feb. 18, 01 h 13 m	Type-3
NO. 7	NO. 8	NO. 9	Feb. 18, 01 h 21 m	Type-2 (tremor)

Table 2.   Characteristic frequencies of the later part of the seismograms associated with the 1973 successive eruptions of Asama Volcano, analyzed by F.F.T. A type in this table corresponds to the same one in Table 1.

Data	Component		
	U-D	N-S	E-W
Type-1 NO. 73, 74, 75	0.22(Hz)	1.84	1.04
NO. 76, 77, 78	1.93	0.70	1.06
NO. 31, 32, 33	0.80	1.68	1.70
NO. 70, 71, 72	0.18	1.68	1.07
NO. 43, 44, 45	0.80	1.00	1.09
NO. 22, 23, 24	1.60	1.68	1.04
Type-2 NO. 28, 29, 30	0.20	1.72	1.09
NO. 19, 20, 21	1.00	1.88	1.07
NO. 25, 26, 27	1.17	1.00	1.15
NO. 37, 38, 39	0.49	1.48	1.07
NO. 61, 62, 63	1.37	1.09	1.06
NO. 7, 8, 9	1.88	1.78	1.76
Type-3 NO. 64, 65, 66	0.90	1.06	1.09
NO. 19, 20, 21	0.98	1.74	1.09
NO. 34, 35, 36	0.20	0.66	1.09
NO. 49, 50, 51	0.80	1.68	1.45
NO. 40, 41, 42	0.80	1.50	1.45
Type-4 NO. 13, 14, 15	0.80	0.68	1.11
NO. 79, 80, 81	0.22	1.82	1.02
NO. 58, 59, 60	0.18	1.47	0.61

structure in the E-W component may be attributed to the seismological structure of the volcano in the radial direction, from the summit crater to the seismic station NAK, provided that the coda waves are backscattering waves from numerous heterogeneities (AKI and CHOUET, 1975) in the volcanic body.

5.   Maximum Amplitude and Its Onset Time

Because of the location of the seismic station NAK, the E-W component gives us the information of P-waves and or SV-type waves, whereas the NS-component reflects SH-type waves. We assume that the seismic waves purely consist of body waves and the maximum amplitude in each component is given by the shear waves which are completely decoupled to SV-waves and SH-waves. Then, we get, $A_V = A_{SV} \sin \varphi$, $A_{EW} = A_{SV} \cos \varphi$, and $A_{NS} = A_{SH}$, where A_V, A_{EW} and A_{NS}, A_{SV} and A_{SH} are the maximum trace maximum amplitudes of the vertical, the E-W and the N-S components, and those of SV-waves and of SH-waves, respectively, and also φ is the incident angle of a seismic ray measured from the vertical line. The maximum amplitude ratio of the horizontal component to the vertical one is given by

$$\frac{A_{EW}}{A_V} = \frac{1}{\tan \varphi}, \quad \text{or} \quad \frac{A_{NS}}{A_V} = \frac{A_{SH}}{A_{SV} \cdot \sin \varphi}.$$

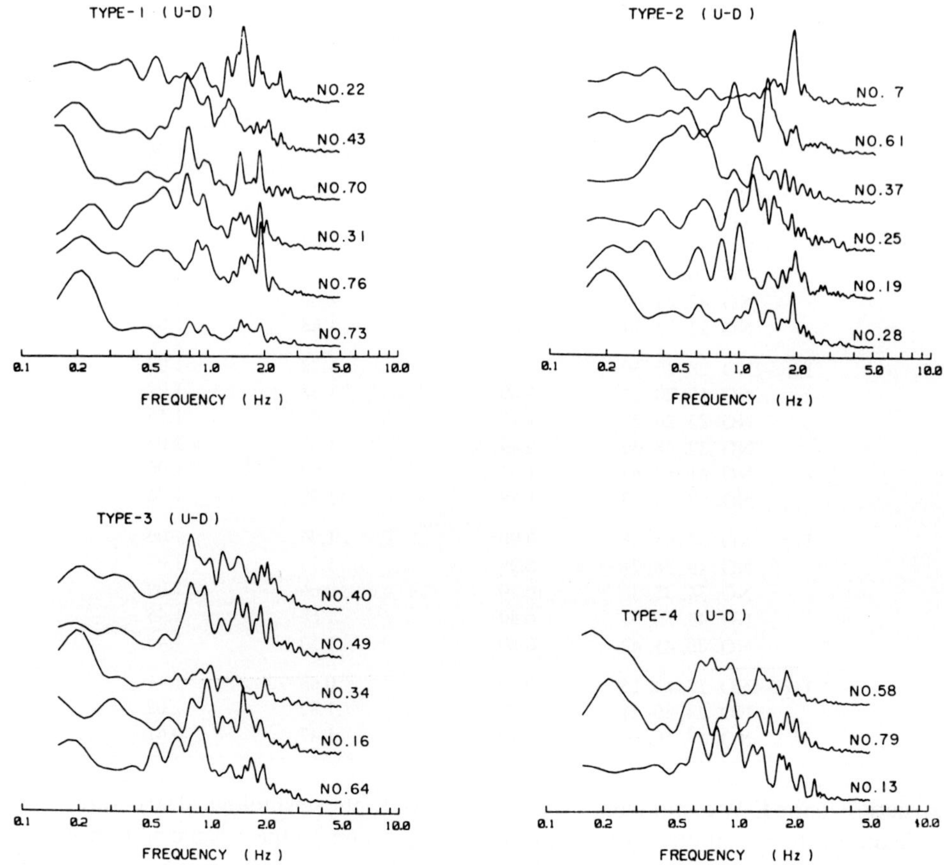

Fig. 5. Normalized power spectra of coda waves (excluded the first 8 sec) of the seismograms in Fig. 4.

Therefore, it is expected that the deeper an earthquake occurs, the larger a value of A/A_v will be, because an incident angle φ becomes smaller according as an earthquake occurring at deeper depth. From this view point, maximum amplitudes and their onset times are compiled as in Fig. 6. A and T represent the maximum trace amplitude and its onset time, respectively. The subscript $_\mathrm{v}$ denotes the datum of the vertical component, whereas A and T without subscripts denote the horizontal ones. The abscissa represents the delay time, $T - T_\mathrm{v}$, of the horizontal maximum amplitude from the vertical maximum amplitude, whereas the ordinate represents the ratio of the horizontal maximum amplitude to the vertical one, A/A_v. T1 in the figure means that the datum corresponds to event Type-1, and similarly for T2–T4. NT represents the datum which does not belong to any of these types. Open and solid circles are the data of the E-W and N-S components, respectively. The estimated errors for the abscissa and the ordinate are roughly 0.3 sec and 3%, respectively. The circles of the same

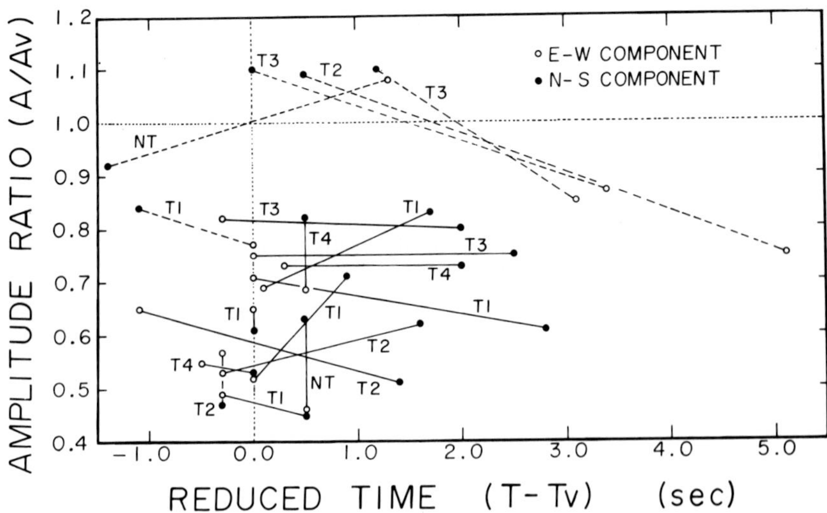

Fig. 6.   Reduced onset time and the maximum trace amplitude of seismic waves associated with successive eruptions.

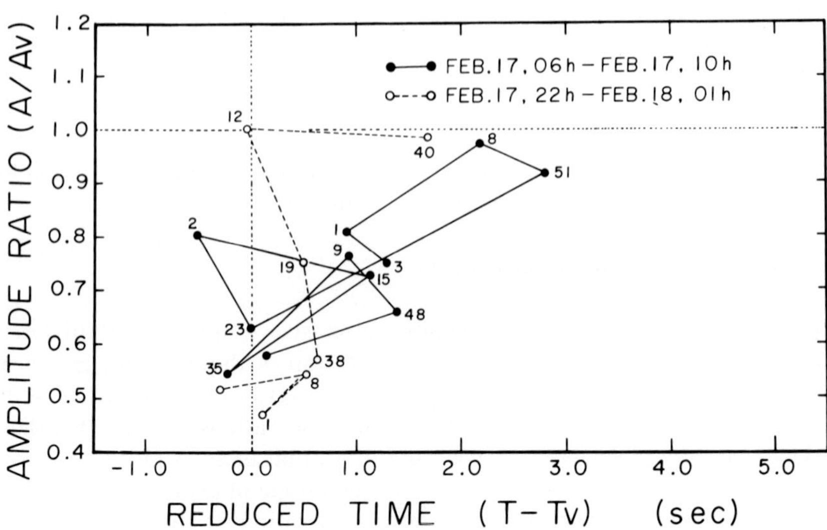

Fig. 7.   Migration of the arrival time of the maximum trace amplitude ratio in the time domain. *A* and *T* are the averages of horizontal components which are found in Fig. 6. The number in the figure is the time of occurrence in minutes of an earthquake after the preceding one.

earthquake are connected by either solid and broken lines, where the solid line means that the E-W component has the maximum amplitude in advance or at the same time as the N-S component, whereas the broken line means that the E-W component has the maximum amplitude delayed from the maximum amplitude of the N-S component.

The temporal sequence of the earthquakes under discussion is examined in Fig. 7, in which a point $(T - T_V, A/A_V)$ is plotted for each earthquakes. The onset time T and the maximum amplitude A are taken as the arithmetic mean of the corresponding values in N-S and E-W components. Numbers in the figure are the time intervals (in minutes) from each preceding earthquake. The circles without a number are the initial data of each active group appearing in the figure. As mentioned previously, the deeper the earthquake occurs, the larger the amplitude ratio will be. The locational migration of circles in Fig. 7, therefore, can be interpreted as the upward migration of an effective magma head in a volcanic conduit, provided that earthquakes occur at the magma head.

6. Implosion Earthquakes

The initial motion of explosion earthquakes is "push" or "upward" in all bearings, which indicates the sudden build-up of pressure in a volcano (probably in a few seconds) before the eruption (e.g. Minakami *et al.*, 1970), although some exceptional evidence was reported by Tanaka (1971) concerning the 1970–1971 Akita-komaga-take (Fig. 1) eruptions.

In contrast, the initial motion of the earthquakes associated with the successive eruptions is "downward" and/or "pull" in all bearings in some cases which could be discriminated. Figure 8 shows an example of seismograms associated with a successive eruption. In the overall data set, there are many seismograms overlapped with tremors or by the coda of seismic waves from a preceding event.

Thus, the earthquakes associated with a single explosive eruption and the earthquakes associated with the successive eruptions are quite different from each other in their source mechanisms. Here, the latter is termed as "implosion earthquakes" and the former, "explosion earthquakes."

7. Model Presentation and Discussion

As mentioned in the previous section, successive eruption earthquakes have several characteristics different from that of explosion earthquakes, for instance, the waveform similarities, the initial motions of "downward" or "pull" in all stations. Moreover, according to the visual and the seismic observations, it seems that the ratio of the kinetic energy of single explosive eruption to that of an eruption of successive eruptions is much larger than the ratio of the magnitude of explosion earthquake to that of successive eruption earthquake. In consideration of the above features, a possible model appropriate for the successive eruptions and the associated earthquakes is proposed on the basis of the idea of two-phase flow consisting of a mixture of liquid magma and gas bubbles.

In general, there are several flow patterns of a liquid-gas mixture, or of a gas-solid

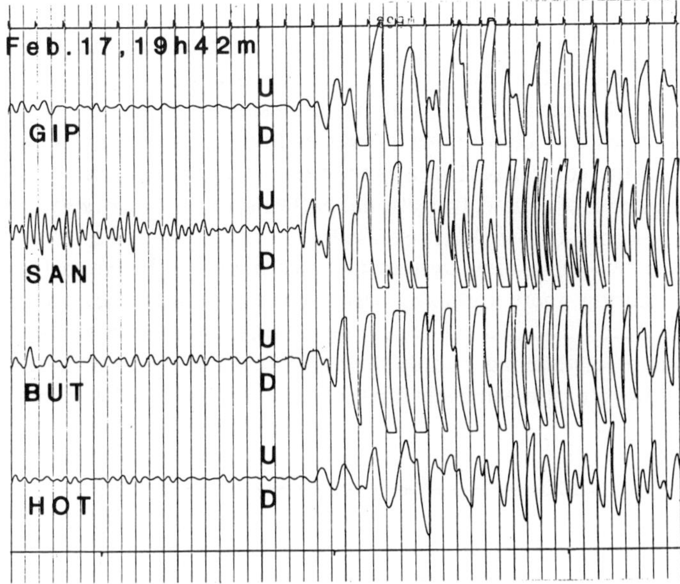

Fig. 8. An example of the seismograms associated with successive eruptions. The time corrections to the original reading are +0.20 sec for GIP, +0.10 sec for SAN, +0.05 sec for BUT, and +0.10 sec for HOT, respectively.

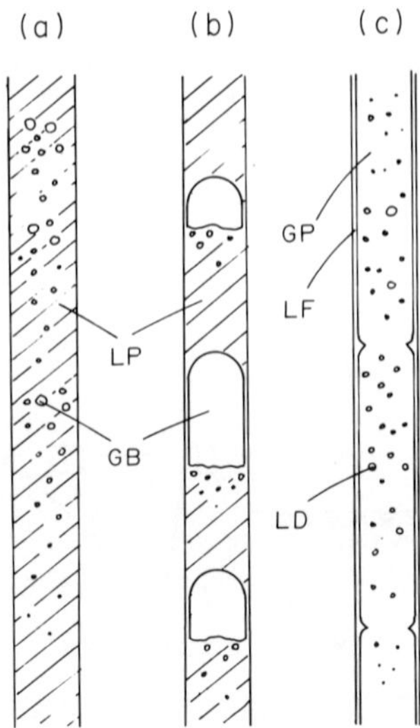

Fig. 9. Schematic illustrations of typical upward flow patterns of two-phase flow in cylindrical pipe: (a) bubbly flow, (b) slug flow, and (c) annular flow. In the figure, liquid-phase and gas-phase are contracted as LP and GP. GB, LF, and LD indicate gas bubbles, liquid film (flow), and liquid droplets, respectively.

mixture through a cylindrical pipe. The typical flow patterns are shown in Fig. 9. The bubbly flow (a) is characterized by discrete gas bubbles suspended in a continuous liquid phase. The slug flow (b) is characterized by a succession of liquid slugs and large gas bubbles which almost fill the available cross section of the flow. The annular flow (c) is characterized by a continuous liquid film flow along the pipe wall, while the gas phase flows in the central part of a pipe, often accompanied by liquid and/or solid droplets. There are, of course, transitional patterns, for instance, between bubbly-slug flow and slug-annular flow (Wallis, 1969).

Since the two-phase flow system includes many problems in the field of heat transfer engineering, there are many papers concerned with the two-phase flow of a liquid-gas mixture through a cylindrical pipe. According to Ozawa *et al.* (1979), the governing factor controlling bubbly flow and slug flow in a cylindrical pipe is the volumetric flow rate of a water-air mixture as shown in Fig. 10. When the flow rate

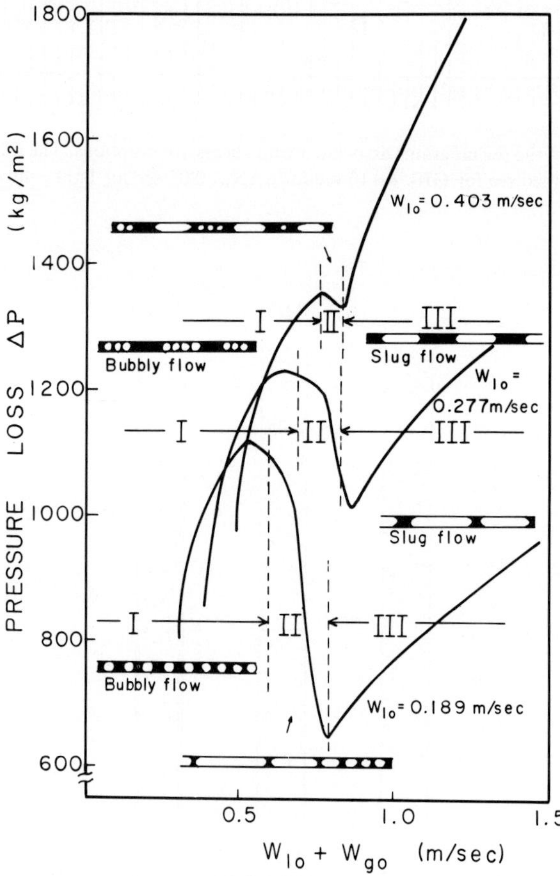

Fig. 10. An experimental result concerned with the pressure loss versus the flow rate of a water-air mixture flowing through a cylindrical pipe (after Ozawa *et al.*, 1979).

increases, frictional loss at the pipe wall increases, however, it decreases markedly at a certain flow rate as shown in the figure. The region can be termed the transition region (II) from a bubbly flow region (I) to a slug flow region (III).

The initial motions of successive eruption earthquakes indicate that implosive events may take place in a volcanic body, probably in a volcanic conduit. A sudden decrease of volume is required for the implosive events. Then, we can attribute the implosions to large gas bubbles collapsing instantanously. Thus, large gas bubbles may be needed in a volcanic conduit in order to interprete the initial motions of successive eruption earthquakes. Therefore, it is believed that the process proceeds from (a) to (c) in Fig. 9 during successive eruptive activities. To make the argument simple, the effect of viscosity, thermal feedback and also chemical reactions are all ignored here. Let us consider how a gas and magma mixture flows in a cylindrical shaped conduit, according to a simplified model as illustrated in Fig. 11 (a)–(e). (a) to (e) in the figure are models of the mixture flow illustrated from the depth to the surface. (a) In the initial stage, gas bubbles are assumed to be nucleated at the exsolution surface (ES). (b) Bubbles grow gradually due to decreasing pressure as they move upward by their buoyancy and large bubbles may be formed by coalescence. There may be decompressions due to magma rising up in a volcanic conduit. The bubbles may become more larger. Thus, such a bubbly flow changes into a slug flow. As mentioned previously, once the transition occurs, the kinetic energy loss by frictional force from the wall is reduced so that the large gas bubbles can rise more easily in viscous liquid magma than before. (d) As soon as a large gas bubble reaches the effective magma head (MH), it disrupts at its tip (topmost portion) and a gravitational downflow of magma film may occur along the conduit wall. Downgoing magma film may produce shear

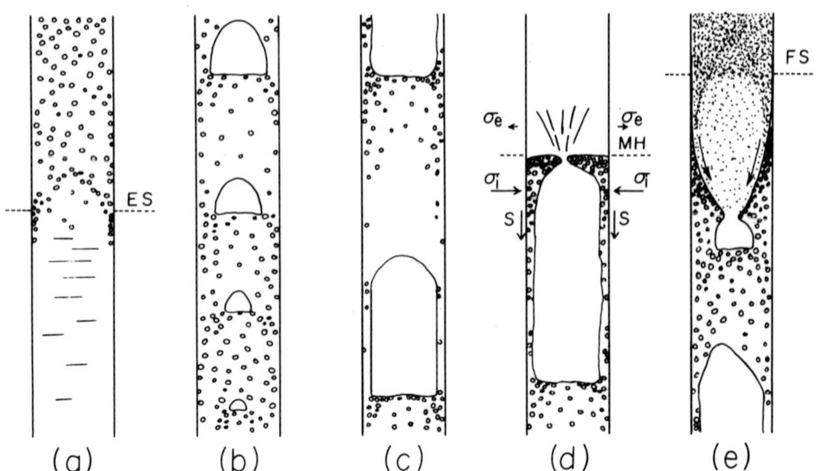

Fig. 11.   Schematic illustrations of a process modelled as a two-phase flow pattern. ES, MH, and FS denote the exsolution surface, the (effective) magma head, and the fragmentation surface, respectively. σ_i, σ_e, and S denote the normal stress acting inward on, the normal stress outward on, and the shear stress along, the volcanic conduit wall, respectively.

H. Imai

stress (S) along the conduit as well as the normal stress (σ_i). In this case, σ_i might be somewhat greater than the normal stress σ_e caused by the released gas itself. (e) The gas squeezed out of the bubble flows up with fragments or fine drops of magma and/or fractured conduit wall rocks as an annular flow. Each component in the annular flow, a gas and fragments in this case, flows up with its own velocity. The velocities are not very fast even at first and decrease immediately because of conduit wall friction, gravitation, etc. Then, they result in forming a gentle and small eruption cloud.

Let us examine how the implosion earthquakes are generated based on the present model. A simple estimation for the potential energy loss by the gravitational downflow of magma film through a volcanic conduit is carried out for the model shown in Fig. 12.

The volume of the magma film, Fig. 12(1), V_0, is given by

$$V_0 = \pi H(R^2 - r^2) \tag{1}$$

where R and r are the radii of the volcanic conduit and the gas bubble, and H is the effective bubble length, respectively. The volume V_0 flows down to form a column having the same volume, thus,

$$V_0 = \pi R^2 h \tag{2}$$

as shown in Fig. 12(2). Therefore, h is given by

$$h = (1 - r^2/R^2)H. \tag{3}$$

On the other hand, if we consider that the magma film having volume, V, flows down to form a cylinder having the same volume as shown in Fig. 12(3 and 4), then V is given by

$$V = \pi(H - h)(R^2 - r^2). \tag{4}$$

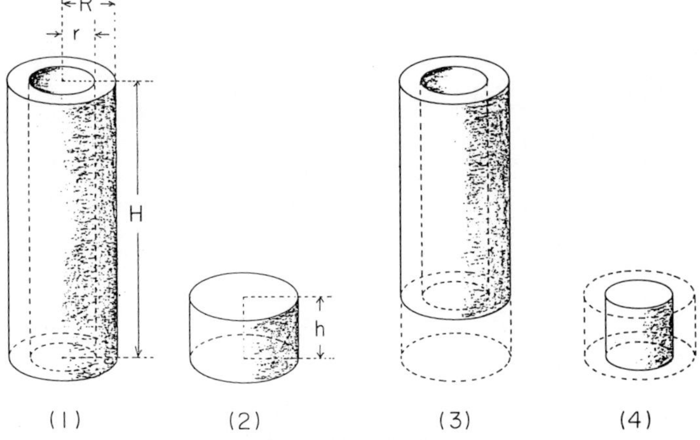

Fig. 12. Schematic model for the calculation of potential energy loss, $-\Delta E$, associated with magma (film) flowing down.

Therefore, the potential energy loss $-\Delta E$ associated with the magma flowing down is given by

$$-\Delta E = \rho V g \left(\frac{H-h}{2} + \frac{h}{2} \right),$$

provided that the total mass is concentrated at the center of gravity. Then the following equation is obtained as

$$-\Delta E = \frac{1}{2} \pi \rho g H^2 r^2 (1 - r^2/R^2), \tag{5}$$

where ρ and g are the density of magma and the acceleration of gravity, respectively. Since the thickness of liquid film of an annular flow is roughly one tenth of the inner diameter of a pipe (ŌYA, 1971) and g is 9.8 m/sec^2, Eq. (5) can be rewritten as

$$-\Delta E = 3.5 \rho H^2 R^2. \tag{6}$$

The bulk density of andesitic magma round at 1,000°C may range roughly 2,500 kg/m^3 (MURASE and MCBIRNEY, 1973) to 500 kg/m^3 owing to volumetric concentration of gas phase exsolved into bubbles in magma at the magma head. The density of 500 kg/m^3 is considered as an extreme case. Eq. (6) yields the relation between $-\Delta E$ and H as shown in Fig. 13, as a function of R with two values of densities, 2,500 kg/m^3 (broken line) and 500 kg/m^3 (dotted line). The vent radius of Asama Volcano is considered to be a minimum of 5 m. Even though $R = 5$ m and $\rho = 500$ kg/m^3, if $H = 20$ m or more, which is one of the most special cases, we can explain the magnitudes of successive eruption earthquakes which roughly range from 1.0 to 1.6 (Fig. 13). Therefore, the model seems to be reasonable and, from energetic view point, downward negative source as discussed above is considered as the source of implosion earthquakes, although there is some uncertainty on the conversion efficiency from potential energy to seismic energy. Taking account of the present model for the mechanism of the implosion earthquake, IMAI (1982) investigated the hypocenters of the earthquakes. According to his result, if the implosion earthquakes occurred roughly at the depth ranging from 0.5 to 1.5 km below the top of the volcano, then the seismic data relative to the initial motion and relative to the onset time difference could be explained satisfactorily.

In the previous section a simple model was presented concerning the successive eruptions by use of the two-phase flow system. The model, however, has a problem concerning the possibility of forming a large gas bubble of which size is 10 m or more in diameter and 10 m or more in length. SPARKS (1978) suggested that a vesiculation theory on the basis of coalescence of bubbles is untenable for bubble growth up to a large size as mentioned above, based on previous works (e.g. MCBIRNEY, 1963, MCBIRNEY and MURASE, 1970, BENNETT, 1974,). The decompressive expansion and/or diffusion as a means of forming the large gas bubbles is also likely to be untenable. In view of the above discussion, we present here an alternative model which is more considered realistic than the previous model. An alternative model introduces the syphonage (see below and Fig. 14). In the initial stage, the peripheral magma cham-

 H. Imai

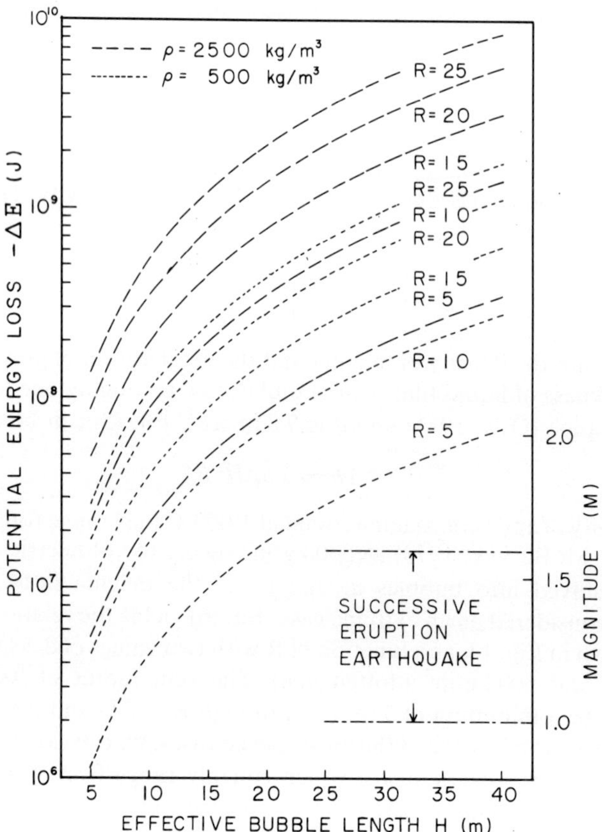

Fig. 13. Potential energy loss $-\Delta E$ versus H as a function of R, where H is the effective bubble length and R is the radius of a volcanic conduit, with the seismic magnitude scale for comparison.

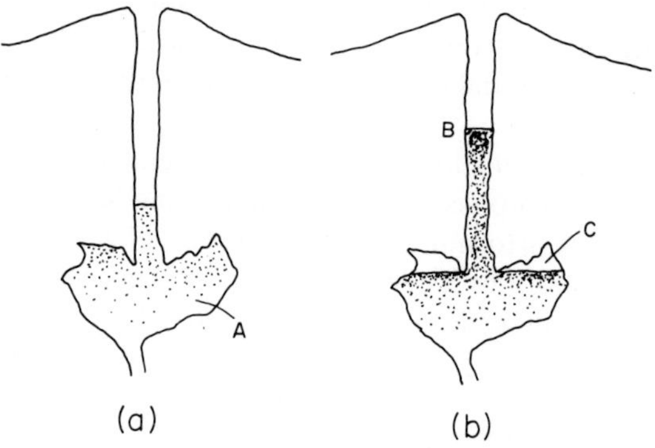

Fig. 14. Schematic illustrations of an alternative model to form large gas bubbles within a volcanic conduit. A, B, and C denote the magma supersaturated by volatile components (mostly, water), the effective magma head and the accumulation of exsolved volatile components.

ber is filled with magma which is supersaturated with water (A in Fig. 14). In the next stage, vesiculation occurs and is accompanied by the magma head (B in Fig. 14) rising; the exsolved gas then accumulates in the upper part of the magma chamber (C in Fig. 14). The highly pressurized gas squeezes magma out of the magma chamber through the volcanic conduit. As soon as the magma surface in the magma chamber reaches the lower end of the volcanic conduit, the syphonage takes place. The syphonage is just the phenomenon commonly seen in a coffee syphon. In a coffee syphon, the water is first boiled to form steam bubbles, alternate slugs of a liquid and vapour then flow up through a cylindrical pipe. The phenomenon is interpreted by use of the slug flow pattern, and is exactly the reason why the concept has just been adopted for the successive eruption mechanism.

8.   Conclusion

A mechanism for the 1973 successive eruptions of Asama Volcano has been investigated by use of the visual and seismic observational data. The apparent feature of these eruptions is the similarity of the focal process as supported by the seismic waveform analysis. The earthquakes associated with the successive eruptions seem to have occurred at shallow depth just under the summit crater of the volcano (IMAI, 1982) and suggest an implosive nature in which the direction of initial motion is "pull" or "downward" in every bearing. Upon consideration of the above results, the qualitative interpretation of the mechanism of successive eruptions has been made by a model introducing two-phase flow consisting of a liquid magma and gas bubble mixture, and by a model introducing the syphonage. Although the models are somewhat speculative and oversimplified, they do account for the visual and the seismic observational data satisfactorily.

The author would like to express his sincere thanks to Prof. D. Shimozuru, Prof. T. Watanabe, and Mr. Kagiyama of the Earthquake Research Institute, The University of Tokyo, for countless discussions throughout the course of this work. The author also thanks Mr. S. McNutt of the Lamont-Doherty Geol. Obs., Columbia University, who reviewed this manuscript.

REFERENCES

AKI, K. and B. CHOUET, Origin of coda waves: source, attenuation, and scattering effects, *J. Geophys. Res.*, **80**, 3322–3342. 1975.

BENNETT, F. D., On the volcanic ash formation, *Am. J. Sci.*, **274**, 648–661, 1974.

HAMAGUCHI, H. and A. HASEGAWA, Recurrent occurrence of the earthquakes with similar waveforms and its related problems, *J. Seismol. Soc. Japan*, **28**, 153–169, 1975.

IMAI, H., N. GYODA, and E. KOYAMA, Explosion earthquakes associated with the 1973 eruptions of Asama Volcano (Part I)—Spectral sudies, *Bull. Earthq. Res. Inst.*, **54**, 161–186, 1979.

IMAI, H., Explosion earthquakes associated with the 1973 eruptions of Asama Volcano (Part II)—The summary of studies on explosion earthquakes and a model of explosive eruptions inferred from seismic data, *Bull. Earthq. Res. Inst.*, **55**, 537–576, 1980.

IMAI, H., Implosion earthquakes associated with the 1973 eruptive activity of Asama Volcano, *Bull. Earthq. Res. Inst.*, **57**, 303–315, 1982.

McBIRNEY, A. R., Factors govering the nature of submarine volcanism, *Bull. Volcanol.*, **26**, 455–469, 1963.

McBIRNEY, A. R. and T. MURASE, Factors govering the formation of pyroclastic rocks, *Bull. Volcanol.*, **34**, 372–384, 1970.

MINAKAMI, T., S. UTIBORI, T. MIYAZAKI, S. HIRAGA, N. GYODA, and T. UTSUNOMIYA, Seismometrical studies of Volcano Asama (Part I). Seismic and volcanic activities of Asama during 1934–1969, *Bull. Earthq. Res. Inst.*, **48**, 235–301, 1970.

OKADA, Hm., H. WATANABE, H. YAMASHITA, and I. YOKOYAMA, Seismological significance of the 1977–1978 eruptions and the magma intrusion process of Usu Volcano, Hokkaido, *J. Volcanol. Geotherm. Res.*, **9**, 311–334, 1981.

ŌYA, T., Upward liquid flow in small tube into which air streams (Second report, pressure drop at the confluence), *Bull. Japan Soc. Mech. Eng.*, **14**, 1330–1339, 1971.

OZAWA, M., K. AKAGAWA, T. SAKAGUCHI, and T. FUJII, Oscillatory flow instabilities in air-water two-phase flow system (First report: Pressure drop oscillation), *Bull. Japan Soc. Mech. Eng.*, **22**, 1763–1770, 1979.

SCRIVEN, L. R., On the dynamics of phase growth, *Chem. Eng. Sci.*, **10**, 1–13, 1959.

SELF, S., L. WILSON, and I. A. NAIRN, Vulcanian eruption mechanism, *Nature*, **277**, 440–443, 1979.

SHIMOZURU, D., S. UTIBORI, N. GYODA, E. KOYAMA, T. MIYAZAKI, T. MATSUMOTO, N. OSADA, and H. TERAO, the 1973 eruptive activity of Asama Volcano: General description of volcanic and seismic events, *Bull. Earthq. Res. Inst.*, **50**, 115–151, 1975.

SHIMOZURU, D., Dynamics of magma in a volcanic conduit—special emphasis on viscosity of magma with bubbles, *Bull. Volcanol.*, **41**, 333–340, 1978.

SHIMOZURU, D., An eruptive mechanism inferred from volcanic earthquakes of Asama Volcano (abstr.), in *Abstracts and Timetable of IUGG*, Canberra, Australia, 1979.

SPARKS, R. S. J., The dynamics of bubble formation and growth in magma: a review and analysis, J. *Volcanol. Geotherm. Res.*, **3**, 1–37, 1978.

TANAKA, Y., The mechanism of the explosion earthquakes, *Bull. Volcanol. Soc. Japan*, **16**, 153–161, 1971.

TSUJIURA, M., The difference between foreshocks and earthquake swarms, as inferred from the similarity of seismic waveform (Preliminary report), *Bull. Earthq. Res. Inst.*, **54**, 309–316, 1979a.

TSUJIURA, M., Mechanism of the earthquake swarm activity in Kawanazaki-oki, Izu Peninsla, as inferred from the analysis of seismic waveforms, *Bull. Earthq. Res. Inst.*, **54**, 441–462, 1979b.

VERHOOGEN, J., Mechanics of ash formation, *Am. J. Sci.*, **249**, 729–739, 1951.

WALLIS, G. B., Bubbly flow, sug flow and annular flow, in *One-dimensional Two-phase Flow*, pp. 243–374, McGraw-Hill, 1969.

Arc Volcanism: Physics and Tectonics, edited by D. Shimozuru and I. Yokoyama, 81–93.
Copyright © 1983 by Terra Scientific Publishing Company (TERRAPUB), Tokyo.

Changes in Water Level and Their Implications to the 1977–1978 Activity of Usu Volcano

Hidefumi WATANABE

Usu Volcano Observatory, Hokkaido University,
Sapporo 060, Japan

Measurements of the water level in a 370 m deep well, which is located 2 km east of Usu volcano, have provided us with valuable information on the 1977–1978 activity of Usu volcano. In December after the pumice eruptions in August 1977, the water level was found to have risen as much as 37 m compared with the level in 1972. Observation of the ground deformation of the summit suggested that this anomalous increase was caused by the eastward thrust of the summit which was remarkable in the earlier period of the present activity.

Continuous measurement of the water level has been made since December 1977, and some anomalous changes were observed also in 1978, associated with the developments of the volcanic activities. Mechanism of these anomalous changes and their implications to the volcanic activities were studied, based on various information obtained by seismic, geodetic and geologic observations. It was shown that the anomalous changes in the water level in 1978 might have been caused by the diffusion of the increased pore pressure which occurred beneath the summit crater.

1. Introduction

Usu volcano (42°32′N, 140°50′E) situated on the southern side of Toya caldera in Hokkaido, is a unique stratovolcano for its dacitic domings in historical activity. Its highest peak is about 730 m above sea level, and its base is about 6 km in diameter. Within the summit crater of about 2 km diameter there are two dacite lava domes, Ko-Usu and Oo-Usu, which were formed in the 1663 and the 1853 eruptions respectively (Fig. 1).

The historical rhyolitic eruptions started in the 17th century. The earlier four eruptions in 1663, 1769, 1822, and 1853 occurred at the summit crater, causing nuée ardentes and formation of lava domes. The succeeding two eruptions in 1910 and 1943–1945 formed respectively Meiji-shinzan (a cryptodome) and Showa-shinzan (a lava dome) at the foot of Usu volcano. The 1910 eruption and the 1943–1945 eruption were thoroughly investigated at the highest level in those epochs by OMORI (1911) and MINAKAMI *et al.* (1951) respectively.

The present eruption began on August 7, 1977 after 32 years of dormancy, and stopped in October 1978. However, the seismic activity and ground deformation are still continuing as of January 1982. The detailed descriptions of the sequence of the

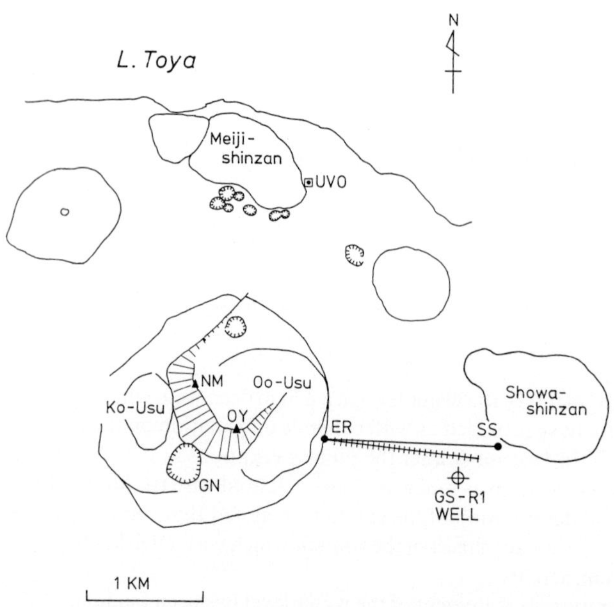

Fig. 1. Topographical sketch map on and around Usu volcano, and location of an observation well GS-R1. Solid line ER-SS is a base line for distance measurement, and hachured line indicates a ropeway. NM = "New Mountain" (cryptodome); OY = Ogariyama (cryptodome); GN = "Gin-numa crater"; UVO = Usu Volcano Observatory.

1977–1978 eruptions and the nature of the ejecta have been reported by KATSUI *et al.* (1978) and NIIDA *et al.* (1980). Geophysical characteristics of the 1977–1978 eruptions and the seismological significance of the doming have been discussed by YOKOYAMA *et al.* (1981) and OKADA *et al.* (1981) respectively.

In relation to the 1977–1978 activities anomalous changes in water level have been observed in a well 370 m deep which is located 2 km east of Usu volcano. In December 1977, after the outburst of the pumice eruption, the water level was found to have increased as much as 37 m compared with the level in 1972. Some anomalous changes in the water level were observed also in 1978, associated with the developments of the volcanic activities. It is a rare opportunity to observe anomalous changes in water level at a well situated very near the active volcano. Mechanism of these anomalous changes and their implications to the volcanic activities are studied in this paper, based on various information obtained by seismic, geodetic and geologic observations.

2. Changes in Water Level in the Usu Volcanic Region

A 370 m deep well (herein referred to as the GS-R1 well), which is located 2 km east of Usu volcano and 1 km south of Showa-shinzan lava dome (Fig. 1), was drilled by Geological Survey of Japan in 1967 originally intending to examine the possible existence of the cryptodome that was suggested by seismic prospecting. Strainers of the well are inserted at the depth of 205–225 m and 335–365 m. This is an important well to observe changes in terrestrial heat flow and water level related to the volcanic activities, as it is situated very closely to the active volcano. Since 1967 measurements of the well water temperature have been repeated by Geological Survey of Hokkaido and Hokkaido University. The temperature changes were very small during 1967–1972 except for the initial recovery after drilling, and the water level was almost constant during the same period.

Measurements of the water level at GS-R1 well have provided us with valuable information on the 1977–1978 activity of Usu volcano. Remarkable changes in the water level have occurred on various time scales; long term, short period, and coseismic changes.

The long-term variations of the GS-R1 water level are indicated in Fig. 2. The water level had been almost constant around 72 m a.s.l. from 1967 till July 1972. But on August 30, 1977 after the pumice eruptions, the water level was higher than 92 m a.s.l. and was found to be 109 m a.s.l. on December 12, 1977. This means that the water level had increased unexpectedly as much as 37 m in relation to the 1977 eruptions of Usu volcano. This was an interesting phenomenon especially from the viewpoint of prediction of volcanic eruptions. It was a regret, however, that we had no data of the GS-R1 water level in the period from July 1972 to August 1977, and could not clarify when the anomalous change had occurred. Continuous measurement of the water level has been made since December 1977. The water level has begun to decrease gradually, after reaching a maximum in December 1977. As of December 1981, the level is almost the same as in 1972, but the decreasing is continuing at a lower rate.

Besides the general trend of decrease, anomalous changes occurred in three periods; March, May–June and August–December 1978 as seen in Fig. 3. These anomalous changes can be seen more clearly in Fig. 4 where the decreasing trend of the water level is subtracted. Among them the rise in March was most remarkable and

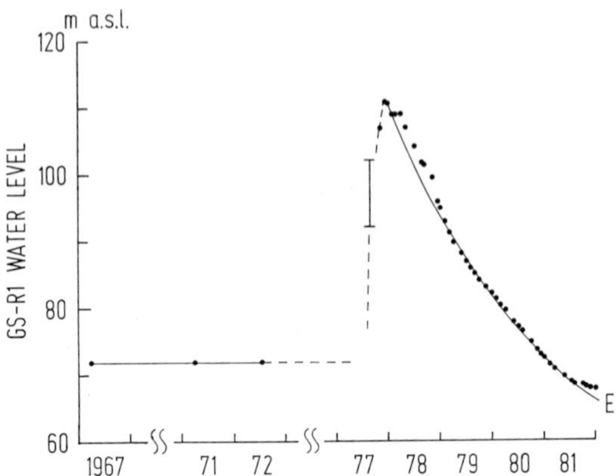

Fig. 2.   Long-term variations of the water level at GS-R1 well. The vertical bar is the uncertainty in the approximate water level on August 30, 1977. The solid curve E indicates the least square fitting of an exponential curve.

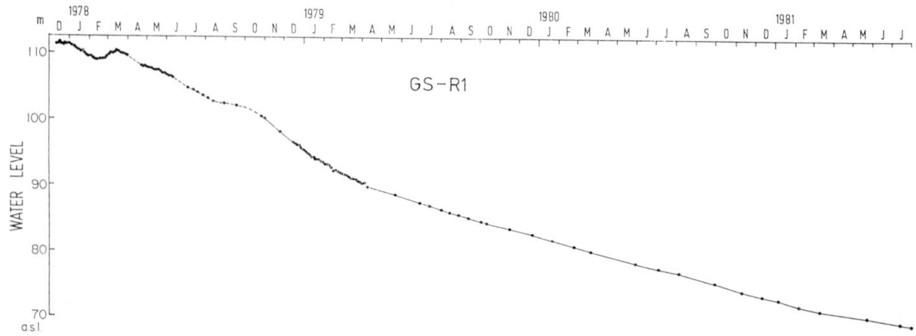

Fig. 3.   Changes in water level at GS-R1 well since December 1977.

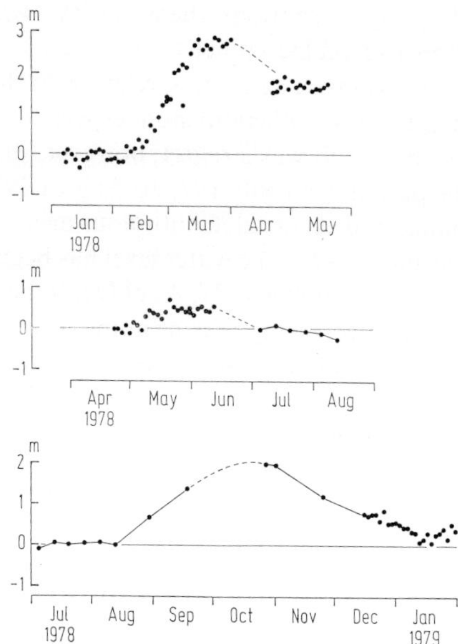

Fig. 4.   Anomalous changes in water level at GS-R1 well in March, May–June, and August–December 1978. They are obtained by subtracting the decreasing trend of the water level in Fig. 3.

amounted to about 2.7 m. The second and third rises were 0.5 m and 2.0 m, respectively.

It was also found that almost all felt shocks of which magnitudes $M > 3$, resulted in offsets of the water level. Here only the brief description of this coseismic phenomena will be made. The offsets coincided with the occurrence of earthquakes and remained for several hours. Both positive and negative offsets were observed and the polarity of the offsets were definitely dependent on the locations of earthquake foci. The amounts of the offsets were closely related to earthquake magnitude. At Usu volcano, large earthquakes occur only in two regions; at the corners of the major U-shaped fault (OKADA *et al.*, 1981). These groups of earthquakes have their own focal mechanisms, which are almost identical within the groups. The offset of the water level is related to a sudden release of the pore pressure when an earthquake occurs.

3.   Interpretation of the Anomalous Changes in the Water Level

Extensive observations and researches of the 1977–1978 eruptions have been made since the outburst of the eruptions (e.g., KATSUI *et al.*, 1978; NIIDA *et al.*, 1980; YOKOYAMA *et al.*, 1981; OKADA *et al.*, 1981). Then the mechanisms of the anomalous water level changes in 1977–1978 can be investigated, based on various kinds of the information obtained by seismic, geodetic and geologic methods.

3.1   Long-term variations

The general features of the long-term variations of the GS-R1 water level (Fig. 2) are characterized by rapid increase in the beginning and gradual decrease after reaching a maximum in December 1977. It is unfortunate that there were no available data in the period from July 1972 to August 30, 1977, because the beginning of the increase is important for considering the cause of the increase in the water level. However, it can be said at least that the beginning of the increase might not be much later than the onset of the pumice eruption on August 7, 1977, judging from the rapid increase in the water level during August 30 to December 12, 1977.

It was suggested (YOKOYAMA *et al.*, 1981) that the ground in the summit and at the eastern foot of Usu volcano had begun to deform a few days before the outburst of the eruption. Recently, other evidence for the onset and developments of the ground deformations in the summit has been found. There is a ropeway (of 1.4 km length) between the east rim of the summit and the eastern foot of Usu volcano (see Fig. 1). Apparent extension of the ropeway has been measured daily since 1965. The results of the measurements are indicated in Fig. 5. Until August 6, 1977, only the secular extension of the cable due to stretching had been observed. But on August 7, about 07:30 (100 min before the eruption), an anomalous extension of 24 cm was noticed, and then it continued to increase as seen in the figure. The rate of the apparent extension, however, decreased gradually and stopped at the end of May 1978. This is important observational data of the ground deformations, because the apparent extension of the cable is a measure of the contraction of the eastern side of the volcano caused by the eastward thrust of the summit.

The remarkable eastward thrust of the summit must have affected the water level at GS-R1 well by compressing the aquifer at the eastern foot of the volcano. Accordingly, one may say that the rapid increase in the water level in the beginning was caused by the thrust of the summit which was remarkable in the earlier stage, and that the water level stopped its increase in December 1977 when the rate of the eastward thrust was significantly reduced.

After reaching a maximum in December 1977, the water level has decreased gradually. The decreasing trend of the water level can be approximated by an exponential curve:

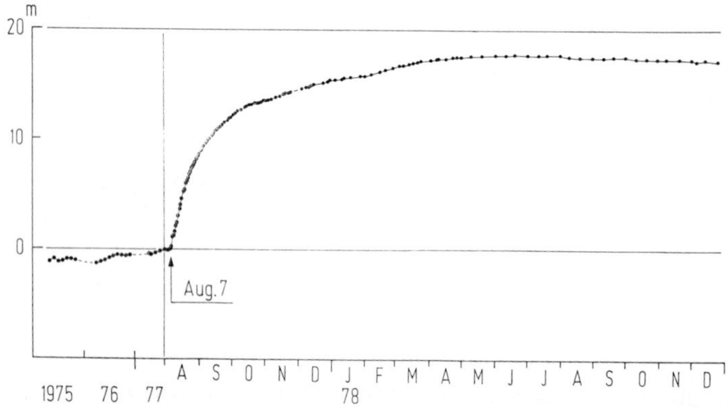

Fig. 5. Apparent extension of a ropeway which is located on the eastern side of Usu volcano (Fig. 1), and was a measure of the contraction between the east rim and the eastern foot of the volcano. Anomalous extension occurred immediately before the onset of the eruption on August 7, 1977.

$$H = (H_0 - H_\infty) \exp\left(-t/T\right) + H_\infty \tag{1}$$

where H is the water level, H_0 and H_∞ are the initial and the ultimate water levels respectively, t denotes time, and T is the characteristic time constant of the exponential decrease. Least square fitting of the exponential curve gave the approximate estimations; $H_\infty = 50$ m and $T = 1,100$ days (curve E in Fig. 2).

The above result suggests that the gradual decrease in the water level at GS-R1 well since 1978 may be caused by the leaking of the aquifer to which GS-R1 well is open. The time constant of the leaking is rather large ($T = 1,100$ days), and consequently it is noted that changes in the water level of which time scales are shorter than half a year, can be analyzed separately from the long-term variations.

3.2 Anomalous changes in 1978

First, the mechanism of the anomalous rise in March 1978 will be discussed. Interpretation of this rise will be simpler than other anomalous changes in 1978, because there occurred no eruption in that period which might have exerted influences on the water level.

In general, the change in water level in a well open to an artesian aquifer is caused not only by the compression of the aquifer but also by the pore pressure change in the aquifer. Then it must be decided which is the main cause of the change in the water level. Length of a base line between the east rim and the eastern foot of Usu volcano (ER-SS in Fig. 1) has been measured repeatedly. Figure 6 shows the rates of shortening in length of the base line ER-SS. It is noted that the maximum rate of shortening occurred in February 1978 and did not coincide with the peak of the water level at GS-R1 well which occurred at the end of March as seen in Fig. 4. This means that the rise of the water level in March 1978 was not caused by the compression of the aquifer.

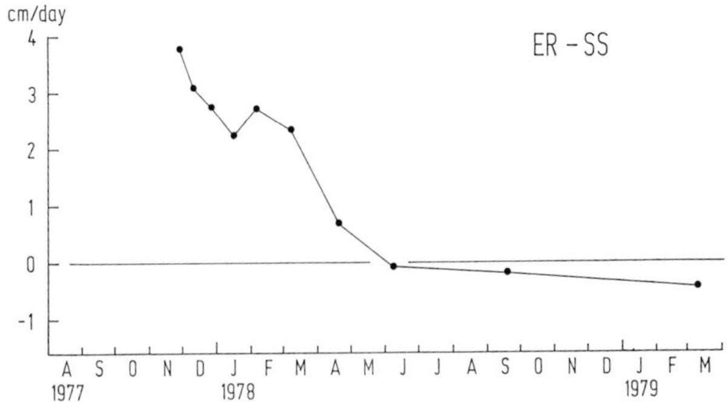

Fig. 6.   Shortening rate of a base line ER-SS between the east rim of the summit and the eastern foot of Usu volcano.

General features of developments in the 1977–1978 activities of Usu volcano can be seen in its temporal variations of energy release (YOKOYAMA *et al.*, 1981). Figure 7 shows the variations of the daily discharge of seismic energy, and of the daily rate of upheaval of a cryptodome Ogariyama. In the figure the variations of discharge of seismic energy are divided into four stages; eruption stage, and first, second, and third post-eruption stages. The first and second stages are bounded by a discontinuity at the end of January 1978. Both the first and second stages were characterized by exponential decreases of seismicity and deformation activity. These facts suggested that the energy source beneath the volcano was isolated from deeper sources during the first and second stages, and that a small amount of energy might have been supplied at the

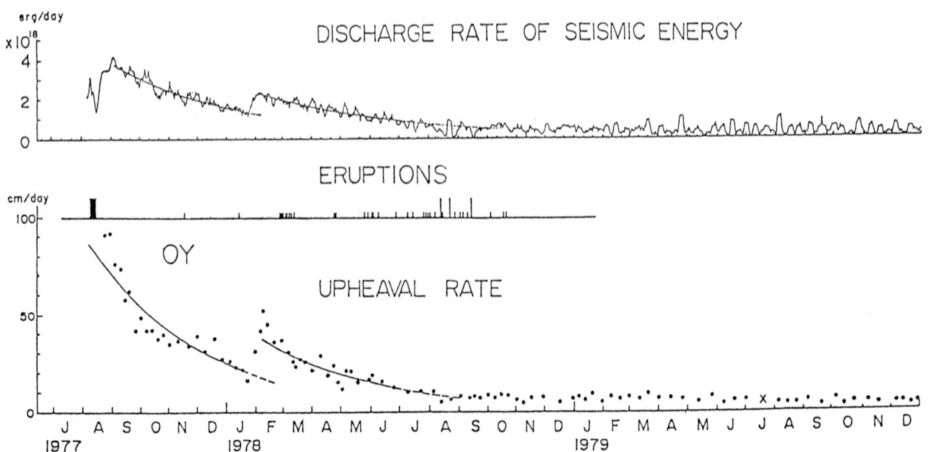

Fig. 7.   Daily discharge rate of seismic energy averaged for every five days, and daily upheaval rate of Ogariyama cryptodome averaged around observation periods (after YOKOYAMA *et al.*, 1981).

end of January 1978 and this resulted in the increased discharge of seismic energy and upheaval of cryptodomes.

Now we take a supposition that magma and/or volcanic gases may have been supplied from depth at the end of January 1978 to increase the pore pressure beneath Usu volcano, and that the increased pore pressure may have diffused to the surrounding medium and eventually caused the rise of the water level at GS-R1 well with a time lag of 50 days. In the following, mechanism of the anomalous change in water level is discussed based on a pore pressure-diffusion model.

The model is shown schematically in Fig. 8. In the figure a spherical pressure source is assumed at depth, $z = 0$; the ground surface, $z = F$. P denotes any observation point of which distance from the center of the spherical source is r.

The governing equation of the pore pressure is as follows:

$$\frac{\partial P}{\partial t} = K\nabla^2 P + S \tag{2}$$

where t, K, and S denote time, hydraulic diffusivity and the pressure source respectively. For simplicity, the hydraulic diffusivity is taken to be constant throughout the half space beneath the surface. It is assumed that the initial pressure is constant at zero both inside and outside the spherical source:

$$P = 0, \, t = 0, \tag{3}$$

and that the pressure at the ground surface is always kept at zero:

$$P = 0, \, z = F. \tag{4}$$

The hydraulic diffusivity and the location and time-function for the pressure source are the unknown parameters to be postulated.

As to the location of the pressure source, we refer to the distribution of hypocenters. Figure 9 shows the distribution of hypocenters in the period of the increased seismic activity which began on January 24, 1978. An earthquake-free zone can be seen clearly in the summit. It is suggested that the dacite magma has been intruding from depth to the earthquake-free zone which underlies the area of doming

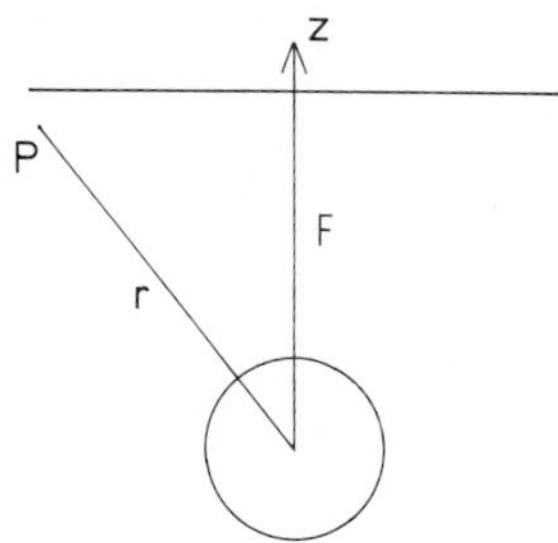

Fig. 8. Schematical illustration of the pore pressure-diffusion model. Spherical pressure source is placed at depth F beneath the surface.

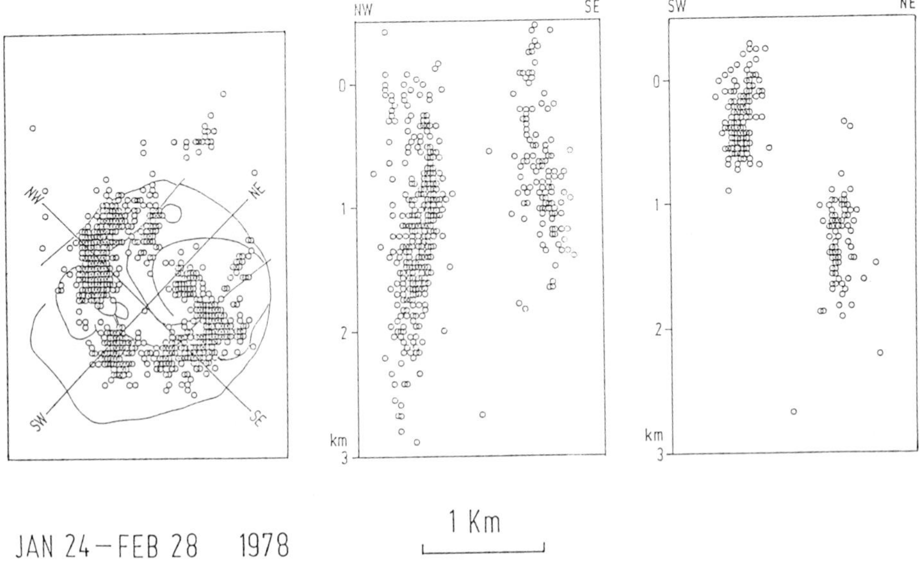

Fig. 9.   Distribution of hypocenters of the earthquake swarm during January 24–February 28, 1978. An earthquake-free zone is clearly seen in the central part of the summit crater where remarkable doming has been occurring.

within the summit crater (OKADA *et al.*, 1981). We may suppose that the pressure increase has taken place at the bottom part of the earthquake-free zone; about 2 km b.s.l. in depth.

The seismic activity began to increase on January 24, 1978, and reached a maximum ten days later and then decreased again. This fact suggests the following time-function for the pressure source:

$$S = \begin{cases} S_0(1 + \sin{(t/T_0 - 1/2)\pi}), & 0 \le t \le 2T_0 \\ 0, & t > 2T_0 \end{cases} \tag{5}$$

where T_0 is taken to be ten days and S_0 is a parameter adjusted to result in the observed changes in the water level at GS-R1 well.

In order to satisfy the boundary condition at the surface (Eq. (4)), we may take the image source, $-S$, which is placed at the symmetrical position in regard to the ground surface. Solutions of the pore pressure can be obtained by the superposition of the solutions which are derived from the source S and its image $-S$, respectively.

Among unknown parameters, only the hydraulic diffusivity exerts significant effects on the time-evolution of the pore pressure. Numerical calculations of Eq. (2) were made by taking various values for the diffusivity. The results are shown in Fig. 10 where the diffusivity (K) is taken as 1, 2, and 3 $\times$ 10^3 cm^2/s, and the depth (F) and the radius (A) are kept constant as $F = 2.5$ km and $A = 200$ m, respectively. In the lower part of the figure solid lines indicate the changes in the water level corresponding to the

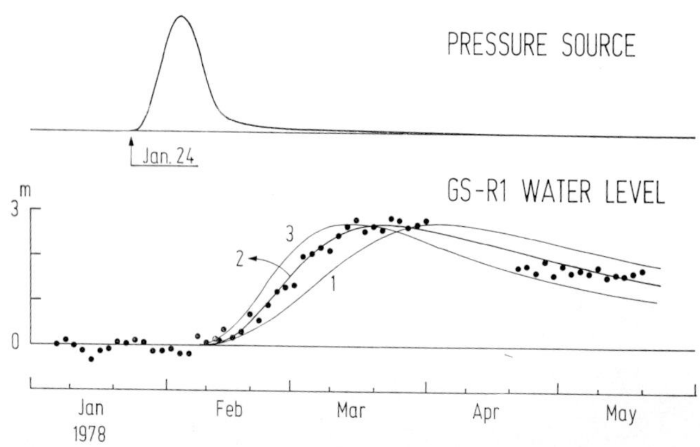

Fig. 10. Time-evolutions of the water level at GS-R1 well which were calculated on the pore pressure-diffusion model, by taking various values for the hydraulic diffusivity (K). Numbers 1, 2, and 3 correspond to the solutions for $K = 1, 2, 3, \times 10^3$ cm^2/s, respectively. Upper part of the figure indicates the pressure change at the center of the spherical source (on an arbitrary scale).

pore pressure changes at GS-R1 well. Observed changes in the water level are well fitted by the theoretical curve based on the pore pressure-diffusion model, if the hydraulic diffusivity is taken as $K = 2 \times 10^3$ cm^2/s. In the upper part of the figure pressure change at the center of the spherical source is also shown on an arbitrary scale.

Magnitude of the pressure in the source depends on both the size and depth of the source. Figure 11 shows the relation of the maximum pressure increase in the source region to the radius and the depth of the spherical source. Suggested ranges of the depth and the radius are indicated by the shaded area.

From Figs. 10 and 11 it can be concluded that the anomalous rise in the water level at GS-R1 well in March 1978 might have been caused by the pore pressure increase

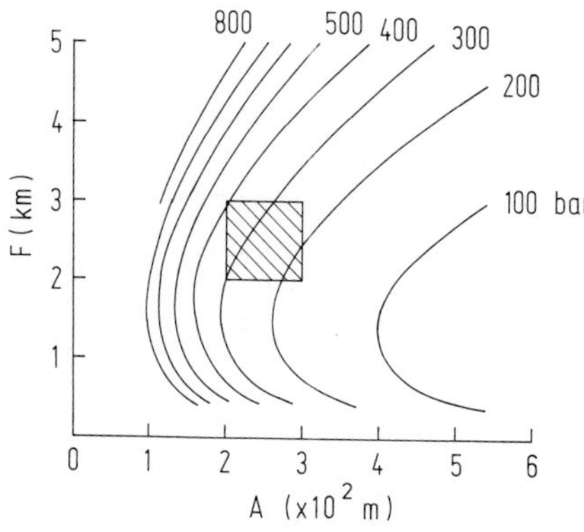

Fig. 11. Relation of the maximum pressure in the source region to the depth (F) and radius (A) of the spherical pressure source.

amounting to about 200–300 bar in the source region beneath the summit. The hydraulic diffusivity of the medium beneath Usu volcano is also estimated to be about 2×10^3 cm^2/s. This is almost the same value, in order of magnitude, as the reported estimations of the crustal diffusivities of the pore pressure (e.g., ANDERSON and WHITCOMB, 1975).

Another anomalous change in GS-R1 water level which occurred during August–December 1978, can also be interpreted by the similar pressure diffusion model. During the second stage, February–October 1978, the discharge of the total energy of volcanic activity decreased exponentially, as already seen in Fig. 7. However, the eruptions became violent during July, August and September 1978, in contrast to the decreasing tendency of the energy discharge. This is because the eruptions are governed also by the depth to the magma which appears to have shallowed considerably in the earlier period of 1978. A large number of eruptions took place at the southern border of the major U-shaped fault, forming one big depression which was called "Gin-numa crater." With progression of the eruptions the degree of vesiculation of the essential ejecta increased generally, reaching to the maximum at the end of August, then rapidly decreased (NIIDA et al., 1980).

From these observations it is possible to suppose that the pressure might have increased beneath Gin-numa crater during July, August, and September 1978. The pressure increase beneath Gin-numa crater is supported also by the active regional seismicity during that period. Figure 12 illustrates the magnitude-time diagram of the earthquakes of which hypocenters are located around Gin-numa crater. It can be seen in the figure that an intense earthquake swarm occurred during June–September 1978.

As for the size and depth of the pressure source, the results of the seismic observations suggest that the depth of the head of the pressure source may be around

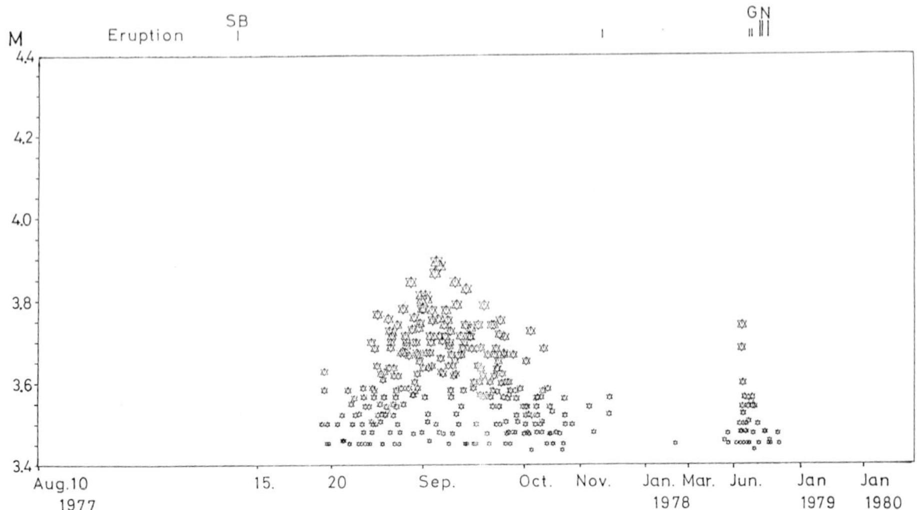

Fig. 12. Magnitude-time diagram of the earthquake swarm which occurred around Gin-numa crater (after UMEHARA, 1981).

the sea level, that is about 400 m below the surface, and the horizontal extent of the source region may be about 300 m. This is supported by the facts that the radial extent of the location of the craterlets J, K, L, and M, which were combined to form Gin-numa crater, is about 300 m, and that the biggest explosions during August and September 1978 must have taken place a few hundred meters below the crater bottom because of their high explosion pressures. Consequently, the depth and radius of the spherical pressure source are taken as 500 m and 150 m, respectively. The sinusoidal time-function for the pressure source was also assumed and the hydraulic diffusivity was taken to be of the same value, $K = 2 \times 10^3$ cm^2/s, as before.

Under these assumptions, the pore pressure changes beneath Usu volcano were calculated. The results of the calculations are summarized in Fig. 13. The lower part of the figure shows the observed changes in GS-R1 well water level. The calculated water level is indicated by the solid curve. Pressure variations in the source region are estimated as shown in the middle row, where the locations of the pressure source are different; at the bottom part of the earthquake-free zone (2 km b.s.l.) for the rise in March, and beneath Gin-numa crater (around the sea level) for the rises in May–June and August–December. Eruptions are shown on a relative scale of intensity in the upper row.

The observed anomalous changes in the water level are well fitted by the calculated curves, based on the pore pressure-diffusion model. This strongly suggests that the observed changes in the water level were caused by the increased pore pressure beneath the summit crater associated with the developments of the volcanic activity in 1978.

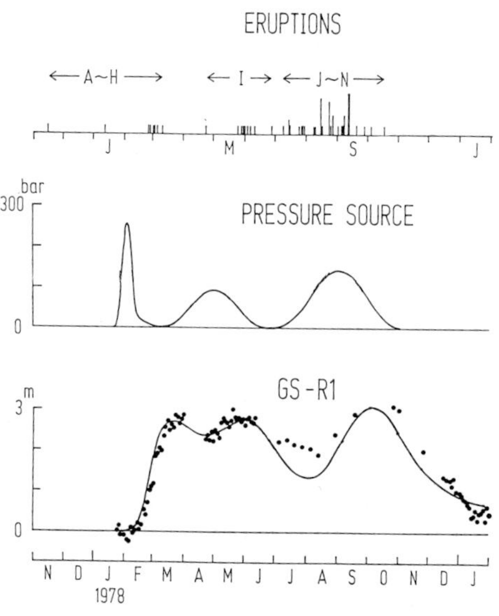

Fig. 13. Relationship between the anomalous changes in water level at GS-R1 well, the estimated pore pressure increases beneath the summit, and the surface activities of Usu volcano in 1978.

4. Conclusion

Anomalous changes in water level have been monitored at an observation well near Usu volcano in relation to the 1977–1978 activities. After the pumice eruptions in August 1977 the water level was found to have increased as much as 37 m compared with the level before the eruptions. An observation of the ground deformation of the summit suggested that the anomalous increase in the water level was caused by the eastward thrust of the summit which was remarkable in the earlier stage of the present activity.

Another anomalous changes were observed in 1978 associated with the developments of the volcanic activities, and this enabled us to estimate the possible pore pressure changes beneath the summit crater and to discuss their relations to the volcanic activities. Coseismic offsets of the water level were also found accompanying the felt shocks.

The present investigations suggest that monitoring of the underground water level may offer another useful informations, independent of those obtained by the seismic and geodetic observations, for understanding the physical processes of the volcanic activities, if some suitable observation wells are available near the active volcano.

I wish to express my sincere thanks to Prof. I. Yokoyama for his valuable advice throughout this work and critical reading of an earlier version of the manuscript. Thanks are also due to Dr. Hm. Okada and H. Yamashita who offered helpful comments, and to Mr. T. Maekawa who helped me in the field observations.

REFERENCES

ANDERSON, D. L. and J. H. WHITCOMB, Time-dependent seismology, *J. Geophys. Res.*, **80**, 1497–1503, 1975.

KATSUI Y., Y. OBA, K. ONUMA, T. SUZUKI, Y. KONDO, T. WATANABE, K. NIIDA, T. UDA, S. HAGIWARA, T. NAGAO, J. NISHIKAWA, M. YAMAMOTO, Y. IKEDA, H. KATAGAWA, N. TSUCHIYA, M. SHIRAHASE, S. NEMOTO, S. YOKOYAMA, T. SOYA, T. FUJITA, K. INABA, and K. KOIDE, Preliminary report of the 1977 eruption of Usu volcano, *J. Fac. Sci., Hokkaido Univ.*, Ser. 4, **18**, 385–408, 1978.

MINAKAMI, T., T. ISHIKAWA, and K. YAGI, The 1944 eruption of Volcano Usu in Hokkaido, Japan, *Bull. Volcanol.*, Ser. 2, **11**, 45–157, 1951.

NIIDA, K., Y. KATSUI, T. SUZUKI, and Y. KONDO, The 1977–1978 eruption of Usu volcano, *J. Fac. Sci., Hokkaido Univ.*, Ser. 4, **19**, 357–394, 1980.

OKADA, Hm., H. WATANABE, H. YAMASHITA, and I. YOKOYAMA, Seismological significance of the 1977–1978 eruptions and the magma intrusion process of Usu volcano, Hokkaido, *J. Volcanol. Geotherm. Res.*, **9**, 311–334, 1981.

OMORI, F., The Usu-san eruption and earthquakes and elevation phenomena, *Bull. Imp. Earthq. Inv. Comm.*, **5**, 1–38, 101–107, 1911.

UMEHARA, H., Study of earthquake families in the 1977–1980 activity of Usu volcano, M.C. Thesis, Faculty of Science, Hokkaido University (in Japanese), 1981.

YOKOYAMA, I., H. YAMASHITA, H. WATANABE, and Hm. OKADA, Geophysical characteristics of dacite volcanism—The 1977–1978 eruption of Usu volcano, *J. Volcanol. Geotherm. Res.*, **9**, 335–358, 1981.

Arc Volcanism: Physics and Tectonics, edited by D. Shimozuru and I. Yokoyama, 95–113.
Copyright © 1983 by Terra Scientific Publishing Company (TERRAPUB), Tokyo.

A Theoretical Model for Dispersion of Tephra

Takeo SUZUKI

Department of Physics, The Institute of Vocational Training,
Sagamihara, Kanagawa 229, Japan

The dispersion of tephra is a function of many factors: total mass, median diameter and standard deviation of material erupted, height of the eruption column, wind velocity, and the nature of particle diffusion from the eruption column. The relationship between these factors and the dispersion of tephra is made clear using a model of two-dimensional diffusion in the atmosphere. It was found that the shape of the curves showing the relationship between logarithmic mass per unit area of deposited tephra and logarithmic area enclosed by iso-mass contours is a function of the standard deviation of material erupted; the inflection-point of the curves is a function of median diameter of the material erupted and height of the eruption column. Variations in wind velocity have an effect upon the shape of the dispersion isopleth curves. The effects of total mass of tephra on dispersal patterns may be expressed by plotting the area enclosed by iso-mass contours as a function of mass per unit area. Variations in total mass cause the resulting curves to shift laterally, but do not change their basic shape.

A quantitative definition for the classification of "bigness" by Walker is proposed for comparison of various volcanic eruptions.

1.   Introduction

The mass of erupted material and the areal extent and thickness of tephra fallout are primary factors in understanding the destructive potential of volcanic eruptions. The former is a function of the total energy of a volcanic eruption, and the latter depends upon the specific expression of this energy; that is, the initial size of ejected tephra, the extent of fragmentation of the tephra, and the height of the eruption column, as well as wind velocity and wind direction. Therefore, tephra fallout patterns may vary greatly in individual eruptions and the tephra takes on a variety of the destruction even though total mass of erupted material may be the same.

The total mass of erupted material is estimated from the accumulation volume of the tephra by empirical methods, since there is no general theoretical basis for computation. The author investigated the relationship between the thickness of air fall deposits and area enclosed by isopach lines and deduced the following formula:

$$Y = A \cdot X^n, \quad X = \log_{10}(h_{max}/h), \quad Y = \log_{10}(S/S_0).$$

S: the area enclosed by isopach of thickness h. S_0:   that of maximum thickness h_{max} (SUZUKI, 1981). This curve was called the "thickness-isopach area" curve which does not conflict with the following assumptions:

(a)   The erupted material consists of a finite quantity of volcanic particles.

(b)   The distribution of the diameter of these particles has a single mode.

(c)   All of the particles fall at terminal velocity and finally accumulate on the ground.

Meanwhile, there is no empirical method proposed for the dispersion of tephra, which depends upon many factors and conditions at the time of an eruption. Their effect on the dispersion of tephra must be investigated. The simulation based on the data gathered during volcanic eruptions at present is a better method to determine the factors and the conditions of ancient tephra eruptions.

The data on air-fall deposits obtained in the field and the laboratory are thickness, particle size distribution, proportion of pumice to lithic fragments, bulk density of deposits, and density of particles. All of these data are a function of sampling sites and yield information on the conditions at the time of eruption, total quantity of erupted material and its particle size distribution, height of the eruption column, velocity and direction of wind, vertical velocity within the eruption column, and diffusion of volcanic particles from the eruption column. Interpretation or estimation of these phenomena of volcanic eruptions from the field data of tephra characteristics is a very useful technique for the study of volcanic activity and estimation of volcanic disaster-potential.

Figure 1 shows the relationship of total mass of tephra and its dispersal area to the destructive potential of volcanoes, and indicates observable or measurable characteristics of eruptions that are necessary for understanding the dispersive activity of an eruption column.

Tsuya (1955) classified the intensity of volcanic activity into ten grades, founding on the volume of erupted material. Walker (1980) pointed out that explosive volcanic eruptions show five kinds of "bigness:" magnitude, intensity, dispersive power, violence, and destructive potential. He defined the magnitude as the total energy released by the eruption, and proposed that the magnitude is expressed by the dense rock equivalent volume (DRE). Alternatively, Yokoyama (1956) calculated the thermal energy founding on the total mass of erupted products. The author prefers to define magnitude of volcanic eruptions by means of total mass of eruptive products (see Appendix).

In this paper dispersion of tephra is investigated using a two-dimensional diffusion model. This model is based on the following factors:

(1)   Diffusion of volcanic particles from an eruption column.

(2)   Horizontal transfer of these particles with the horizontal movement of the atmosphere.

(3)   Horizontal diffusion of the particles due to turbulency of the atmosphere.

(4)   Sinking of the particles in the atmosphere.

2.   Dispersion of Tephra in the Atmosphere

Since the movement of air mass is random in time and space due to the interaction of many eddies, movement of dispersing particles in the atmosphere is also random. Small particles diffuse in the atmosphere in both vertical and horizontal directions;

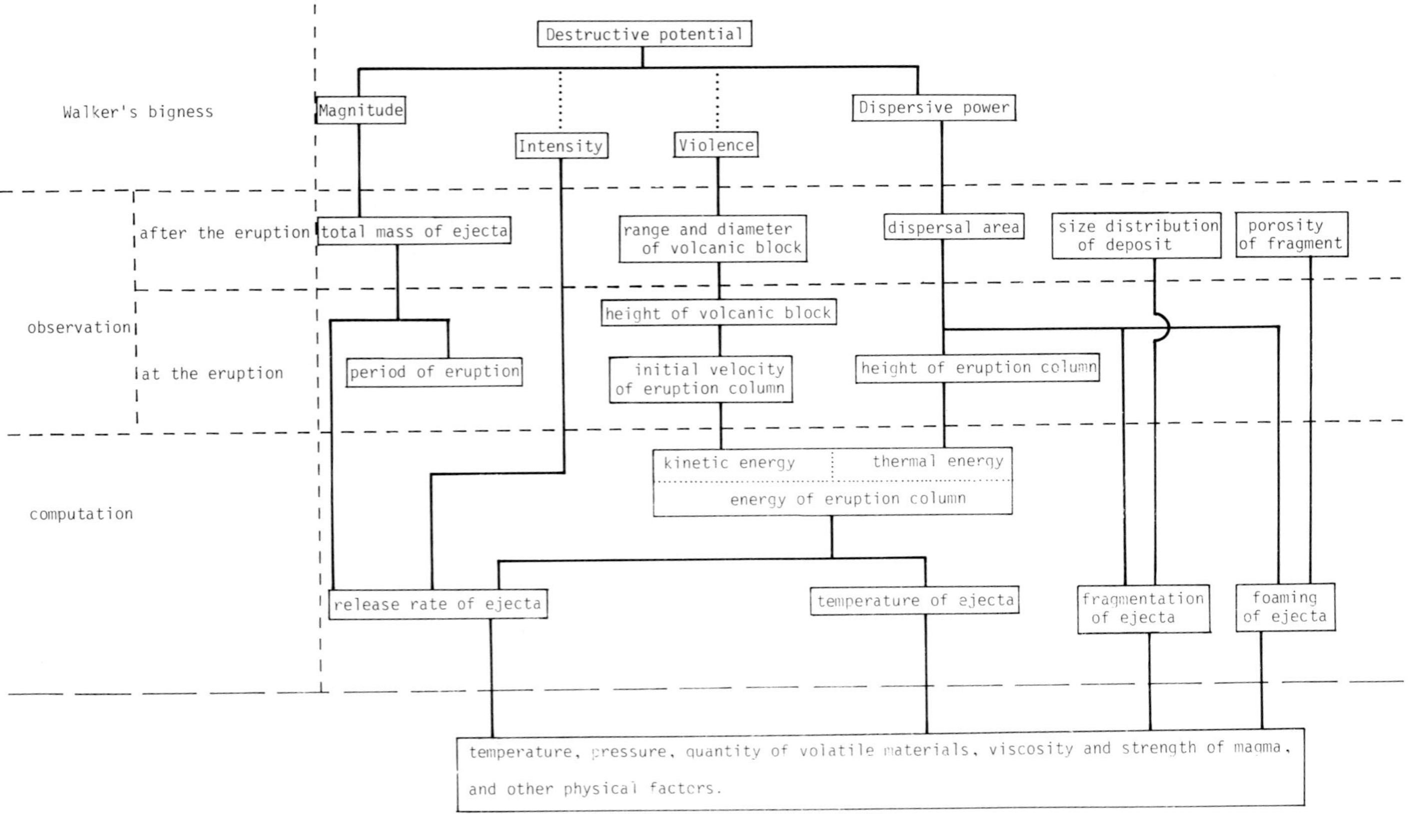

Fig. 1.   Block diagram showing the relationship of the factors concerned with volcanic eruptions.

however, the scale of horizontal turbulence is much greater than that of vertical turbulence. In this paper, therefore, particle dispersion is investigated using a two-dimensional diffusion model in which only horizontal turbulent diffusivity is considered.

The two-dimensional differential equation for diffusion in wind of uniform velocity, taking x leeward (CSANADY, 1973), is as follows

$$\frac{\partial \chi}{\partial t} = - u \frac{\partial \chi}{\partial x} + \frac{\partial}{\partial x}\left(K \frac{\partial \chi}{\partial x} \right) + \frac{\partial}{\partial y}\left(K \frac{\partial \chi}{\partial y} \right), \tag{1}$$

where χ is the concentration of the diffusing substance, K is the eddy diffusivity, and u is the velocity of wind. The eddy diffusivity $K(r, t)$ in turbulent diffusion is a function of the distance r from the center of diffusion and the diffusion time t, where $r = \{(x - u t)^2 + y^2\}^{1/2}$.

The function form of $K(r, t)$ has been expressed in various ways. Comparison of the solution of Eq. (1) with published data was made by using the following variance σ_r^2 and apparent eddy diffusivity A_L.

$$\sigma_r^2 \equiv \int_0^\infty r^2 \cdot \chi(t,r) \cdot 2\pi r \, dr \bigg/ \int_0^\infty \chi(t,r) \cdot 2\pi r \, dr, \quad A_L \equiv \sigma_r^2/4t.$$

For the typical $K(r, t)$, σ_r^2 and A_L are as follows:

Fick type diffusion	$K = \text{const.};$	$\sigma_r^2 \propto t,\ A_L = K$
JOSEPH and SENDNER (1958)	$K \propto r;$	$\sigma_r^2 \propto t^2,\ A_L \propto L$
OZMIDOV (1958)	$K \propto r^{4/3};$	$\sigma_r^2 \propto t^3,\ A_L \propto L^{4/3}$

where $L \equiv 3\,\sigma_r$.

On the basis of observations, the following relations have been reported.

$$A_L \propto L^{4/3} \qquad \text{(RICHARDSON, 1926)}$$
$$\sigma_r^2 \propto t^2 \sim t^3 \qquad \text{(BATCHELOR, 1950)}$$
$$A_L \propto L^{1.1}, \ \sigma_r^2 \propto t^{2.3} \ \text{(OKUBO, 1971)}$$

The above stated K is expressed as a function of r only. In the turbulent diffusion of air-fall particles, however, the diffusion time is the fall time of particles. The eddy diffusivity K may, therefore, also be a function of the fall time of particles. The scale and time effects of turbulent diffusivity are caused by superposition of the effects of many eddies, and the effect of large-scale eddies may be equivalent to that of plolonged time in the random movements of dispersing particles. The solution of Eq. (1) is obtained by writing $K = C\, t^{3/2}$

$$\chi = \frac{5q}{8\pi C\, t^{5/2}} \exp\left[- \frac{5\{(x - ut)^2 + y^2\}}{8C\, t^{5/2}} \right], \tag{2}$$

where C is a constant and q is the amount of material. Equation (2) gives the variance and the apparent eddy diffusivity as follows:

$$\sigma_r^2 = \left(\frac{8C}{5} \right) t^{5/2}, \ A_L = 0.08073\, C^{2/5}\, L^{6/5}.$$

The assumption for $K = Ct^{3/2}$, therefore, satisfies Eq. (1) and is compatible with the observed data of Richardson, Batchelor, and Okubo. The data on diffusion in the atmosphere shown in Fig. 2 gives

$$A_L = 0.887\, L^{6/5},$$

where A_L is given in cm^2/s and L in cm, and hence $C = 400$.

3.   Atmospheric Settling Velocity of Pyroclastic Particles

WALKER *et al.* (1971) determined the terminal fall velocity of pyroclastic particles by laboratory measurements, and calculated the sea-level terminal fall velocity of particles with various sizes and densities. WILSON (1972) estimated the fall time of pyroclastic particles from great height using Walker *et al*'s method. WILSON and HUANG (1979) founding on the laboratory measurement of terminal fall velocities for pumice, glass shards and feldspar crystals of ash size, proposed the following formula on the relationship between the drag coefficient C_a, Reynold's number R_a and shape parameter F,

$$C_a = \frac{24}{R_a} F^{-0.828} + 2\sqrt{1.07 - F}, \tag{3}$$

where

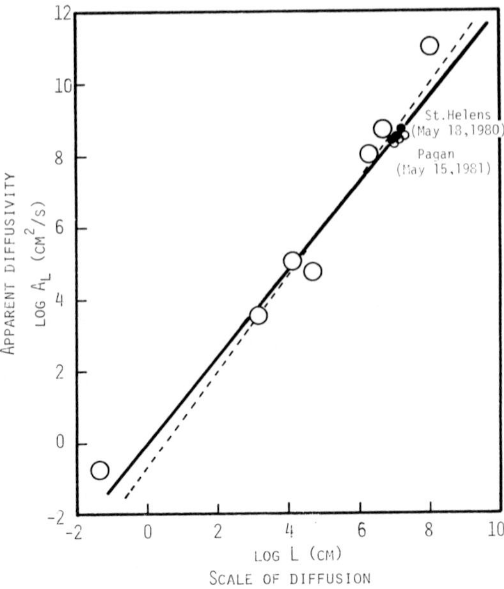

Fig. 2.   Diffusion diagram for apparent diffusivity vs. scale of diffusion in the atmosphere. The large open circles and dotted line are from RICHARDSON (1926), the solid line is $A_L = 0.887\, L^{6/5}$.

$$C_a = \frac{4\psi_p g \, d}{3\psi_a V_t^2}, \quad R_a = \frac{\psi_a V_t d}{\eta_a}, \quad F = \frac{b+c}{2a}, \quad d = \frac{a+b+c}{3}$$

(d is the mean diameter of particles with principal axes a, b, c (a is the longest)), V_t is the terminal velocity of the particle, η_a, ψ_a are the viscosity and density of air, ψ_p is the density of particles, and g is the acceleration due to gravity.

Equation (3) gives the terminal fall velocity V_t as a function of the mean diameter d and shape parameter F of particles:

$$V_t = \frac{\psi_p g d^2}{9\eta_a F^{-0.828} + \sqrt{81\eta_a^2 F^{-1.656} + \frac{3}{2}\psi_a \psi_p g \, d^3 \sqrt{1.07 - F}}}. \tag{4}$$

For particles having a mean diameter less than 0.01 cm, however, the terminal fall velocity for pumice estimated by Eq. (4) does not agree with the experimental data obtained by Wilson and Huang (1979). Equation (3) depends heavily on the data for glass shards. They used 2.40 (g/cm^3) for the density of all glass-shards, and ascribed the change of C_a to the shape parameter F. However, the assumption of uniform 2.40 (g/cm^3) density for glass-shard is questionable, because the effects of density and the shape parameter on C_a cannot be separated. On the basis of experimental measurements of pumice and feldspar crystals, Eq. (3) is corrected as follows:

$$C_a = \frac{24}{R_a} F^{-0.32} + 2\sqrt{1.07 - F}, \tag{3'}$$

and Eq. (4) is corrected to

$$V_t = \frac{\psi_p g d^2}{9\eta_a F^{-0.32} + \sqrt{81\eta_a^2 F^{-0.64} + \frac{3}{2}\psi_a \psi_p d^3 \sqrt{1.07 - F}}}. \tag{4'}$$

Figure 3 shows the experimental data and the results using Eqs. (4) and (4)′ for comparison.

On the other hand, in order to calculate the fall time (T) from the high altitude, the above expression (4)′ is inconvenient for the successive integration during the passage of particles. Therefore, let us derive a simple formula to calculate the fall time of particles, assuming that the change of fall velocity depends only on the density of surrounding air. The effect of the density change on V_t is indicated in the following equation for terminal fall velocity:

$$V_t = \left(\frac{mg}{C_a A \psi_a}\right)^{\frac{1}{2}},$$

where m is the mass of particle, A is the effective cross-sectional area, C_a is drag coefficient and g is the acceleration due to gravity; then

$$V_z/V_0 \fallingdotseq (\psi_{a0}/\psi_{az})^{\frac{1}{2}},$$

where V_0 and V_z are the terminal fall velocity of particle at sea-level and height z, and

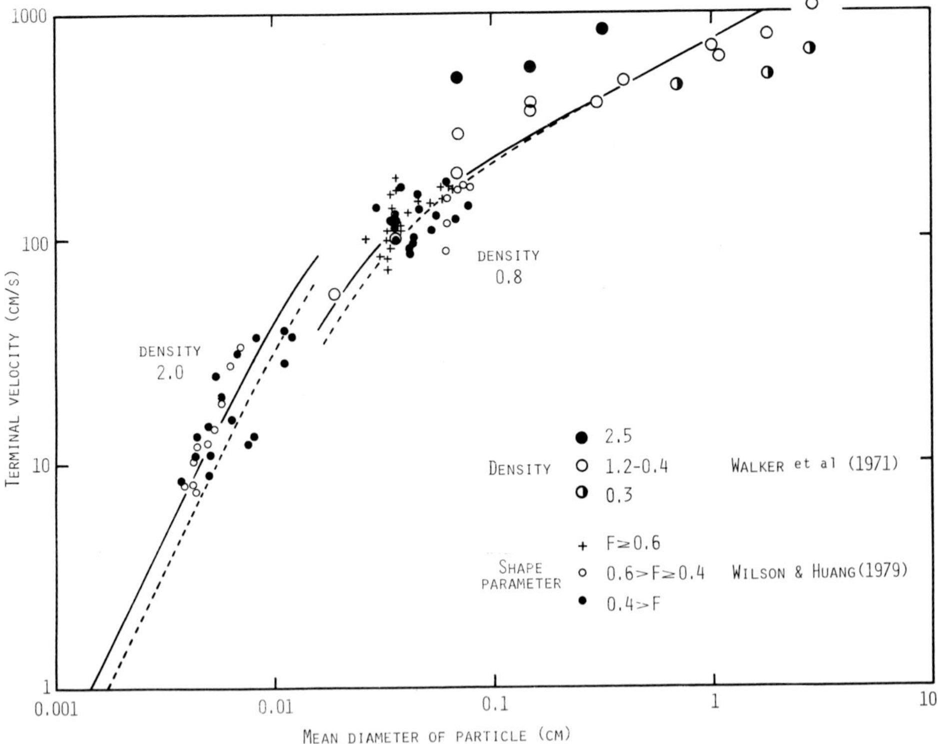

Fig. 3.   Terminal fall velocity of volcanic particles vs. mean diameter of the particle at sea-level. Dotted curves; Eq. (4), solid curves; Eq. (4)'. All experimental data are from WALKER *et al.* (1971) and WILSON and HUANG (1979).

ψ_{a0} and ψ_{az} are the density of air at sea-level and height z. Assuming constant temperature of air, Boyle's law gives

$$\psi_{az}/\psi_{a0} = \exp\{(-\psi_{a0}/P_0)gz\},$$

where P_0 is the atmospheric pressure at sea-level; then

$$V_z = V_0 \exp(0.0625\, z(\text{km}))$$

and the fall time from height z is

$$T = \int_0^z \frac{\mathrm{d}z}{V_z} = \frac{1 - \exp(-0.0625\, z(\text{km}))}{0.0625\, V_0}. \tag{5}$$

WILSON (1972) computed the effect of height on the fall time using the Runge-Kutta method for $\eta_a(z), \psi_a(z)$, and $g(z)$. The fall time obtained by Eq. (5) agrees roughly with that of Wilson. Figure 4 shows the fall time deduced from WILSON (1972) compared with that calculated from Eq. (5). In order to reduce the slight disagreement, the following corrections should be done:

$$t = 0.752 \times 10^6 \left[\frac{1 - \exp(-0.0625z)}{V_0} \right]^{0.926}, \tag{5}'$$

where t is given in seconds, z in km, and V_0 in cm/s. The corrected relation is shown by a solid line in Fig. 4.

4.   Diffusion of Volcanic Particles from the Eruption Column

The size distribution of erupted pyroclastic material may be explained by Rosin's law, or logarithmic probability distribution. In this paper the latter distribution is adopted. The function of probability density is given by

$$f(\log_{10} d) = \frac{1}{\sqrt{2\pi}\,\sigma_d} \exp\left\{ - \frac{(\log_{10} d/d_m)^2}{2\sigma_d^2} \right\},$$

where σ_d is the standard deviation of size distribution and d_m is the median diameter. The quantity dq of particles having diameters between d_j and d_{j+1} for the total quantity of erupted material (Q) is

$$dq = \frac{Q \log_{10}\left(\dfrac{d_{j+1}}{d_j}\right)}{\sqrt{2\pi}\,\sigma_d} \exp\left\{ - \frac{(\log_{10} d_j/d_m)^2}{2\sigma_d^2} \right\}. \tag{6}$$

Previous study of the diffusion of particles from an eruption column has left the problems of diffusion unsolved. Morton *et al.* (1956) and others studied the vertical velocity of a thermal plume by theoretical and experimental methods, and Seino (1969), Blackburn *et al.* (1979), and others have studied the vertical velocity of volcanic eruption columns. The theories are complicated and results differ from each other.

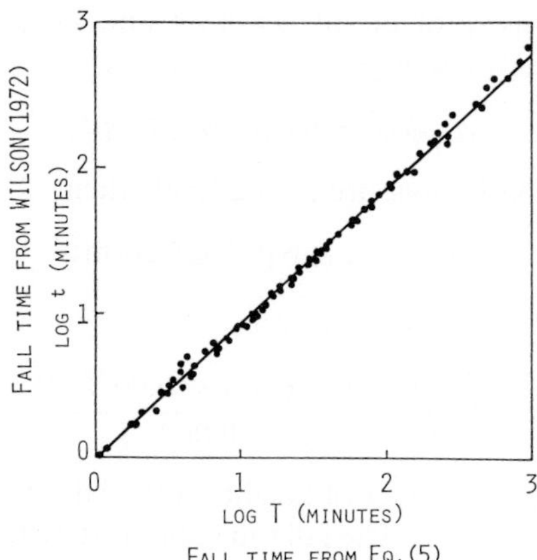

Fig. 4.   Comparison of tephra fall times deduced from Wilson (1972) and from Eq. (5). Solid line is $t = T^{0.926}$.

In this paper, therefore, a convenient model in which the vertical velocity of a volcanic eruption column is maximum at the vent and zero at the top of the column is adopted, and the diffusion of volcanic particles from an eruption column is calculated on the following assumptions:

(1)   The vertical velocity of a volcanic eruption column $W(z)$ is a function of height z.

$$W(z) = W_0\left(1 - \frac{z}{H}\right)^\lambda,$$

where W_0 is the initial velocity at $z = 0$, H is the maximum height of the eruption column, and λ is a constant.

(2)   The diffusion parameter of the eruption column Y is a function of the vertical velocity of column $W(z)$ and the terminal fall velocity of particle at sea-level V_0 as follows:

$$Y(z) = \frac{\beta\{W(z) - V_0\}}{V_0},$$

where β is a constant.

(3)   The probability density of diffusion $P(z)$ is a function of parameter Y as given by the equation

$$P(z) = A\,Y\exp(-Y),$$

where A is given by

$$\int_0^H P(z)\mathrm{d}z = 1.$$

Under these assumptions, we get the probability density of diffusion $P(z)$ for $\lambda = 1$ as follows:

$$P(z) = \frac{\beta W_0 Y \exp(-Y)}{V_0 H\{1 - (1 + Y_0)\exp(-Y_0)\}}, \tag{7}$$

where

$Y_0 = \beta(W_0 - V_0)/V_0.$

The eruption column is fan-shaped upwards; the diffusion source is generally not a point source. The radius of the eruption column is assumed to be $0.5\,z$, where z is the height, and is equal to three times the standard deviation σ_r; hence $3\,\sigma_r = 0.5\,z$. From Eq. (2) the variance $\sigma_r^2 = (8/5)Ct^{5/2}$ and the time t_s for $\sigma_r = 0.5\,z/3$ is given as follows:

$$t_s = \left(\frac{5z^2}{288C}\right)^{2/5}.$$

This t_s is added only to the diffusion time in Eq. (2), and not to the horizontal transfer time.

5. The Total Mass of Erupted Material and Its Aerial Distribution

The quantity dq of particles having diameters between d_j and d_{j+1} is given by Eq. (6). The quantity $q(d, z)$ of the particles diffused at height $z \sim z + dz$ is

$$q(d, z) = dq\, P(z)\, dz.$$

By substituting q into Eq. (2), the quantity of accumulation at the point (x, y) on the earth's surface, when χ in Eq. (2) is integrated over all sizes of particles and all heights of the eruption column, is given as follows:

$$X(x, y) = \int_{d=0}^{d=\infty} \int_0^H \frac{5P(z)}{8\pi Ct^{5/2}} \exp\left[-\frac{5\{(x - ut)^2 + y^2\}}{8Ct^{5/2}} \right] dq\, dz. \tag{8}$$

The diffusion time t is given by Eq. (5)′, and V_0 in Eq. (5)′ is given by Eq. (4)′ as a function of the diameter of particles d.

6. Results of Numerical Computation

On the basis of the previous arguments, we shall simulate the dispersion of tephra from the eruption column. The quantity $Q = 10^{15}$ gram is used in the present computation for the total quantity of erupted material, This quantity is nearly equal to 1 km³ of tephra. As shown in Eq. (6), the quantity of accumulation at any point on the surface is proportional to the total quantity of erupted material, if other parameters are constant, as is obvious from Eq. (6) and Eq. (8).

The range of values used in the present computation is 0.01–10 cm for median diameter d_m, 0.1–2.0 for the standard deviation of size distribution σ_d, 3–30 km for the height of eruption column H. For the simplicity, the horizontal wind velocity is constant through the concerned altitude, which is assumed to be 5–30 m/s.

A typical size distribution of erupted material is shown in Fig. 5. The particle diffusion from the eruption column is almost controlled by the value of βW_0 in Eq. (7). The mass diffusion from the eruption column as a function of β is shown in Fig. 6(a), where 10^4 cm/s is used for the value of W_0 (initial velocity in the eruption column); 0.1 cm is used for d_m, 0.4 for σ_d and 10 km for H. Fig. 6(b) shows iso-mass distribution maps and Fig. 6(c) shows variations of mass per unit area on the axis of mass distribution as a function of distance from the vent. In the case of large β, mass diffusion takes place in the upper part of the eruption column as shown in Fig. 6(a) and the distribution has a maximum mass per unit area at some distance from the vent (Fig. 6(b), $\beta = 0.5$). Such bimodal tephra distribution is reported in some cases, but quite seldom. The value of β takes, therefore, 0.1, and hence, $\beta W_0 = 10^3$ cm/s.

For $\beta = 0.1$ in Fig. 6(a), the variation of the mass diffusion from the eruption column with d_m, σ_d, and H is shown in Fig. 7(A), (B), and (C), respectively. In these figures, f, m, and c show fine, medium, and coarse particle, respectively. Figure 8 shows the distribution area enclosed by the 10 gram per unit area estimated by using Fig. 7. Figures 8(A), (B), and (C) correspond to Figs. 7(A), (B), and (C) for $u = 10$ m/s, respectively. Figure 8(D) shows the variation of the distribution area with u for $\beta = 0.1$ in Fig. 6(a). Figure 9 shows the variation of mass per unit area on the mass distribution

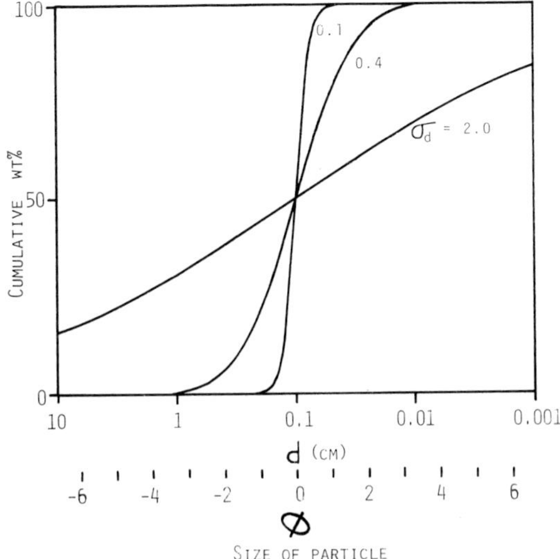

Fig. 5. Size-distribution of material erupted. Median diameter $d_m = 0.1$ cm and standard deviation σ_d = 0.1, 0.4, and 2.0.

axis with distance from the vent shown in Fig. 8, and Fig. 10 shows the variation of the area enclosed by isopleths of mass per unit area, the area enclosed by iso-mass contours, with the mass per unit area.

For tephra having small median diameter, mass diffusion takes place in the upper part of eruption column, as shown in Fig. 7(A) (a), and the dispersion area is broad at a distance, as shown in Fig. 8(A). The accumulated mass is small near the vent and is commonly maximum at a distance, as shown in Fig. 9(A). The iso-mass lines enclose a large area of thin tephra accumulation, as shown in Fig. 10(A). For large median diameter, on the other hand, mass diffusion takes place in the lower part of the eruption column. The dispersion area is a small oval near the vent, and the accumulated mass is greatest near the vent. The area enclosed by iso-mass lines is small and the ash accumulation thick.

Variation of σ_d has a remarkable effect on the mass diffusion, as shown in Fig. 7(B). The shape of the dispersion area is an oval lying inside the vent for large, σ_d, and is spread in a fan-shape from the vent for small σ_d, as shown in Fig. 8(B). The accumulated mass decreases linearly with distance for large σ_d, decreases sharply at a distance with the decrease of σ_d and is maximum at a distance for small σ_d, as shown in Fig. 9(B). The area enclosed by iso-mass lines decreases linearly with increasing mass per unit area for large σ_d, but for small σ_d, it remains almost constant at a certain value of the mass per unit area and then decreases sharply, as shown in Fig. 10(B).

As to the relationships between mass diffusion and height (Fig. 7(C)), the shape of the dispersion area (Fig. 8(C)), the accumulated mass (Fig. 9(C)), and the area enclosed by iso-mass (Fig. 10(C)), each of them has similar shapes for the variation of the height

T. Suzuki

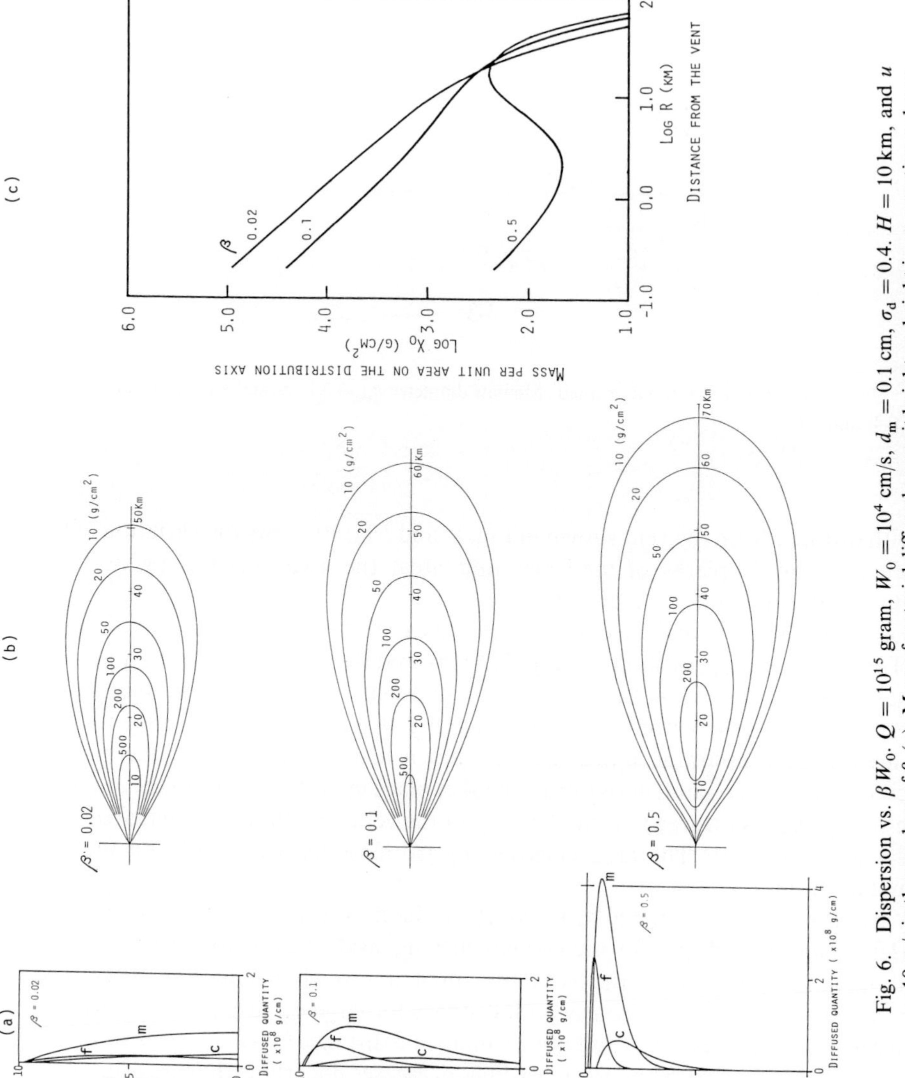

Fig. 6. Dispersion vs. βW_0. $Q = 10^{15}$ gram, $W_0 = 10^4$ cm/s, $d_m = 0.1$ cm, $\sigma_d = 0.4$. $H = 10$ km, and $u = 10$ m/s in three values of β. (a): Mass of material diffused per unit height vs. height in an eruption column. (b): Iso-mass maps (maps of isopleths of mass in gram/cm²). (c): Variation of mass per unit area on the distribution axis with distance from the vent.

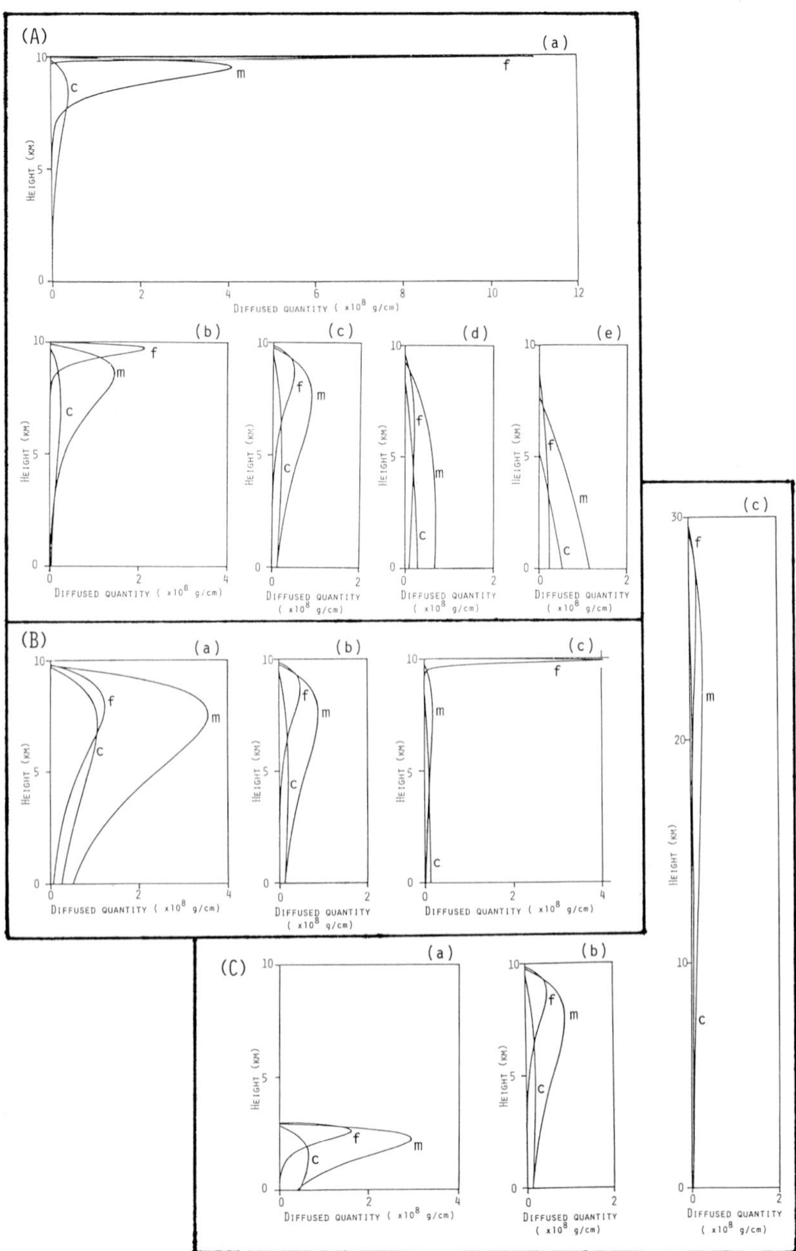

Fig. 7. Mass of material diffused per unit height vs. height in an eruption column. (A): (a) $d_m = 0.01$ cm, (b) $d_m = 0.03$ cm, (c) $d_m = 0.1$ cm, (d) $d_m = 1$ cm, (e) $d_m = 10$ cm; $\beta W_0 = 10^3$ cm/s, $\sigma_d = 0.4$, $H = 10$ km, (B): (a) $\sigma_d = 0.1$, (b) $\sigma_d = 0.4$, (c) $\sigma_d = 2.0$; $\beta W_0 = 10^3$ cm/s, $d_m = 0.1$ cm, $H = 10$ km. (C): (a) $H = 3$ km, (b) $H = 10$ km, (c) $H = 30$ km; $\beta W_0 = 10^3$ cm/s, $d_m = 0.1$ cm, $\sigma_d = 0.4$. f: fine size particles. m: medium size particles ($= d_m$). c: coarse size particles.

T. Suzuki

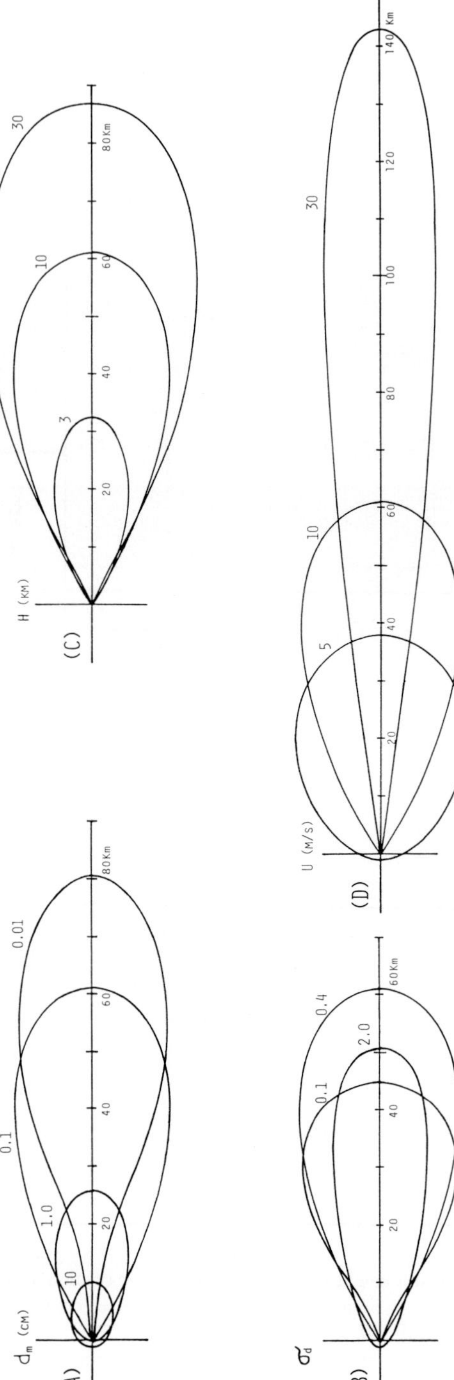

Fig. 8. Area enclosed by isopleths of 10 gram/cm² for $Q = 10^{15}$ grams. The parameters are chosen as follows: (A) mediam diameter of material erupted; (B) standard deviation of material erupted; (C) height of eruption column; and (D) velocity of wind.

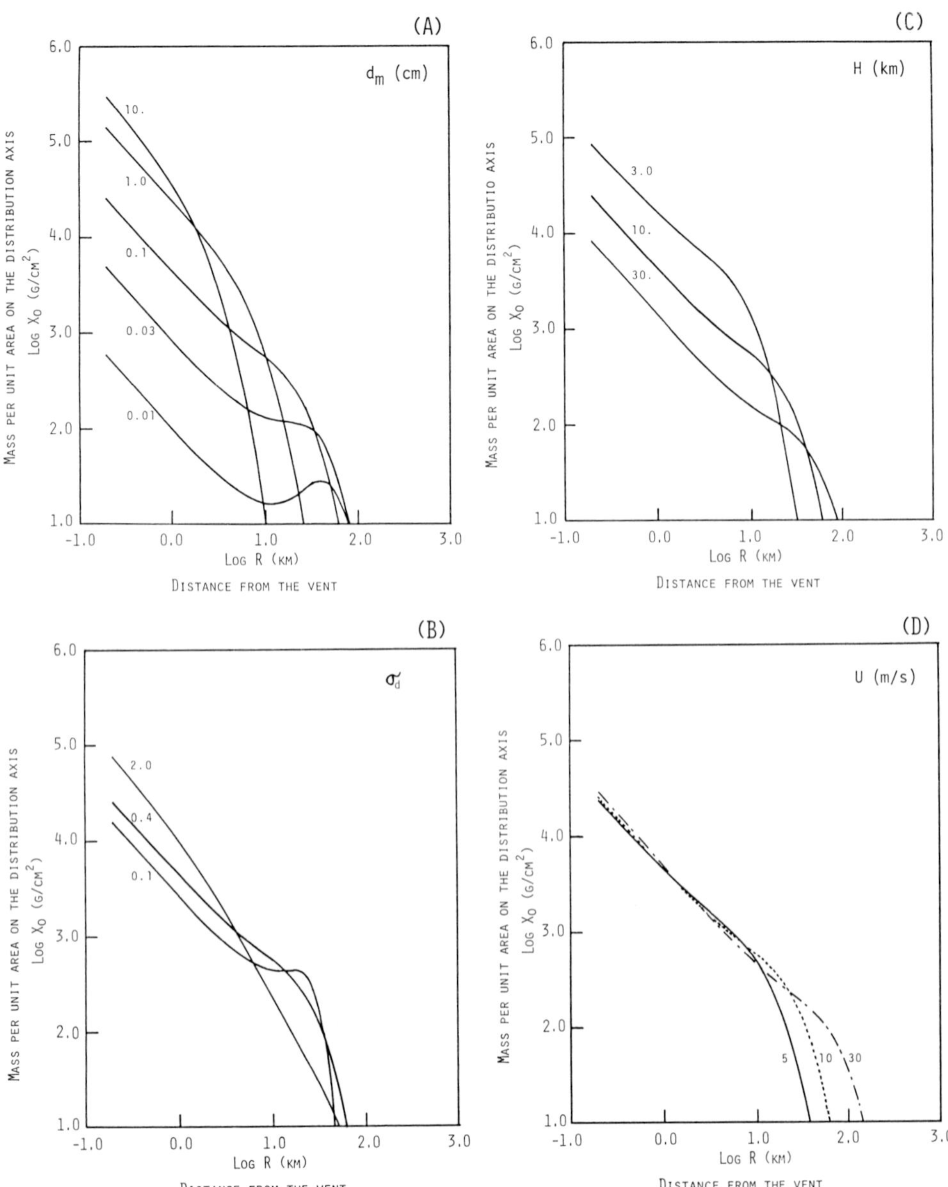

Fig. 9.   Variation of mass per unit area on the distribution axis with distance from the vent. The parameters are shown in the top right-hand corner: (A) median diameter of material erupted; (B) standard deviation of material erupted; (C) height of eruption column; and (D) velocity of wind.

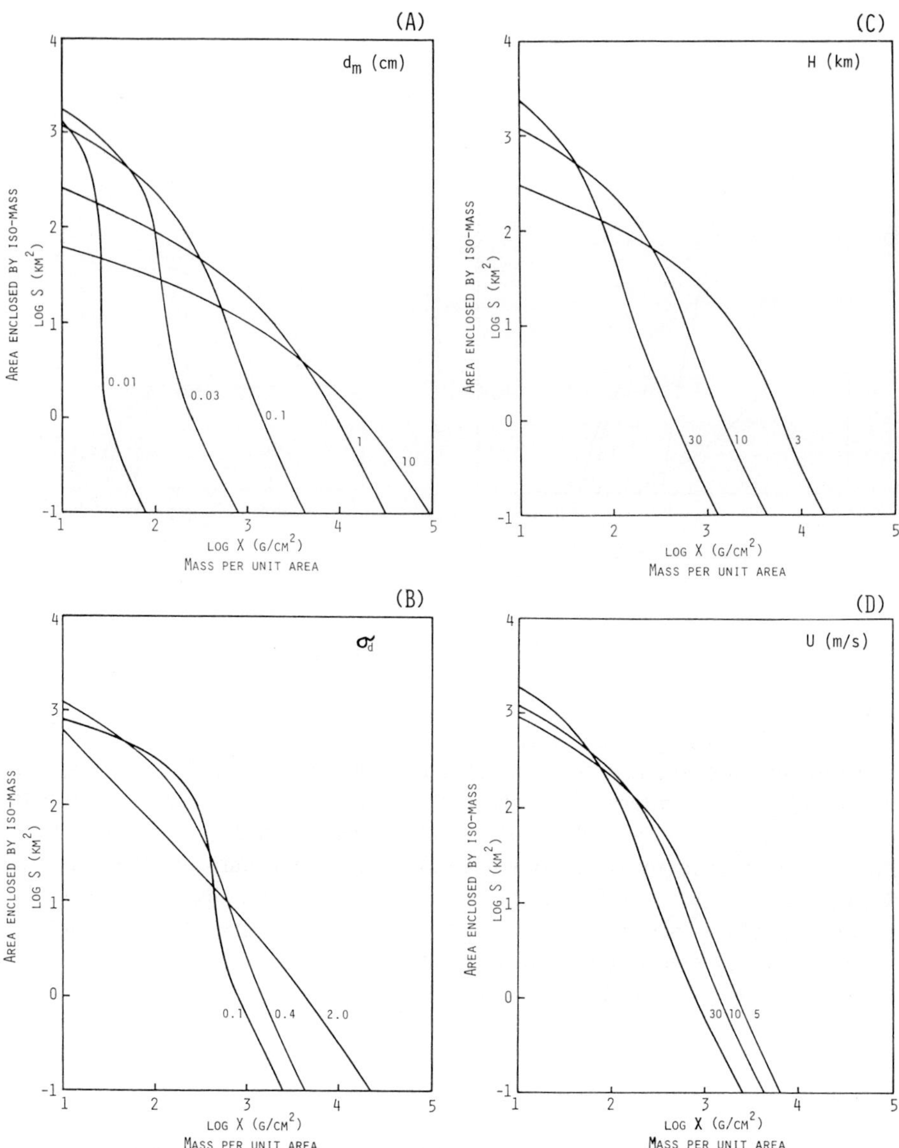

Fig. 10.   Mass per unit area vs. area enclosed by isopleths of mass per unit area. The parameters are shown in the top right-hand corner: (A) median diameter of material erupted; (B) standard deviation of material erupted; (C) height of eruption column; and (D) velocity of wind.

of eruption column, if other parameters are constant. The higher the eruption column, the larger the dispersion area is and the less the accumulated mass near the vent.

The dispersion area surrounding the vent is oval for low wind velocity and is elongated leeward for large wind velocity, as shown in Fig. 8(D). In the area near the vent, the accumulated mass decreases outward at an almost constant rate, and the distance from the vent where the mass decreases sharply depends on the wind velocity as shown in Fig. 9(D). The effect of the wind velocity on the area enclosed by iso-mass lines is small, as shown in Fig. 10(D).

7.   Conclusions

The results of numerical and graphical analysis of tephra diffusion and dispersion in the atmosphere lead to the following conclusions:

(1)   The variations in the total mass of tephra do not affect the dispersal pattern, and the mass per unit area in the dispersion area is in proportion to the total mass.

(2)   For an eruption column with ejecta of large median diameter and at low wind velocity, the iso-mass contour lines of tephra accumulation are almost a circle with the center located at the eruptive vent.

(3)   For a high eruption column with ejecta of small median diameter and small standard deviation of size distribution, the dispersion area is the broadest and has a maximum mass per unit area at some distance from the vent.

(4)   For an eruption column with large median diameter and large standard deviation of material erupted, the tephra dispersion curves show that the mass per unit area of tephra decreases at a logarithmically linear rate with increasing distance from the vent, and the area enclosed by iso-mass contours decreases almost logarithmic-linearly with increasing mass per unit area.

Comparison of the results of computational model studies of tephra dispersion with the dispersion pattern of tephra in the field, based on observation of actual eruption columns, will help clarify the mechanisms of ash diffusion from eruption columns. There are many aspects of volcanic eruptions and characteristics of eruption products that can be studied by empirical methods to shed additional light on these mechanisms. Study of total mass of tephra and dispersion patterns of the material erupted, lithic fragment-pumice ratios, porosity of pumice, and others will accelerate analysis of the dynamic mechanisms of volcanic eruptions.

The author will discuss the relationship between the size distribution of tephra deposits and the factors of the dispersion in later papers, using this model of tephra dispersion.

The author wishes to thank Prof. Tsutomu Murase of our institute for his valuable advice and encouragement.

APPENDIX

The classification of "bigness" by WALKER (1980) is very useful for the investigation of the relationships between the destructive potential of volcanoes and these factors. His classification, however, is qualitative and a quantitative definition of "bigness" is necessary for more rigorous analysis of volcanic eruptions. A quantitative definition of "bigness" is proposed as follows:

(1)   Magnitude, M_g

"Magnitude" is defined by the total mass of erupted material, because the total energy of a volcanic eruption is calculated from the total erupted mass:

$$M_g = \log_{10} M \text{ (kg)},$$

where M is total mass of ejecta. If the erupted material is all essential material, the total energy of a volcanic eruption is given by $\log_{10} E \text{ (J)} = 6.2 + M_g$, where J is Joule, (Nakamura, 1965).

(2)   Intensity, I_t

"Intensity" is defined by the mass emission rate.

$$I_t = \log_{10} \dot{M} \text{ (kg/s)},$$

where $\dot{M}$ is the mass emission rate; $M = \int_0^T \dot{M}\, dt$, and T is the duration of eruption. The energy emission rate, therefore, is $\log_{10} \dot{E} \text{ (J/s)} = 6.2 + I_t$.

(3)   Dispersive power, D_p

"Dispersive power" is defined by the characteristic area of tephra dispersal. This area was discussed by Suzuki (1981) and is a function not of total mass erupted, but of the height of the eruption column and fragmentation of ejecta:

$$D_p = \log_{10} S_c \text{ (m}^2\text{)},$$

where S_c is defined by the area at the inflection point of the "thickness-isopach area" curve of tephra.

(4)   Violence, V_l

"Violence" is defined by the vacuum range (Wilson, 1972), because the real range of an ejected volcanic block is a function of the size and density of block. The vacuum range is related to the vertical velocity of the eruption column at the vent.

$$V_l = \log_{10} L_v \text{ (m)},$$

where L_v is the range in the vacuum expressed in meter.

(5)   Destructive potential, D_t

"Destructive potential" is defined by the area enclosed by the 10 cm isopach of tephra, because the crops on farm-land are completely destroyed by the air-fall deposit 10 cm thick.

$$D_t = \log_{10} S_{10\,cm} \text{ (m}^2\text{)},$$

where $S_{10\,cm}$ is the area enclosed by 10 cm isopach of tephra.

REFERENCES

Batchelor, G. K., The application of the similarity theory of turbulence to atmospheric diffusion, *Quart. J. R. Met. Soc.*, **76**, 133–146, 1950.

BLACKBURN, E. A., L. WILSON, and R. S. J. SPARKS, Mechanisms and dynamics of strombolian activity, *J. Geol. Soc. London*, **132**, 429–440, 1976.

CSANADY, G. T., *Turbulent Diffusion in the Environment*, D. Reidel Publishing Co., Boston, 1973.

JOSEPH, J. and H. SENDNER, Über die horizontale Diffusion im Meere, *Dt. Hydrogr. Z.*, **11**, 49–77, 1958.

MORTON, B. R., Sir G. TAYLOR, and J. S. TURNER, Turbulent gravitational convection from maintained and instantaneous sources, *Proc. R. Soc. London*, A, **234**, 1–23, 1956.

NAKAMURA, K., Energies dissipated with volcanic activities-classification and evaluation, *Bull. Volcanol. Soc. Japan*, **10**, 81–90, 1965.

OKUBO, A., Oceanic diffusion diagrams, *Deep-Sea Research*, **18**, 789–802, 1971.

OZMIDOV, R. V., On the calculation of horizontal turbulent diffusion of the pollutant patches in the sea, *Doklady Akad. Nauk, SSSR*, **120**, 761–763, 1958.

RICHARDSON, L. F., Atmospheric diffusion shown on a distance-neighbour graph, *Proc. R. Soc., London*, A, **110**, 709–737, 1926.

SEINO, M., Motion of volcanic smoke (1), *Bull. Volcanol. Soc. Japan*, **14**, 111–126, 1969.

SUZUKI, T., "Thickness-isopach area" curve of tephra, *Bull. Volcanol. Soc. Japan*, **26**, 9–23, 1981.

TSUYA, H., Geological and petrological studies of volcano, Fuji, 5. On the 1707 eruption of volcano Fuji, *Bull. Earthq. Res. Inst., Univ. Tokyo*, **33**, 341–384, 1955.

WALKER, G. P. L., The Taupo pumice: product of the most powerful known (ultraplinian) eruption? *J. Volcanol. Geotherm. Res.*, **8**, 69–94, 1980.

WALKER, G. P. L., L. WILSON, and E. L. G. BOWELL, Explosive volcanic eruptions-1. The rate of fall of pyroclasts, *Geophys. J. R. Astron. Soc.*, **22**, 377–383, 1971.

WILSON, L., Explosive volcanic eruptions-2. The atmospheric trajectories of pyroclasts, *Geophys. J. R. Astron. Soc.*, **30**, 381–392, 1972.

WILSON, L. and T. C. HUANG, The influence of shape on the atmospheric settling velocity of volcanic ash particles, *Earth* Planet. Sci. Lett., **44**, 311–324, 1979.

YOKOYAMA, I., Energetics in active volcanoes. 2nd paper, *Bull. Earthq. Res. Inst., Univ. Tokyo*, **34**, 75–97, 1956.

TECTONICS

Arc Volcanism: Physics and Tectonics, edited by D. Shimozuru and I. Yokoyama, 117–140.
Copyright © 1983 by Terra Scientific Publishing Company (TERRAPUB), Tokyo.

Thermal Process in Subduction Zones
—A Review and Preliminary Approach on the Origin of Arc Volcanism—

S. HONDA and S. UYEDA

Earthquake Research Institute, University of Tokyo,
Yayoi 1–1–1, Bunkyo-ku, Tokyo 113, Japan

Origin of arc volcanism is difficult to understand because subduction of cold oceanic plate intuitively disfavors local heating and generation of magma. Present views on the general thermal state under subduction zones are briefly reviewed and importance of transport of heat associated with wedge flow induced by subduction is recognized. Simple corner flow model has been tested for various subduction zones as a preliminary approach. These models can explain regional features of surface heat flow and over-all thermal structures under subduction zones. To produce arc magma and to explain more localized heat flow anomalies, however, more localized treatments as suggested by ANDREWS and SLEEP (1974) and BODRI and BODRI (1978) must be included. Finally, some comments are made on potentially important factors in future model studies of subduction zones.

1. Introduction

Major volcanic activities on the earth are observed at divergent and convergent plate boundaries. It is easy, at least intuitively, to understand the origin of the volcanism at divergent boundaries, since it appears to be a natural consequence of upward movements of hot mantle materials, resulting in the pressure-release melting. On the contrary, volcanism associated with convergent boundaries, which we will call arc volcanism hereafter, is much more difficult to understand. Downward movements of oceanic plates will generally make the mantle around them cool, so that there arises a problem of heat sources which are needed to generate and maintain arc volcanism. Also there is a problem of the origin of the volcanic materials. There seems to be two alternative material sources for arc volcanism, that is, mantle wedge overlying the slab and the slab itself. Despite many geophysical, geochemical and geological studies, there are no definite answers to these problems.

In this paper, we will briefly review the previous studies of the thermal structure of the subduction zones, paying special attention to some of the above mentioned problems of arc volcanism. Next, we will consider these problems by using the simple corner flow model of BATCHELOR (1967) (also see McKENZIE (1969)). The reason why we adopt this model is because this model can easily parameterize the effects of such important factors as rheology, lateral heat source and viscous dissipation. Finally we will discuss the future model studies of subduction zones.

2. Problems

As described in the previous section, there are two major unresolved problems associated with arc volcanism.

(1) What is the heat source of arc volcanism?
  How can heat be carried to the magma source?
(2) What is the material source of arc volcanism?
  How can that material be carried to the earth's surface and be differentiated into a variety of rock types?

In addition to these problems, (3) evaluation of the role of water or other volatile elements is important, because (a) they will lower the melting point, (b) they will act as negative heat sources (i. e. dehydration reactions), and (c) they will carry the heat efficiently, if they exist as free phases. These problems (1), (2), and (3) undoubtedly are connected with each other. In this paper, however, main attention will be paid on their physical aspects.

Two alternatives are proposed for the heat source of arc volcanism, the mechanical work (shear or viscous heating) and heat transported laterally or upwards by the induced convection in the mantle wedge. The subducting cold slab will work against the resistance of the overlying mantle and this work will be dissipated in the form of thermal energy. The total dissipated energy is approximately expressed as,

$$q \simeq u\tau, \tag{1}$$

where, q is the heat flux per unit area and unit time produced by the mechanical work, u the rate of subduction, and τ is the shear stress. Although u is relatively well constrained by geophysical observations, the magnitude of shear stress τ is still in much debate. If we assume that $u = 10$ cm/yr, we obtain the heat flux dissipated $q = 320$ mW/m^2 $\simeq 7.5 \times 10^{-6}$ cal/cm^2 sec (for $\tau = 1$ k bar) and 32 mW/m^2 $\simeq 0.75 \times 10^{-6}$ cal/cm^2 sec (for $\tau = 100$ bars). These values suggest the importance of the shear heating as a heat source. However, since much of the generated heat will be conveyed downward by the descending slab, and the shear stress will be strongly dependent on the temperature through constitutive relations, simple application of Eq. (1) is not meaningful in estimating the temperature at depth. It is necessary to consider the movement of the slab and rheology of the slip zones. Studies along this line were performed analytically (Turcotte and Oxburgh, 1968; McKenzie and Sclater, 1968; McKenzie, 1969; Turcotte and Schubert, 1973; Yuen et al., 1978) and numerically (Minear and Toksöz, 1970; Hasebe et al., 1970). Turcotte and Schubert (1973), assuming constant shear stress and constant frictional coefficient at the slip zone, concluded that the shear stress of about 1.4 kbar would be necessary to melt the crust of the subducting lithosphere beneath the volcanic lines. Minear and Toksöz (1970) and Hasebe et al. (1970) conducted a numerical simulation of the subduction zone, in which the slab having a prescribed angle and speed kinematically subducts into essentially conductive mantle, and obtained a similar conclusion that the shear stress of a few kilobars is necessary to melt the subducting plate or the overlying wedge mantle. These studies assume predetermined amount of shear heating. Turcotte and Oxburgh (1968) and Yuen et al. (1978) assumed that the rheology of the slip zones, in which two half space blocks move with a constant relative velocity, is described by the Newtonian constitutive relations of which viscosity is strongly dependent on the temperature,

$$\eta \propto T \exp\left(\frac{E^*}{kT}\right), \tag{2}$$

where, η is the viscosity, T temperature, E^* activation energy, and k is the Boltzmann constant. Equation (2) implies that, the higher the temperature, the lower the viscosity. The lowering of the shear stress and also of viscous heating may be expected when the temperature rises. TURCOTTE and OXBURGH (1968) who studied the steady state and YUEN et al. (1978) who included the time dependent term of this problem found that the maximum temperature attained in the slip zones depends only on the relative motion and the creep properties of the rocks. YUEN et al. (1978) concluded that the maximum temperatures are always less than those required for partial melting. However, it is well known that these problems are quite sensitive to the boundary conditions (YUEN and SCHUBERT, 1979). For example, constant shear stress condition may lead to the thermal runaway in which temperature in the slip zones becomes great enough to melt the rocks (LOCKETT and KUSZNIR, 1982) (note that the amount of viscous dissipation per unit time and unit volume σ can be expressed as $\sigma \sim \tau^2/\eta \propto \tau^2/T \exp(-E^*/kT)$). Also, it is clear that the thermal properties of slip zones such as thermal diffusivity are important in this problem. At the present stage, the boundary conditions of the slip zones especially at the subduction zones are not well known and the experimental data on the rheological properties of rocks seem to be inadequate to constrain the models sufficiently. We may summarize the previous studies on the shear stress heating as follows:

(1)   Several kilobars of shear stress is necessary to generate the melt along the slip zone.

(2)   Strong temperature dependent viscosity may control the slip zone temperature which generally would be lower than that required for partial melting, although with some special conditions and rheology, melting may be expected.

The magnitude of the shear stress can be estimated from several different lines of geophysical arguments such as those of seismic stress drop, bending and loading of elastic part of the plate, heat flow anomaly at major fault zones, in-situ measurement of tectonic stress, and petrological observations (e.g. HANKS and RALEIGH, 1980 for references). There is a considerable diversity in the estimate which ranges from a few bars to a few kilobars depending on the problems studied. Generally, if the medium is treated as elastic, it will give several kilobars of stress. On the contrary, if it is treated as a viscous fluid, it will give several hundred bars. This suggests that the diversity of the estimated magnitudes of tectonic stresses is originated from our ignorance on the appropriate rheology. Tectonic stress can also be estimated from the tectonic setting. For example, BIRD (1978a) estimated the shear stress acting on the slip zone of the subducting plate at Mariana and Tonga subduction zones by assuming simple force balance on the landward plate. His estimation was about 200 bars which would provide negligible contribution to the heat source of arc volcanism. BIRD (1978a) concluded that the lateral heat transport from the landward side of the mantle wedge is necessary to produce arc volcanism. Bird (1978b) also conducted a similar force balance calculation at continent-continent convergence zones, and derived the same conclusion, that is, negligible contribution of shear heating. He explained the volcanism observed at collision zones by the delamination model in which the mantle part of the lithosphere is detached from the crustal part of the lithosphere because of the negative buoyancy.

The difficulty with the shear heating hypothesis for arc magmatism suggests that the second hypothesis of the lateral or upward heat transport by induced convection in the mantle wedge may be worth further consideration. It may be easily inferred that the moving slab will induce a secondary convection in the mantle wedge both mechanically (Sleep and Toksöz, 1971) and thermally (Rabinowicz et al., 1980). Supplement to this hypothesis, the role of dehydration in the subducted crust may become important. Since the material flowing with the induced convection will be colder near the subducting plate than at upwelling position, in order to produce arc magmas, lowering of melting point may be required near the subduction zone. Otherwise, there would be more melts far behind the arc which seems to be unrealistic because, except for the cases where active back arc spreading is in progress, like in the Mariana system, the amount of the volcanic materials is greatest at the volcanic front (Sugimura et al., 1963). It is well known that the water lowers the solidus of rocks (Kushiro et al., 1968) and the subducted crust contains hydrated minerals produced by the hydrothermal circulation in the crust near the ridge (Anderson et al., 1976). As the subduction proceeds, dehydration of crust will release its water into the overlying mantle wedge, and the solidus of the mantle wedge will be lowered. It would thus be important to assess the significance of this effect in the generation of arc magmas.

At this point, we may comment on the relation between the high heat flow in marginal seas and arc volcanism. Sometimes arc volcanism and high heat flow in the back arc basins were assumed to be of the same origin (e.g. Sugimura and Uyeda, 1973; Hasebe et al., 1970). However it seems that high heat flow is not always observed in the back arc region (Watanabe et al., 1977). On the contrary, arc volcanism seems to be more universally associated with subduction. These observations suggest that the mechanisms of arc volcanism and high heat flow in the back arc basins are not the same. We now believe that the origin of high heat flow in the back arc basins is related with the process of back arc spreading (Anderson, 1980; Sclater et al., 1980).

Andrews and Sleep (1974) suggested the importance of the ablation process on the arc volcanism, based on numerical simulation of the Newtonian viscous fluid having a temperature and depth dependent viscosity. The subducting plate will remove the material near the surface of the plate by shear. The mantle material to fill this gap comes from the lower mantle, because the side part of the mantle which is cold and hard is too rigid to move. This mechanism will carry the heat effectively to the corner region of the wedge mantle and encroach the lower part of the landward plate under the volcanic zone. This phenomenon is called the ablation by Andrews and Sleep (1974). Bodri and Bodri (1978) also conducted similar calculations and confirmed the importance of the ablation process. However, these studies are still rather qualitative.

Anderson et al. (1978, 1980) calcualted the thermal structure of the subduction zones including the effect of dehydration. They found that the reaction heat, amounting to about 50 cal/g (ANDERSON *et al.*, 1976), may be important as a heat sink. They proposed that the wedge part of the mantle will be melted by the introduction of water in the case of ocean-ocean convergence plate boundary. For the case of ocean-continent convergent boundary, they suggested that melting may take place both in continental lower crust and wedge mantle. Importance of the dehydration reaction has been already noted by HASEBE *et al.* (1970). Since low heat flow is generally observed from the trench to the aseismic front (YOSHII, 1975) and possibly to the volcanic front, HASEBE *et al.* (1970) had to assume a heat sink which cancels the frictional heating from the surface to the depth of 60 km along the slip zone.

It is usually assumed that the released water will migrate directly upward. However, OXBURGH and TURCOTTE (1976) suggested that because of the temperature reversal (high temperature top and low temperature beneath), released water may go down into the slab. They also found that the water carries a great amount of heat from the slab, if it comes out from the slab. These problems are very important, but the fate of the released water and its effect on the temperature have not been sufficiently discussed.

There seems to be only few studies on the significance of heat transported laterally or upwards producing the arc magmatism. In the next section, we will discuss this problem by using a simple fluid dynamical model proposed by MCKENZIE (1969), because his model can easily parameterize such important properties, as viscous heating, lateral heat source, viscosity, and convection. Also we will discuss the deficit and merit of the simple model.

3.   Simple Model Study of Subduction Zones

3.1   Model

Figure 1 shows the two-dimensional subduction zone model that we have studied. The oceanic plate having a prescribed dip angle, speed and age (t1) subducts kinematically beneath the landward plate having a prescribed age (t2) and thickness. We assume that the trench position is fixed with respect to the deep mantle. If the penetration depth of the slab is long enough, this assumption may be valid (UYEDA and KANAMORI, 1979). Absolute plate motion is given on both subducting and overlying

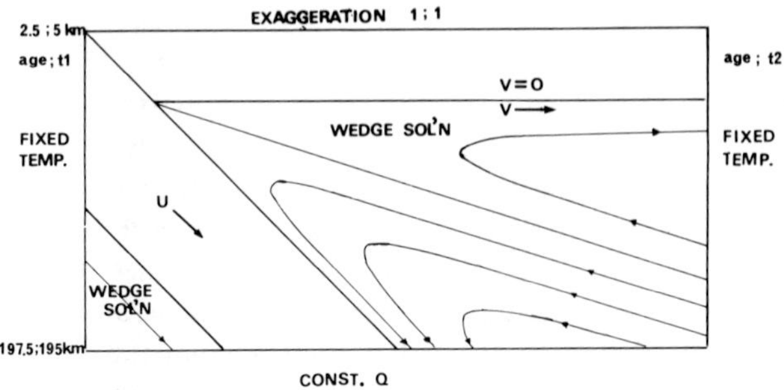

Fig. 1.   Sketch of subduction model. The range of depth is from 5 km to 195 km for coarser mesh and from 2.5 km to 197.5 km for finer mesh. See text for detail.

plates. However, in actual calculations, we assume a stationary conducting layer having a constant thickness for the landward plate, and give the plate velocity at the base of the conducting layer to represent the effect of motion of the landward plate. Although this assumption of stationary conducting lithosphere seems artificial and unrealistic, the heat flow distribution near the trench may support this assumption. If there is no conductive layer, calculated heat flow becomes too high near the trench, and contradicts the observation. The conductive layer near the trench may represent the arc lithosphere. Further back in the back arc region, the assumption of fixed lithosphere with a moving bottom surface would not be too bad an approximation either. To specify the temperature distribution in the subducting and overlying plates, we employed the goetherm of the following form at the side boundaries,

$$T(z) = T_{\mathrm{m}} \operatorname{erf}\left(\frac{z}{2\sqrt{\kappa t}}\right), \tag{3}$$

where $T(z)$ is the temperature at depth z, T_{m} the temperature at the base of plate, κ the thermal diffusivity and t is the age of plate. This geotherm is known as the solution of the cooling half space model. This geotherm may not be applicable for the oceanic plates older than 80 Ma. However, the error caused by this is most probably insignificant, considering the uncertainty in other parameters. When the overlying plate is continental, the error may be greater. The top surface is supposed to be at a constant temperature $(0^\circ\mathrm{C})$ and constant heat flux (Q) condition is assumed for the bottom surface of the whole system. The mantle wedge is represented by a Newtonian viscous fluid having a constant viscosity with infinite Prandtl number. Basically the flow field should be obtained by dynamical calculation with appropriate boundary conditions. However, in this paper, instead of doing so, we adopt the corner flow solution of Batchelor (1967). In geological terms, the corner flow solution neglects the buoyancy forces associated with thermal expansion or chemical inhomogeneity and only includes the kinematic conditions of plate motions, that is, the flow is driven by the plate motion (see McKenzie (1969) for detail). In our case, we considered the motion of landward plate also, assigning a velocity at the bottom of it as explained above. When we get the flow field, we can calcualte the "pseudo-evolution" of the temperature field with the aid of energy conservation equation,

$$\rho C_{\mathrm{p}}\left(\frac{\partial T}{\partial t} + v_1\frac{\partial T}{\partial x_1} + v_2\frac{\partial T}{\partial x_2}\right) = k\left(\frac{\partial^2}{\partial x_1^2} + \frac{\partial^2}{\partial x_2^2}\right)T + H, \tag{4}$$

where ρ is density, C_{p} specific heat, t time, x_1 horizontal axis, x_2 vertical axis, v_1 horizontal velocity of the mantle, v_2 vertical velocity of the mantle, k thermal conductivity, and H is the inernal heating term. H includes the heat generation by radioactivity, shear stress, reaction heat, and so on. Instead of incorporating all the heat sources, in this preliminary approach, we only include the viscous heating term, in order to examine the relative importance of shear heating and lateral heat transport. The viscous heating term can be obtained from the flow field according to the equation (McKenzie, 1969),

$$H = \frac{1}{2\eta}S'_{ij}S'_{ij}, \tag{5a}$$

$$S'_{ij} = \eta\left(\frac{\partial v_i}{\partial x_j} + \frac{\partial v_j}{\partial x_i}\right) \qquad (i \neq j), \tag{5b}$$

$$S'_{ii} = 0, \tag{5c}$$

$$(i = 1, 2)$$

where Einstein summation convention is used. Direct application of this formula gives a difficulty near the corner, because, as is well known, the solution diverges at the corner. To avoid this problem we assume that, if the stress becomes greater than some critical value, the stress is kept at this critical value when we calculate H. This assumption will apparently violate the equation of motion. We think, however, the error caused by this assumption is not so great. In reality, at the corner, mantle would not behave as a viscous fluid, instead, more complicated rheology which we do not know yet would be operative. At the thrust zone between the subducting and overlying conductive layer, no heat generation is assumed. There is no clear explanation of this assumption. Considering the low heat flow distribution in the area of trench-arc gap, however, this assumption may not be so bad. If the destruction (earthquake) frequently occurs at the major thrust zones, the slip zone may become weak and, as a result, there may be only a small amount of shear heating. Dehydration reactions may also play some role in cancelling out the effect of shear heating (HASEBE *et al.*, 1970). Thus, this assumption may be probably taking care of heat sink effect of dehydration. The calculation of the temperature was done by the finite difference method similar to the one described by TOKSÖZ and HSUI (1978). Most of the calculations were done by using a mesh of 10 km $\times$ 10/tan α km where α is the dip angle of subducting plate. In some cases, calculation was made with 5 km $\times$ 5/tan α km grids. Finer and coarser grids produce some discrepancies in the results. Usually, the calculated temperature in the subducting plate is colder (about 10%) for the finer mesh than for the coarser mesh. On the contrary, in the mantle wedge, temperature is warmer (about 10%) for the finer mesh. However, overall characteristic of the temperature field agrees in the cases of coarser and finer meshes. For example, the heat flow profiles on the earth's surface do not differ much, and there seems no spuriorius heat source in both meshes. Major conclusions obtained in this study will not be substantially affected by these calculation errors. The physical properties used in this study are listed in Table 1 which are taken

Table 1.   Physical properties used in this study.

Thermal conductivity	k	7.5×10^{-3} cal/cm sec °C
Thermal diffusivity	κ	8.0×10^{-3} cm²/sec
Temperature at the base of plate	T_m	1333°C
Heat flux coming from the lower mantle	Q	0.8×10^{-6} cal/cm² sec

from Scalter *et al.* (1980). In our model, the viscosity only affects the viscous heating term, since the buoyancy term in the momentum equation is neglected. Also, note that, in our model the viscous heating term is proportional to the viscosity. The viscosity of underlying mantle especially near the trench is unknown. Thatcher *et al.*, (1980) proposed an elastic lithosphere, 30 km thick, overlying a 10^{20} poise visco-elastic mantle beneath the Tohoku region, based on the model study of post-earthquake crustal movements.

In most cases, we took the value of 2×10^{20} poise that is consistent with the values derived from the study of post-glacial uplift (Cathles, 1975) and dynamic balance of the subducting slab (Yokokura, 1981a). The critical stress which limits the viscous heating near the corner was usually selected to be 1 kbar.

Calculations were started at time $t = 0$ by abruptly imposing the movement of two plates and the corner flow solution. The initial temperature distribution was set to be the one appropriate for the overlying landward plate. Results represented in this paper are snap shots taken at time $t = 20$ Ma after the initiation of the subduction. Since the motions are set to the steady state situation from $t = 0$, the "time evolution" of the temperature field has no real significance. The time of 20 Ma has, therefore, no meaning. To obtain the fairly stable state, especially near the transition zone, of heat flow from the trench low to the back arc high, we adopted the value of 20 Ma. Usually, the temperature in the mantle wedge is stabilized within a few million years. The thermal time constant of the system is mainly dominated by the diffusion time of the overlying lithosphere. In some cases, 20 Ma is not long enough to attain the stable temperature. However, we found that 20 Ma was appropriate to study the general thermal structure near the subduction zones.

3.2 *Application to the Tohoku subduction zone, northern Honshu, Japan*

Tohoku district is one of the most extensively studied subduction zones in the world. Yoshii (1979) compiled geophysical data obtained around the Tohoku subduction zone. He provided the cross section of those data perpendicular to the Japan trench (see Fig. 2). Well defined seismic plane dipping with an angle of about 30° is observed to the depth of about 600 km. Heat flow is low (~ 1 HFU) on the subducting Pacific plate and high (~ 2 HFU) in the Japan Sea area. It seems that the transition of low heat flow region to the high heat flow region occurs near the volcanic front. However, a gap of heat flow data in the zone from the aseismic front, which is situated near the edge of the land, to the volcanic front does not allow us to derive definite conclusions on the detailed features of heat flow distribution at this interesting plate boundary. Recently, Ishida (1981) found high heat flow values (more than 3 HFU) near the volcanic front which are shown in Fig. 2 by circles. These observations, taken sufficiently away from presently active volcanoes, suggest the existence of a narrow zone of very high heat flow near the volcanic front. The heat in the peak zone may have been transported by some deep rooted process such as magmatic ascent, although some of this high heat flow may be originated from the redistribution of heat by geothermal activity (Matsubayashi *et al.*, 1982).

Our modeling of the subduction zones requires the age and the thickness of the subducting and overlying plates. The age is obtained from the map of Sclater *et al.*

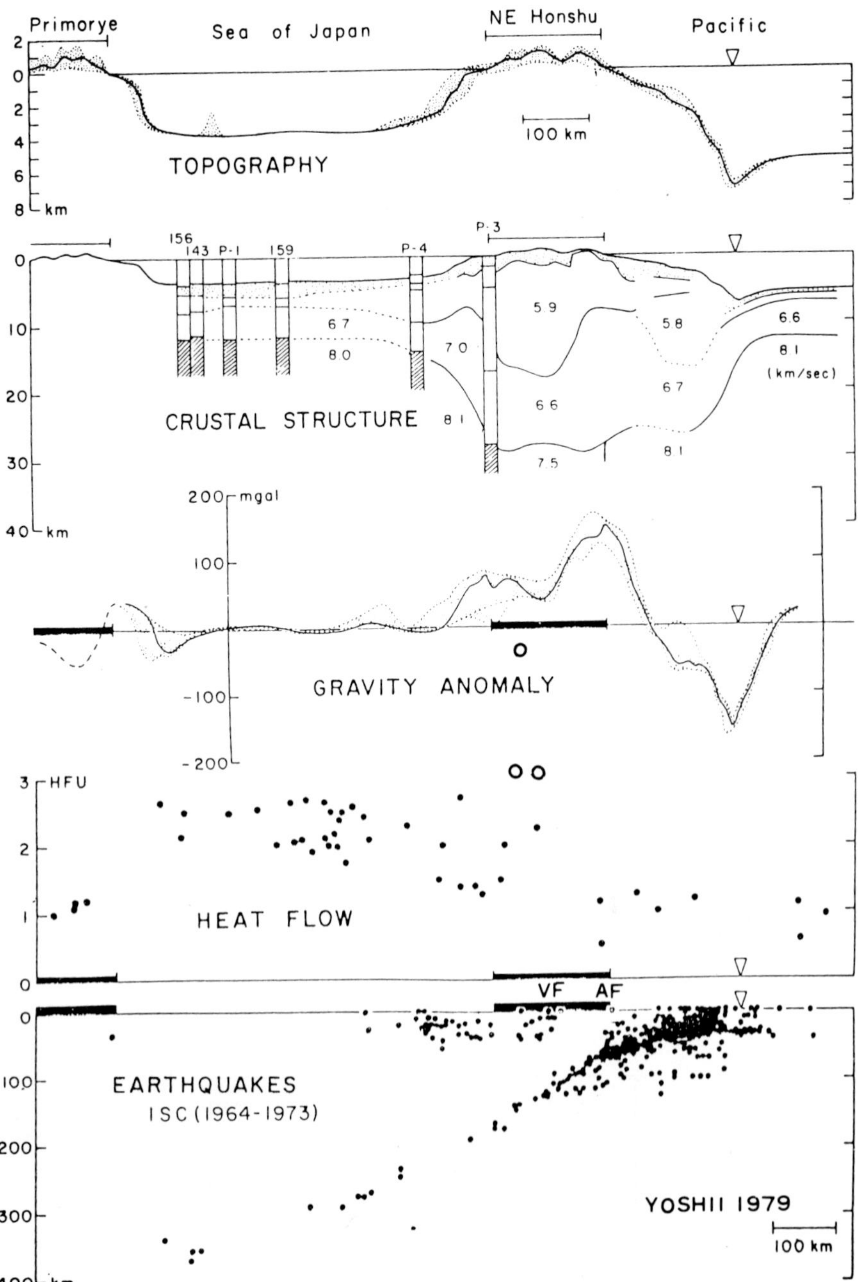

Fig. 2. Geophysical data across the Tohoku subduction zones modified from YOSHII (1979). Circles in the heat flow section are from ISHIDA (1981). These values may suggest a 'peak' of heat flow near the front of volcanoes. However, high heat flow greater than 5 HFU is probably associated with the redistribution of heat in geothermal areas (MATSUBAYASHI et al., 1982).

(1980). For the subducting plate, the thickness was estimated by the thickness-age relationship of Molnar *et al.* (1978). In the map of Sclater *et al.* (1980) the age of the Japan Sea, which is still unknown, is estimated to be 20–35 m.y. from the same relationship as that between heat flow and age for the normal ocean floor. We adopted 25 m.y. for t2 to calculate the temperature using Eq. (3). On the other hand, the thickness of conductive lithosphere of the landward plate was chosen to be 30 km based on the surface wave studies of Abe and Kanamori (1970) and seismic loading studies of Thatcher *et al.* (1980), although thickness of 50 km would fit better the normal age-thickness relation (see next section).

In Fig. 3A which uses the finer grid, the results for the case of 2×10^{20} poise mantle with critical stress of 1kbar are shown. On the temperature structure beneath the Tohoku region, Tatsumi *et al.* (1982), who studied the phase relations of primary magma of arc volcanism which coexists with the upper mantle materials, proposed that a region of about 1,400°C temperature exists between the slab surface and the 70 km depth beneath the volcanic zone. Our model in Fig. 3A does not predict such a high temperature region. The 2×10^{20} poise mantle may be too soft to generate sufficient viscous dissipation. Our model studies show that a mantle having viscosity as high as 10^{22} poise, which may be the upper limit of the normal mantle considering the

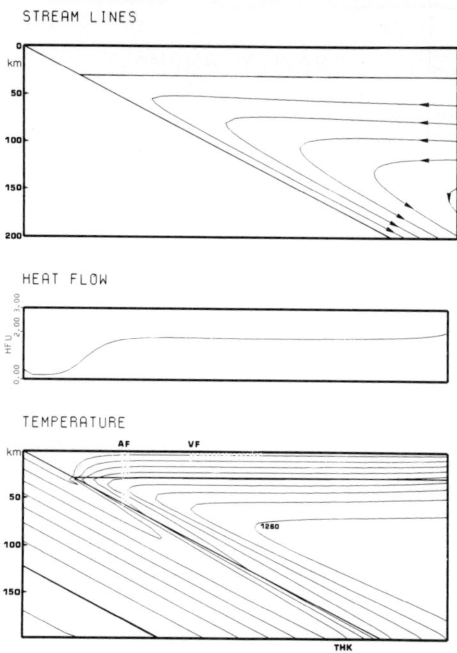

Fig. 3A.   A model showing the temperature structure beneath the Tohoku district. Viscosity of 2×10^{20} poise and critical stress of 1 kbar are assumed. Numerals in the figure of the temperature show degrees in centigrade. Contour interval is 140°C. Stippled line beneath AF shows the aseismic front. VF shows the volcanic front. Horizontally stippled area near VF shows the approximate width of volcanic belt. Finer mesh was used.

postglacial uplift, is required to produce 1,400°C under the volcanic zone. Even in that case, we must grade up the critical stress to a few kbar. Example of such a case is shown in Fig. 3B in which a 2×10^{22} poise mantle with a critical stress of 6 kbar is assumed. The shaded area in Fig. 3B shows the area in which the temperature exceeds 1,400°C. However, this model may not be appropriate, because the stress appears to be too high. In addition, the viscosity is also too high to adopt, because it is well known that the greater the temperature becomes, the smaller the viscosity becomes. If the process of the production of magma is episodic as suggested by OXBURGH and TURCOTTE (1970), the difficulty of viscosity might be removed. In their model, heating is assumed to continue until melting occurs and after the melting, little heating will occur till the liquid is removed.

More self-consistent model can be obtained by raising the temperature at the side boundary of the landward plate, thereby providing more laterally transported heat. Raising the temperature was done by changing T_m (usually 1,333°C) to 1,450°C in Eq. (3). In Fig. 3C, we show the result of such a case for a 2×10^{20} poise mantle with a critical stress of 1 kbar. Although this model cannot produce the 1,400°C isotherm just beneath the volcanic zones, it seems quite possible that additional factors such as ablation, which are not considered in the present model, would help further raising the local temperature beneath the volcanic zone. In fact, preliminary results which include the buoyancy force and temperature dependent viscosity predicted such isotherms (to be published elsewhere). The question is if such a high general temperature under the overlying back arc plate as assumed in the model in Fig. 3C, which gives 1,400°C at the depth of 70 to 100 km, is reasonable or not. Such a high temperature model may be consistent with that of SCHUBERT et al. (1978). They studied the two-dimensional viscous boundary layer model of the mantle circulation including a realistic mantle rheology. They concluded that to fit the observed topography of the sea floor, the temperature is about 1,500°C at the depth 100–200 km for dry olivine rheology or about 1,400°C at similar depths for wet olivine rheology. Other information on the

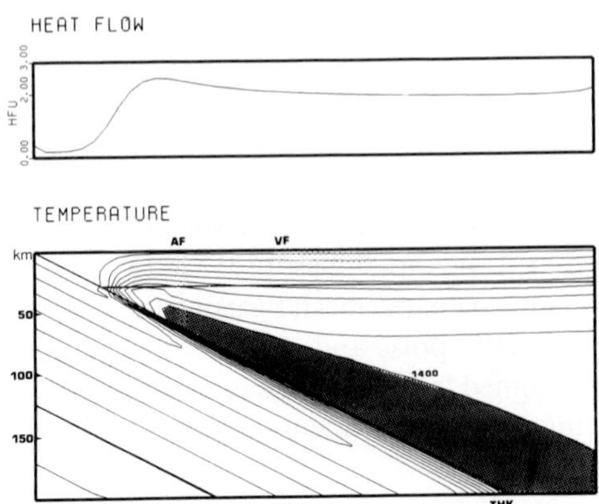

Fig. 3B.   Same as Fig. 3A except for the value of viscosity and critical stress. Viscosity of 2×10^{22} poise and critical stress of 6 kbar are assumed. Finer mesh was used.

                                   S. Honda and S. Uyeda

HEAT FLOW

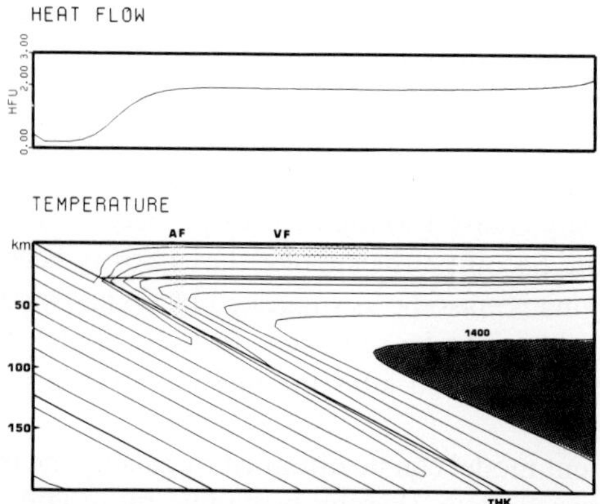

Fig. 3C.   Same as Fig. 3A except for the value of T_m in Eq. (3). T_m is assumed to be 1,450°C instead of 1,333°C. Finer mesh was used.

temperature in the mantle comes from the pyroxene geothermometry (MERCIER, 1980) and seismic velocity distribution in the mantle (ANDERSON, 1981). Pyroxene geotherm shows that a similar temperature distribution exists near the mid-oceanic ridges (MERCIER, 1980). Considering all these, we suspect that the temperature distribution assumed at the right hand side wall in Fig. 3C may be too high but not drastically too high. We expect, by further examination on the effects of buoyancy force and temperature dependent viscosity, that we may come up with a more reasonable model.

It may be noted here that the heat flow distribution near the aseismic front is too high in all the models. In the next section, we will return to this problem.

3.3   Application to other subduction zone

Our simple model was applied to other subduction zones also by adjusting the involved parameters. Figure 4 and Table 2 show the locations of the studied areas and associated parameters. The parameters are mainly referred to YOKOKURA's table (1981b). With respect to the Nankai trough region (NNK), southwestern Japan, we referred to the seismicity map given by NAKANISHI (1980) to obtain the dip angle and to SENO (1977) to obtain the relative plate motion. The dip angles of the subducting plate generally vary with depth. We re-examined the seismicity map tabulated in Yokokura's reference list and determined the dip angles for 0–200 km depth. All the calculations except for NNK, were made on coarse mesh by using the parameters listed in Tables 1 and 2 with viscosity of 2×10^{20} poise and critical stress of 1 kbar. Basically, age and plate thickness were determined from SCLATER et al. (1980) and MOLNAR et al. (1978). However, if the overlying plate is continental, we set plate thickness and age to be 100 km and 150 m.y.

In the following, we briefly review these results. In Figs. 5A-5J, obtained results of temperature and heat flow distribtuion are shown. Stream lines in each region are very similar to Tohoku (Fig. 3A) so that we do not show them except for Mariana (MAR).

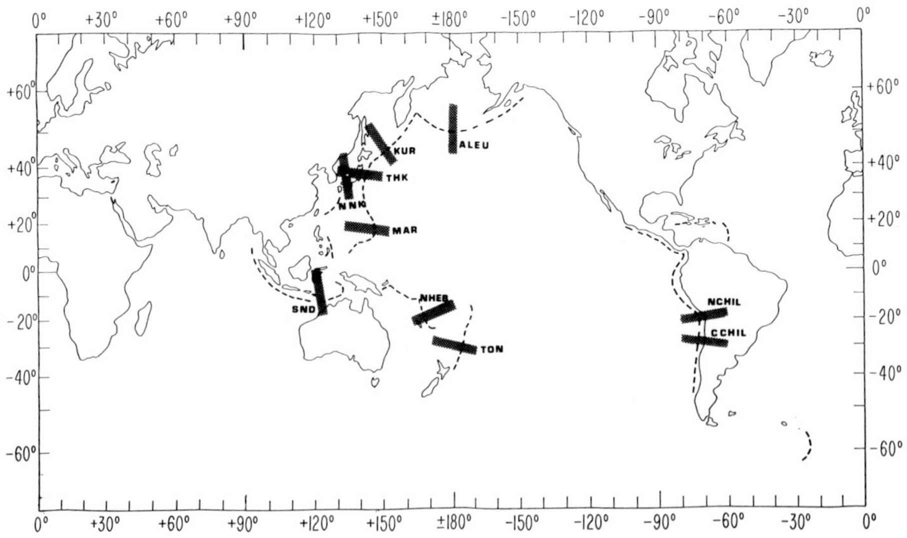

Fig. 4.   The subduction zones studied in this paper. Stippled area shows the approximate positions. Abbreviations are; CCHIL, Central Chile; NCHIL, North Chile; ALEU, Aleutian; KUR, Kurile; THK, Tohoku; NNK, Nankai Trough; MAR, Mariana; SND, Sunda; NHEB, New Hebrides; and TON, Tonga.

Generally, overall regional distributions of heat flow appear to be explained by these simple models, that is, the low heat flow near the trench and gradual increase toward the landward plate. The models predict arc to arc variations in heat flow distribution, namely high heat flow in back arc regions of the Kuril, Tohoku, Nankai, Mariana, New Hebrides, and Tonga arcs and lack of such back arc high heat flow in Chile, Aleutian and Sunda arcs.

Lack of broad high heat flow zones in certain back arc regions has not been sufficiently tested by observation, although there are some indications for it. In the case of the Chilean arc (UYEDA et al., 1978a, b; UYEDA and WATANABE, 1982), low heat flow in the trench-arc zone and high heat flow zone associated with volcanic zone have been observed but a regional high heat flow in the back arc region has not been found. When the back arc region is continental as South America, contribution from crustal radioactive heat sources may be substantial. The observations are too scarce to make detailed discussion on this point worthwhile. In the Sunda arc, high heat flow basins were discovered in the back arc of the Sumatra (CARVALHO et al., 1980; UYEDA, 1980), but for the Java and eastern Sunda arc, which we modeled, no regional back arc high heat flow zone (except obvious potential high heat flow associated with the volcanic zone) has been found (BOWIN et al., 1980; UYEDA et al., 1982). As to the Aleutian arc, the average heat flow in the Aleutian Basin has been reported to be 1.3 HFU, again with some higher values along the arc (LANGSETH et al., 1980).

Also our models show the correlation between the distance from the trench to the transition zone in heat flow distribution and the thickness of the conducting layer in the

Table 2. Subduction zone parameters.

Site	Dip (deg)	U	V (cm/yr)	t1 (m.y.)*	t2 (m.y.)**		Figure***
Central Chile (CCHIL)	15	8.5	−1.9	50 (70 km)	continent		5A (C)
North Chile (NCHIL)	26	9.0	−1.8	50 (70 km)	continent		5B (C)
Aleutian (ALEU)	40	5.6	−0.1	55 (80 km)	80–110	95 (100 km)	5C (C)
Kurile (KUR)	45	7.7	−1.1	100 (100 km)	20–35	25 (50 km)	5D (C)
Tohoku (THK)	26	8.6	−0.8	125 (110 km)	20–35	25 (50, 30 km)	3A (F) 3B (F) 3C (F) 5E (C)
Mariana (MRN)	43	9.1	5.5	160 (110 km)	0–4	2 (20 km)	5G (C)
New Hebrides (NHEB)	70	3.3	−6.3	25 (50 km)	0–10	5 (20 km)	5I (C)
Tonga (TON)	50	8.2	−0.8	100 (100 km)	0– 14	7 (20 km)	5J (C)
Sunda (SND)	40	5.8	−0.8	100 (100 km)	−	130? (100 km)	5H (C)
Nankai Trough (NNK)	15	2.0	−1.3	20 (50 km)	20–35	25 (30 km)	5F (F)

*Brackets indicate the thickness of lithosphere.

**The age and thickness of the continental lithosphere are selected to be 150 m.y. and 100 km respectively. The range of estimated age, adopted values of age and lithosphere thickness are included.

***(C) and (F) mean coarser mesh and finer mesh.

landward plate. Compare the result of THK of Fig. 5E where the thickness of the conducting layer is assumed to be 50 km and that of Fig. 3A in which 30 km thick conductive layer was used. The former shows much more gradual transition. These characteristics are found to be independent of grid size. As noted in the previous section, the thickness of the landward plate at THK is probably about 30 km, while the observed low heat flow near the aseismic front (Fig. 2) require thick lithosphere (greater than 50 km). This discrepancy may be resolved if we assume the existence of a conductive triangular region at the corner to simulate the arc and back arc structures (Honda, in preparation).

Models of CCHIL (Fig. 5A) and NCHIL (Fig. 5B) are selected, because it is well known that there are volcanic activities in the latter region and no activities in the former. Our models show only small difference in the thermal structure under the subduction zones, again suggesting that more sophisiticated models are needed to explain the volcanism. Mariana subduction zone (MAR) shows generally high temperature in the mantle wedge (Fig. 5G). This is because the retreating of the landward plate gives rise to the upwelling of the deep hot mantle. This hot upwelling should accompany the back-arc spreading in the Mariana trough. New Hebrides (NHEB, Fig. 5I) and Nankai (NNK, Fig. 5F) subduction zones show generally high temperature within the subducting plate compared to the other subduction zones. This occurs because of the slow rate of subduction and high temperature in the landward mantle. The depth of the seismic plane just over which volcanic line lies is usually of the order of 100–150 km (SUGIMURA, 1960). However, in the NHEB region, it is about 180 km. There is only rare volcanic activity in the Chugoku district, southwest Japan, which is associated with the subduction of Philippine Sea plate at Nankai trough. I.S. Sacks (private communication, 1982) proposed that the subduction of the Philippine Sea plate is not straight as we have modeled, instead, there is a nearly horizontal part beneath Shikoku island which is situated about 200 km north of the Nankai trough. This type of subduction is proposed beneath Peru also by HASEGAWA and SACKS (1981). KANAMORI (1972) proposed that the present subduction of the Philippine Sea plate started only at about 2 m.y. ago while KARIG (1975) believes that it started about 4 m.y. ago. Obviously, the present models are not able to deal with these factors.

4.   Future Model of Subduction Zones

Present paper reports some of the results of our preliminary attempt to understand the thermal state under subduction zones. We have made calculations using the simplest corner flow model and showed that even with such a simple model, features of regional heat flow may be explained. However, if we want to deal with the origin of arc volcanism, more sophisticated models are needed.

To solve the problems of subduction zones more rigorously, it is required to consider the dynamical forces caused by the density difference such as thermal volumetric expansion and chemical inhomogeneity. RABINOWICZ et al., (1980), based on numerical work, reported the importance of the effect of cold subducting slabs in making convection strong enough to carry the heat effectively from the continent to the ocean. JURDY and STEFANICK (1982) indicated that the velocity of the convection

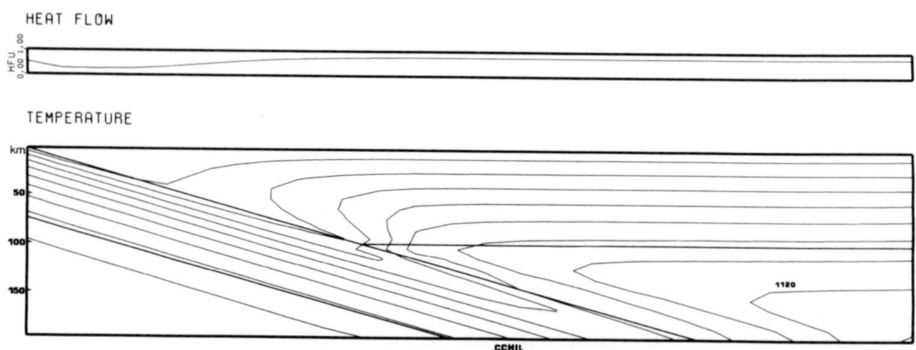

Fig. 5A.   Subduction zone of Central Chile (CCHIL).

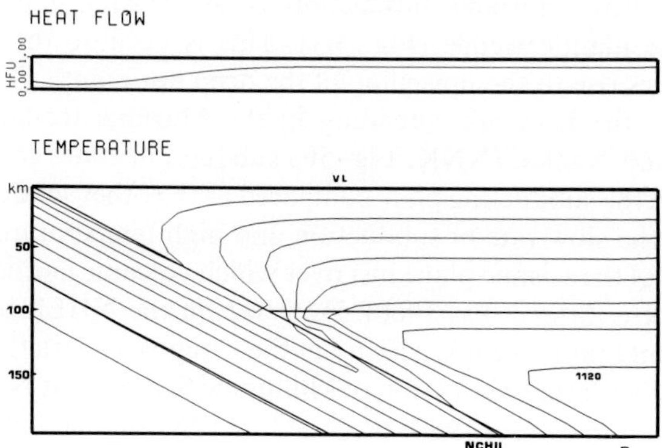

Fig. 5B.   Subduction zone of North Chile (NCHIL).

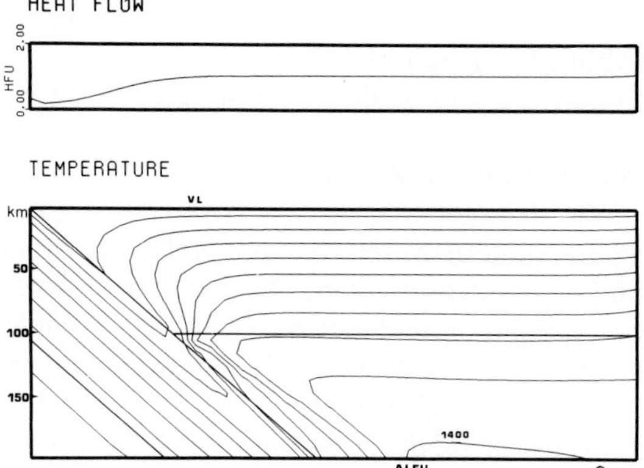

Fig. 5C.   Subduction zone of Aleutian (ALEU).

Fig. 5.   Model for various subduction zones. Numerals in the figure of the temperature show degrees in centigrade. The contour interval is 140°C. Abbrviations are the same as Fig. 4 and Fig. 3. VL implies volcanic line.

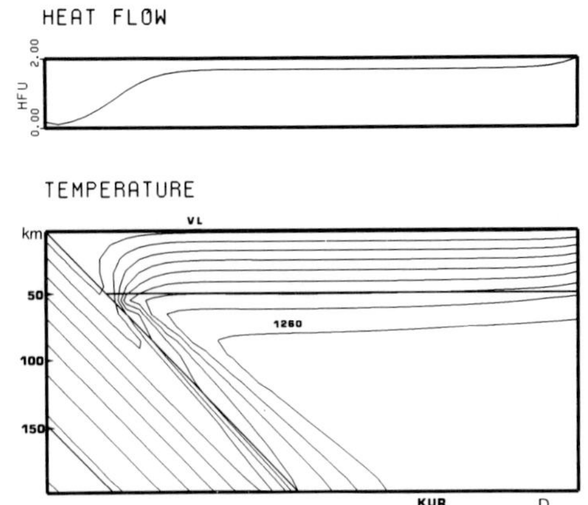

Fig. 5D.   Subduction zone of Kurile (KUR).

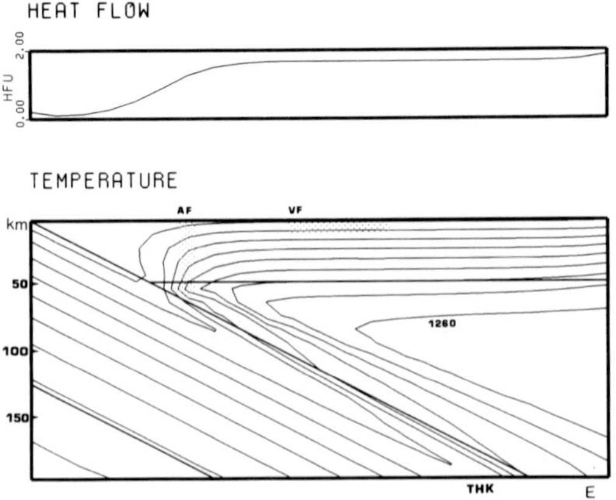

Fig. 5E.   Subduction zone of Tohoku (THK).

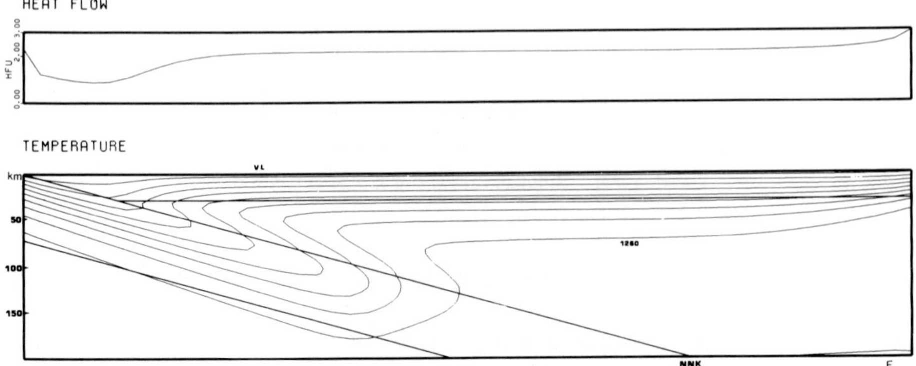

Fig. 5F.   Subduction zone of Nankai Trough (NNK).

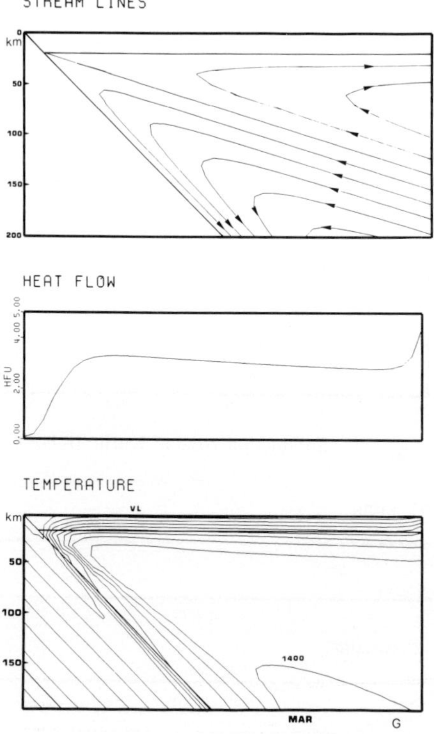

Fig. 5G.   Subduction zone of Mariana (MRN).

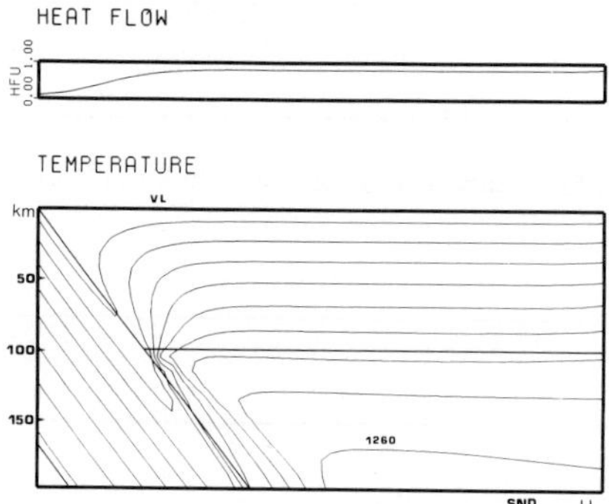

Fig. 5H.   Subduction zone of Sunda (SND).

produced in the wedge mantle can be higher than the subducting plate velocity. These induced convection currents may transport the heat more efficiently and rapidly than our model. Also, time dependence of the convection which generally shows stronger flow in the earlier phase of the subduction may be important.

The effect of the temperature dependent viscosity with an appropriate non-linear

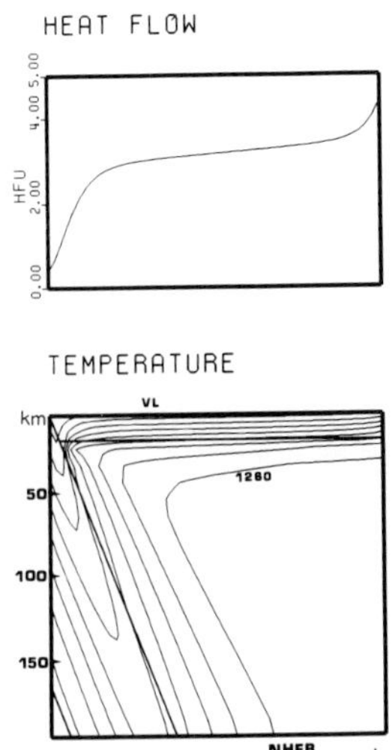

Fig. 5I.   Subduction zone of New Hebrides (NHEB).

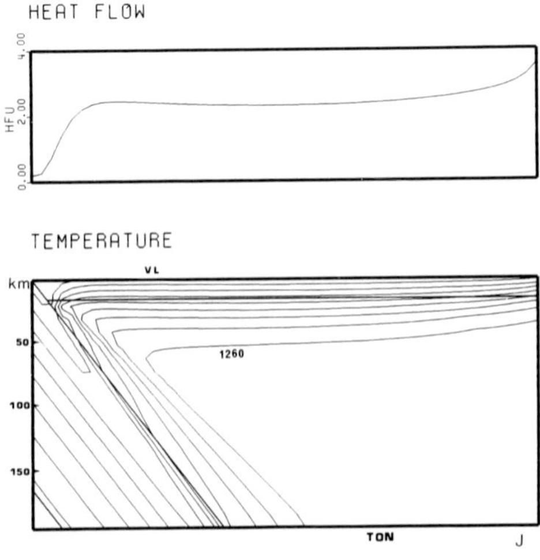

Fig. 5J.   Subduction zone of Tonga (TON).

characteristics must also be included in the models, because the variable viscosity will introduce the ablation which may be the cause of the peak heat flow near the volcanic front and more importantly arc volcanism itself (ANDREWS and SLEEP, 1974; BODRI and BODRI, 1978). The difference in chemical composition between the mantle and crust may play an important role as shown by our study, that is, the role of a conductive layer. Future model must include such an inhomogeneity also. If the melt is produced somewhere in the mantle, it will convey the heat appreciably (TURCOTTE, 1981) depending on the amount of melt. We must manage to model such a complex

                    S. Honda and S. Uyeda

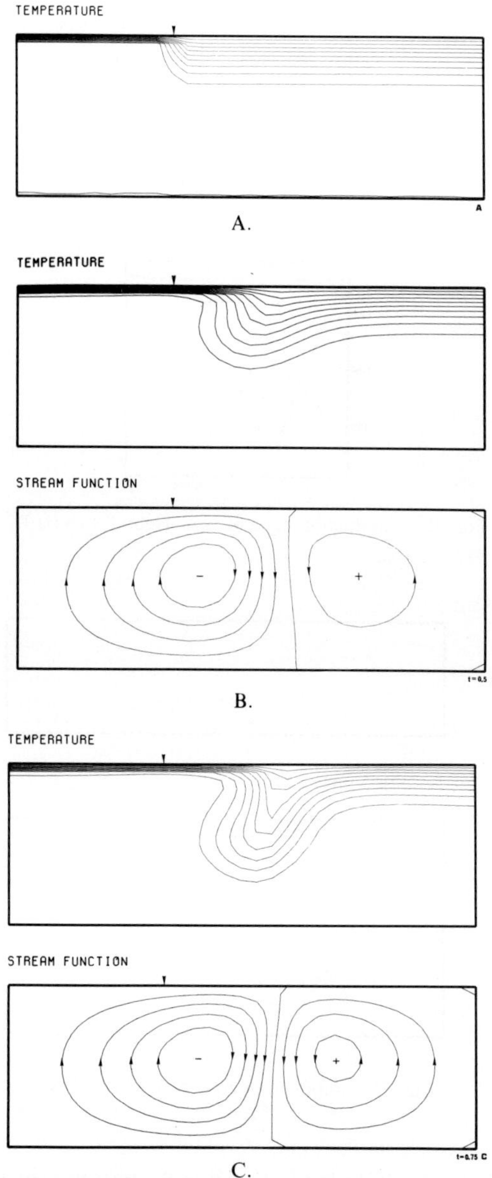

Fig. 6.   An example of time dependent convection which shows subduction-like geotherms. At the beginning, hot mantle having an error function type temperature contacts the cold mantle which also has an error function type temperature (A). Such a temperature structure may occur at the transform fault. Subsequent evolution of the temperature, normalized by an appropriate temperature, shows a subduction-like structure (B and C). Arrows in the upper side of the figure shows the position where the initial contact surface of the hot mantle and cold mantle existed. This model may suggest a retreat of the subducting slab and subsequent spreading of marginal sea. Note the relation between the stream function, which shows the instantaneous motion of the mantle, and temperature. After some time, the descending flow approaches (D and E) the symmetric pattern. (constant viscosity fluid).

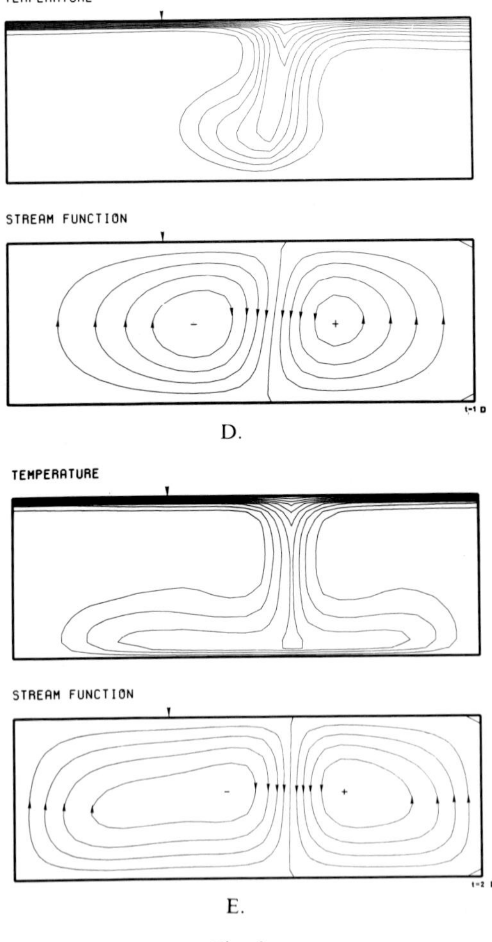

Fig. 6

two phase flow as was suggested by SLEEP (1974) for the oceanic ridge system. It is evident that the subducting lithosphere has a considerable amount of hydrated minerals (ANDERSON *et al.*, 1976), so the dehydration, its reaction energy and the transportation of heat by the released water must be considered properly in the future (OXBURGH and TURCOTTE, 1976). The effect of varying slab angle may also be important. SYDORA *et al.* (1978) modeled such a complex geometry by the method similar to MINEAR and TOKSÖZ (1970), although their model is essentially a conductive one. Finally, we must note that the model treated until recently is more or less kinematic. This is partly because of our ignorance of the past geometry of the subduction zones and partly because we could not model an inclined and one sided subduction zones by using a fully dynamical model. Simple fluid dynamical models usually predict an almost vertical and two sided subduction. Probably inclusion of non viscous rheology such as elastic-plastic rheology and chemical inhomogeneity such as the effect of light crust may help solving these basic problems. Honda (unpublished calculation) suggested that inclined subduction may be modeled when we consider a transitional stage of convection (see Fig. 6 and its caption). Similar mechanism was pointed by Richter (1978) also. Fully dynamical model, including its initiation, evolution and termination, is our final goal to understand the nature of the subduction zones.

5.　Conclusions

We have reviewed some important problems associated with arc volcanism and temperature structure of the subduction zones. The modeling of the subduction zones by using a simple fluid dynamical model of McKenzie (1969) is shown to be helpful, to some extent, to understand the general nature of the subduction zones. We recognized that, as long as the critical stress is kept at 1 kbar level, advective transport of heat plays more important role than shear heating in bringing the wedge mantle temperature sufficiently high to explain arc magmatism. However, unless advective transport of heat is upward like in the case of the Mariana zone (Fig. 5G), upper mantle temperature in back arc region away from the subduction zone must be higher than under the volcanic front. Such a model might present another problem, because volcanoes are confined close to the volcanic front. In order to get around this difficulty, we suspect that, under volcanic zones, there may be some mechanism to generate magmas more efficiently and/or to make its ascent easier. Volatile elements released from the slab and mechanical weakening or stress conditions in the region of magma ascent may be the candidates for such a mechanism. To understand the arc volcanism fully, therefore, more of such factors will have to be included in the model also. Application of such models to many subduction zones with different characteristic parameters will be useful in assessing their relative importance.

Discussions with Drs. I. Kushiro, M. Sakuyama, H. Fukuyama and Mr. Y. Tatsumi prompted us to initiate the present study. Miss. N. Sugi is also thanked for her helpful suggestion and discussions. Dr. D. Shimozuru's encouragement for publishing our work is also acknowledged.

REFERENCES

Abe, K. and H. Kanamori, Mantle structure beneath the Japan Sea as revealed by surface waves, *Bull. Earthq. Res. Inst.*, **48**, 1011–1021, 1970.

Anderson, O. L., Temperature profiles in the earth, in *Evolution of the Earth, Geodynamics Ser.*, **5**, edited by R. J. O'Connell and W. S. Fyfe, pp. 19–27, Am. Geophys. Union, Washington, D. C., 1981.

Anderson, R. N., 1980 Update of heat flow in the East and Southeast Asian seas, in *The Tectonic and Geologic Evolution of Southeast Asian Seas and Islands, Geophys. Monogr.*, **23**, edited by D. E. Hayes, pp. 319–326, Am. Geophys. Union, Washington D. C., 1980.

Anderson, R. N., S. E. Delong, and W. M. Schwarz, Thermal model for subduction with dehydration in the downgoing slab, *J. Geology*, **86**, 731–739, 1978.

Anderson, R. N., S. E. Delong, and W. M. Schwarz, Dehydration, asthenospheric convection and seismicity in subduction zones, *J. Geology*, **88**, 445–451, 1980.

Anderson, R. N., S. Uyeda, and A. Miyashiro, Geophysical and geochemical constraints at converging plate boundaries—Part I: Dehydration in the downgoing slab, *Geophys. J. R. Astron. Soc.*, **44**,

333–357, 1976.

ANDREWS, D. J. and N. H. SLEEP, Numerical modelling of tectonic flow behind island arcs, *Geophys. J. R. Astron. Soc.*, **38**, 237–251, 1974.

BATCHELOR, G. K., *An Introduction to Fluid Dynamics*, pp. 224, Cambridge University Press, Cambridge, Mass., 1967.

BIRD, P., Stress and temperature in subduction shear zones: Tonga and Mariana, *Geophys. J. R. Astron. Soc.*, **55**, 411–434, 1978a.

BIRD, P., Initiation of intracontinental subduction in the Himalaya, *J. Geophys. Res.*, **83**, 4975–4987, 1978b.

BODRI, L. and B. BODRI, Numerical investigation of tectonic flow in island-arc areas, *Tectonophysics*, **50**, 163–175, 1978.

BOWIN, C., G. M. PURDY, C. JOHNSTON, G. SHOR, L. LAWVER, H. M. S. HARTONO, and P. JEZEK, Arc-continent collision in Banda Sea region, *Am. Assoc. Pet. Geol. Bull.*, **64**, 868–915, 1980.

CARVALHO, Humberto da Silva, Ir. PURMOKO, Ir. SISWOYO, Ir. THARMIA, and V. VAQUIER, Terrestrial heat flow in the Tertiary basin of Central Sumatra, *Tectonophysics*, **69**, 163–188, 1980.

CATHLES, L. M., *Viscosity of the Earth's Mantle*, pp. 386, Princeton University Press, New Jersey, 1975.

HANKS, T. C. and C. B. RALEIGH, The conference on magnitude of deviatoric stresses in the earth's crust and uppermost mantle, *J. Geophys. Res.*, **85**, 6083–6085, 1980.

HASEBE, K., N. FUJII, and S. UYEDA, Thermal processes under island arcs, *Tectonophysics*, **10**, 335–355, 1970.

HASEGAWA, A. and I. S. SACKS, Subduction of the Nazca plate beneath Peru as determined from seismic observation, *J. Geophys. Res.*, **86**, 4971–4980, 1981.

ISHIDA, T., Measurements of heat flow utilizing boreholes, Grad. Thesis, Chiba Univ., Chiba, Japan, pp. 62., 1981 (in Japanese).

JURDY, D. M. and M. STEFANICK, Models for back-arc spreading, submitted to *Tectonophysics*, 1982.

KANAMORI, H., Tectonic implication of the 1944 Tonankai and the 1946 Nankaido earthquakes, *Phys. Earth Planet. Inter.*, **5**, 129–139, 1972.

KARIG, D. E., Basin genesis in the Philippine Sea, in *Initial Rep. Deep Sea Drilling Project, Leg. 31*, U.S. Government Printing Office, pp. 857–879, 1975.

KUSHIRO, I., Y. SYONO, and S. AKIMOTO, Melting of a peridotite nodule at high pressure and high water pressures, *J. Geophys. Res.*, **73**, 6023–6029, 1968.

LANGSETH, M., M. A. HOBART, and K. HORAI, Heat flow in Bering Sea, *J. Geophys. Res.*, **85**, 3740–3750, 1980.

LOCKETT, J. M. and N. J. KUSZNIR, Ductile shear zones: Some aspects of constant slip velocity and constant shear stress models, *Geophys. J. R. Astron. Soc.*, **69**, 477–494, 1982.

MATSUBAYASHI, O., T. ISHIDA, S. HONDA, and H. FUJISAWA, Heat flow measurements by using the densely distributed boreholes (In the case of intermediate depth exploring boreholes in the geothermal area), *Programme and Abstracts*, The Seismological Society of Japan, **1**, 27, 1982 (in Japanese).

MCKENZIE, D. P., Speculations on the consequences and causes of plate motions, *Geophys. J. R. Astron. Soc.*, **18**, 1–32, 1969.

MCKENZIE, D. P. and J. G. SCLATER, Heat flow inside the island arcs of the northwestern Pacific, *J. Geophys. Res.*, **73**, 3173–3179, 1968.

MERCIER, J. -C. C., Single-pyroxene thermobarometry, *Tectonophysics*, **70**, 1–39, 1980.

MINEAR, J. W. and M. N. TOKSÖZ, Thermal regime of a downgoing slab and new global tectonics, *J. Geophys. Res.*, **75**, 1397–1419, 1970.

MOLNAR, P., D. FREEDMAN, and J. S. F. SHIH, Lengths of the intermediate and deep seismic zones and temperatures in downgoing slabs of lithosphere, *Geophys. J. R. Astron. Soc.*, **56**, 41–54, 1979.

NAKANISHI, I., Precursors to ScS phases and dipping interface in the upper mantle beneath southwestern Japan, *Tectonophysics*, **69**, 1–35, 1980.

OXBURGH, E. R. and D. L. TURCOTTE, Thermal structure of island arcs, *Geol. Soc. Am. Bull.*, **81**, 1665–1688, 1970.

OXBURGH, E. R. and D. L. TURCOTTE, The physico-chemical behaviour of the descending lithosphere, *Tectonophysics*, **32**, 107–128, 1976.

RABINOWICZ, M., B. LAGO, and C. FROIDEVAUX, Thermal transfer between the continental asthenosphere and the oceanic subducting lithosphere: Its effect on subcontinental convection, *J. Geophys. Res.*, **85**, 1839–1853, 1980.

RICHTER, F. M., Simple numerical models for subduction, *EOS, Trans. AGU*, **59**, 1194–1195, 1978.

SCHUBERT, G., D. YUEN, C. FROIDEVAUX, L. FLEITOUT, and M. SOURIAU, Mantle circulation with partial shallow return flow: Effects on stresses in oceanic plates and topography of the sea floor, *J. Geophys. Res.*, **83**, 745–758, 1978.

SCLATER, J. G., C. JAUPART, and D. GALSON, The heat flow through oceanic and continental crust and heat loss of the earth, *Rev. Geophys. Space Phys.*, **18**, 269–311, 1980.

SENO, T., The instantaneous rotation vector of the Philippine Sea plate relative to the Eurasian plate, *Tectonophysics*, **42**, 209–226, 1977.

SLEEP, N. H., Segregation of magma for a mostly crystalline mush, *Bull. Geol. Soc. Am.*, **85**, 1225–1232, 1974.

SLEEP, N. H. and M. N. TOKSÖZ, Evolution of marginal basins, *Nature*, **33**, 548–550, 1971.

SUGIMURA, A., Zonal arrangement of some geophysical and petrological features in Japan and its environs, *J. Fac. Sci. Univ. Tokyo, Sec. II*, **12**, 133–153, 1960.

SUGIMURA, A. and S. UYEDA, *Island Arcs, Japan and its Environs*, pp. 247, Elsevier, Amsterdam, 1973.

SUGIMURA, A., T. MATSUDA, K. CHINZEI, and K. NAKAMURA, Quantitative distribution of late Cenozoic volcanic materials in Japan, *Bull. Volcanol.*, **26**, 125–140, 1963.

SYDORA, L. J., F. W. JONES, and R. St. J. LAMBERT, The thermal regime of the descending lithosphere: The effect of varying angle and rate of subduction, *Can. J. Earth Sci.*, **15**, 626–641, 1978.

TATSUMI, Y., M. SAKUYAMA, H. FUKUYAMA, and I. KUSHIRO, Generation of arc basalt magmas and thermal structure of the wedge mantle in subduction zones, submitted to *J. Geophys. Res.*, 1982.

THATCHER, W., T. MATSUDA, T. KATO, and J. B. RUNDLE, Lithospheric loading by the 1896 Riku-u earthquake, northern Japan: Implications for plate flexure and asthenospheric rheology, *J. Geophys. Res.*, **85**, 6429–6435, 1980.

TOKSÖZ, M. N. and A. T. HSUI, Numerical studies of back-arc convection and the formation of marginal basins, *Tectonophysics*, **50**, 177–196, 1978.

TURCOTTE, D. L., Magma migration, *Ann. Rev. Earth Planet. Sci.*, **10**, 397–408, 1982.

TURCOTTE, D. L. and E. R. OXBURGH, A fluid theory for the deep structure of dip-slip fault zones, *Phys. Earth Planet. Inter.*, **1**, 381–386, 1968.

TURCOTTE, D. L. and G. SCHUBERT, Frictional heating of the descending lithosphere, *J. Geophys. Res.*, **78**, 5876–5886, 1973.

UYEDA, S., Review of heat flow studies in the Eastern Asia and Western Pacific region, *UN ESCAP, CCOP/SOPAC, Tech. Bull.*, **3**, 151–169, 1980.

UYEDA, S. and H. KANAMORI, Back arc opening and the mode of subduction, *J. Geophys. Res.*, **84**, 1049–1061, 1979.

UYEDA, S. and T. WATANABE, Terrestrial heat flow in Western South America, *Tectonophysics*, **83**, 63–70, 1982.

UYEDA, S., T. WATANABE, and F. VOLPONI, Report of heat flow measurements in San Juan and Mendoza, Argentine, *Bull. Earthq. Res. Inst.*, **53**, 165–172, 1978a.

UYEDA, S., T. EGUCHI, S. KAMAL, and S. MADJO, Preliminary study on geothermal gradient and heat flow in Java, *Tech. Bull., ESCAP/CCOP*, **15**, in press, 1982.

UYEDA, S., T. WATANABE, E. KAUSEL, M. KUBO, and Y. YASHIRO, Report of heat flow measurements in Chile, *Bull. Earthq. Res. Inst.*, **53**, 131–163, 1978b.

WATANABE, T., M. LANGSETH, and R. N. ANDERSON, Heat flow in back arc basins of the Western Pacific, in *Island Arcs, Deep Sea Trenches and Back-Arc Basins, Mawrice Ewing Ser.*, **1**, edited by M. Talwani and W. C. Pitman III, pp. 137–161, Am. Geophys. Union, Washington, D. C., 1977.

YOKOKURA, T., Viscosity of the earth's mantle: Inference from dynamic support by flow stress, *Tectonophysics*, **77**, 35–62, 1981a.

YOKOKURA, T., On the subduction dip angles, *Tectonophysics*, **77**, 63–77, 1981b.

YOSHII, T., Proposal of the "aseismic front," *Zisin 2*, **28**, 365–367, 1975 (in Japanese).

YOSHII, T., A detailed cross-section of the deep seismic zone beneath northeastern Honshu, Japan, *Tectonophysics*, **55**, 349–360, 1979.

YUEN, D. A. and G. SCHUBERT, On the stability of frictionally heated shear flows in the asthenosphere, *Geophys. J. R. Astron. Soc.*, **57**, 189–207, 1979.

YUEN, D. A., L. FLEITOUT, G. SCHUBERT, and C. FROIDEVAUX, Shear deformation zones along major transform faults and subducting slabs, *Geophys. J. R. Astron. Soc.*, **54**, 93–119, 1978.

Volcano Spacing and Subduction

D. Shimozuru and N. Kubo

Earthquake Research Institute, University of Tokyo,
Yayoi 1-1-1, Bunkyo-ku, Tokyo 113, Japan

In this paper, volcano spacing in trench-arc system with its relation to the dip angle of subduction of oceanic plates is discussed. Three kinds of volcano spacing have been calculated. They are (1) distance between the nearest active volcanoes (Spacing 1), (2) distance between the neighbouring projection of location of active volcanoes to the volcanic front (Spacing 11), and (3) number of active volcanoes per 100 km arc length (Linear Density).

Histogram of spacing widely spreads for Tonga, New Hebrides, Mariana, and Ryukyu, while, Central America, Honshu, Kurile, Kamchatka, Java, and Sumatra show much narrow spacing. In order to demonstrate the mechanism of such difference in spacing for these arcs, dip angle of Wadati-Benioff zone just beneath the volcanic front of each arc were measured from the published data. The results are (1) linear density of active volcanoes and dip angle of Wadati-Benioff zone is in positive correlation, and (2) two types of trench-arc system (Chilean type and Mariana type) seems to be clearly separated. The general trend of the linear density-dip relation could be interpreted as follows. Increase of dip angle of decoupling plate implies the increase of degree of decoupling which reflects the decrease of compressional stress. If the compressional stress is low, magma rise through not only the preexisting channel, but also rise through forming new channels. Consequently, the volcano spacing becomes large.

1. Introduction

The one of the most outstanding features of island and continental arc volcanism is the narrow alignment of volcanic centers along the so called volcanic front. Some of them are segmented, specially recognized in the Aleutian arc (Marsh, 1979a). Volume of the volcanic products as well as the spatial density of volcanic centers are dominated at the volcanic front and markedly decreases towards their back-arcs. In some island arcs, volcanic centers are spaced regularly and their population is large. In other island arcs, location of volcanoes is much diversed. Many investigations appeared to debate why and how are they spaced in island and continental arcs.

Fedotov (1975) suggested that, in Kamchatka, magma feeding channels are spaced at a distance about 30 km speculated from the spacing of volcanoes along the Pacific coast of south of Kamchtka. Volcano spacing of both hot-spot type and island arc type was argumented by Vogt (1974) who demonstrated that this spacing is close to the thickness of the lithosphere and increases with increasing thickness of the lithosphere. The mechanism is postulated by him so that the volcanism is controlled primarily by fracture pattern of the lithosphere and they occur at fracture joints.

Interesting studies on island-arc volcanism, elaborate and instructive, are due to Marsh and Carmichael (1974) and March (1979a, b) whose basic idea depends on the theoretical studies on salt dome dynamics by Daneš (1964), Salig (1965) and Biot (1966). According to them, along the Wadati-Benioff zone directly beneath the volcanoes, a long, narrow belt of magma exists and sinusoidal deformation of the upper surface of the gravitationally unstable layer may occur due to a certain hypothetical perturbation. Eventually, the perturbation would generate diapir which spaced along the subducting slab. According to Selig, the spacing of the above diapiric conduit is

$$\lambda = \frac{2\pi h_2}{2.15}\left(\frac{\eta_1}{\eta_2}\right)^{1/3},$$

where λ, h_2 are the wavelength of diapir (volcano spacing) and the thickness of the buoyant layer (magma) and η_1, η_2 are, respectively, the viscosity of the upper and lower layer substances. After such a formation of diapiric conduit, according to Marsh, the same conduit might be occupied by magma for upward migration over a considerable span of time. On putting possible values to the above equation, Marsh obtained the average spacing of diapir as about 70 km.

On the other hand, alternative approach on arc volcanism related with kinetics of subduction zone was done by Acharya (1981) who examined volcanic activity in the circum-Pacific region related with the convergence rate of plate motion. He found the linear relationship between the number of eruptions and aseismic slip rate. Where the degree of decoupling is high, volcanic activity along the subduction zone is high. Where the two plates are strongly coupled, the volcanic activity is low.

It was stressed by Kanamori (1977) that coupling and decoupling of plates play an essential role in the evolution of island arcs and marginal seas. Therefore, volcano spacing may also be influenced by the degree of coupling of the two plates. In this paper, spacing of volcanoes in the circum-Pacific area is discussed in view of the morphology and convergence rate of the Wadati-Benioff zone.

2.   Spacing of Volcanoes

In this kind of study, spacing of volcanoes should be clearly defined. Vogt (1974), taking the fracture system, measured the spacing as the distances to the nearest volcanoes on an adjacent fracture line provided the longer distance is not more than twice the lesser. Thus, he obtained 58 km for island arcs and smaller for hot spots.

March and Carmichael (1974) took the distance from one volcanic crater to its next nearest neighbour. In this paper, we calculated three types of volcano spacing. Data base is referred to "*List of the World Active Volcanoes*" edited by Katsui (1971). Among the 830 active volcanoes in the List, we chose 594 active volcanoes which are associated with subduction zone.

Spacing I   Distance between the nearest volcanoes were calculated based on the given longitude and latitude of each volcano. This is defined as Spacing I as shown in

Fig. 1. But, there is ambiguity to take the nearest distance when the location of volcanoes is complicated.

Spacing II   This is defined as the distance of the vertical projection of volcanoes to the volcanic front. The mean spacing is calculated as

$$S = L/N - 1,$$

where L is the length of the volcanic front and N is the number of volcanoes along the said front.

Linear density (*L.D.*)   To avoid the complexity and uncertainty, population of active volcanoes along the front is considered. Number of active volcanoes along the front length of 100 km is defined as linear density (*L.D.*) expressed as

$$L.D. = \frac{N-1}{L} \times 100 = \frac{1}{S} \times 100.$$

The similar idea was postulated by KAIZUKA *et al.* (1976). however, they took L as the length between the arc junction. Therefore, L includes the part of the arc in which active volcanoes are not observed like the western part of the Aleutian arc. They took the length of the Andes arc as 8,000 km, however, as pointed out by ISACKS and BARAZANGI (1977), segmentation is developed in this arc and should be considered separately.

Example of histograms of the volcano spacing, thus calculated, are shown in Fig. 2. Histogram widely spreads for Mariana, Tonga, New Britain, Solomon, and New Hebrides, while Honshu, Kurile, Kamchatka, Aleutian, and Central America show

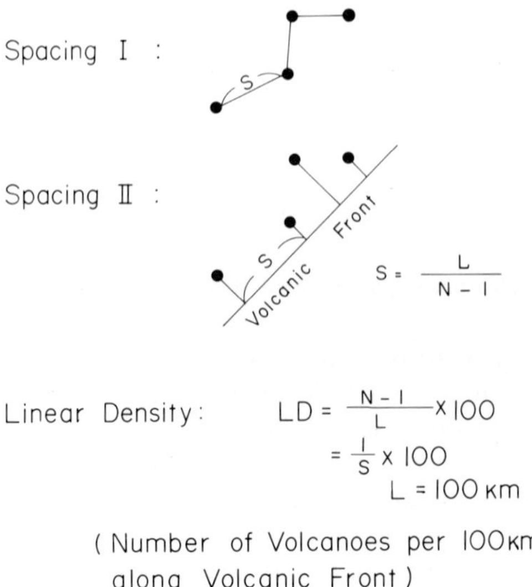

Fig. 1. Method to calculate the three kinds of volcano spacing. Solid circles indicate the location of volcanic centers.

much narrow spacing. Spacing I, II, and *L.D.* are listed in Table 1 together with other geophysical parameters. As shown in the figure, volcano spacing is classified into two types, Type I and Type II. Type I is characterized by the wide volcano spacing and Type II shows narrow volcano spacing. Well developed island arcs and continental arcs are involved in Type II. Even belongs to continental arc, volcano spacing of Cascade and Mexico seems to belong Type 1. These arcs are not considered to be the typical trench-arc system. Especially, in Mexico, volcano alignment is not paralled to the trench axis.

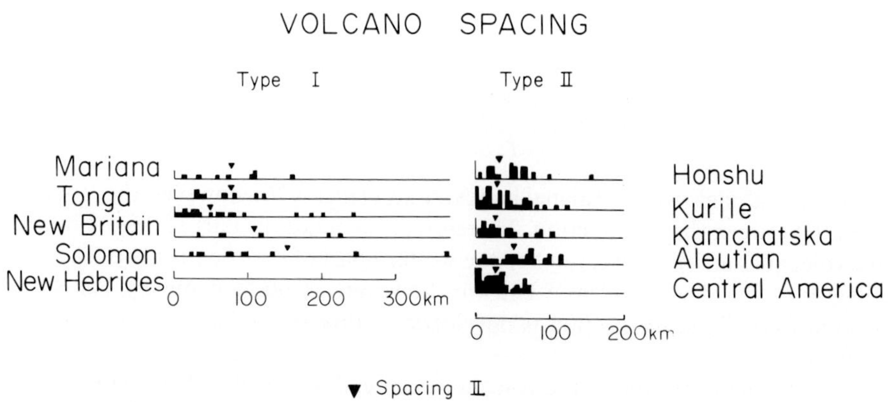

Fig. 2.   Example of the histogram of volcano spacing. Inverse triangles are average of Spacing II.

As listed in Table 1, Spacing II is always smaller than Spacing I. New Hebrides is the only exception which may be due to the difficulty to measure the distance of nearest volcano. If the difference (Spacing I–Spacing II) is large, volcanoes are distributed at the backside of the front with the area of some extent, i.e., the width of volcanic belt is large as illustrated in Fig. 3. If Spacing I is close to Spacing II, volcanoes are located almost linearly along the volcanic front. As clearly seen in Table 1, arcs classified as Type I show small difference in Spacing I and II, which means the linear alignment of volcanoes. But, in arcs of Type II, the difference ranges between 0–20 km.

3.   Geometry and Convergence Rate of Wadati-Benioff Zone

Since island arc magmatism seems to be related with kinetics of subduction of oceanic plate, we try to find any relationship between volcano spacing and dip and convergence rate of subducting plate. We have already clarified that volcano spacing is not always the same in island and continental arcs, but, it is characteristic to each arcs. On affirming the diapiric conduit hypothesis, we have to prove why volcano spacing is different for different arcs. Parameters which specify the configuration and kinetics of subducting plate are dip angle and convergence rate. The upper surface of the downgoing oceanic plate is represented by Wadati-Benioff zone and usually shows low angle in the shallower depth and large angle with increasing depth. In this paper, we defined the "dip angle" of Wadati-Benioff zone as the dip angle just beneath the

Table 1. Volcano spacing and other geophysical parameters of continental and island arcs.

Arc	Dip (deg)	Conv. rate (cm/y)	Spacing I (Km)	Spacing II (Km)	Linear density	Length of volcanic front (Km)	No. of volcanoes
1 Scotia	70	2.0	62.1	54.9	1.82	439	9
2 Lesser Antilles	63	2.0	47.3	43.7	2.29	699	17
South Chile	—	—	185.7	172.4	0.58	348	3
3 Central Chile	22	11.1	60.8	58.1	1.72	1103	20
4 N. Chile-S. Peru	25	10.7	59.7	47.8	2.09	1114	33
Colombia	—	7.7	57.8	47.3	2.12	898	21
5 Central America	63	8.0	28.5	25.3	3.96	1062	43
6 Alaska	42	5.9	55.3	47.4	2.11	900	20
7 Aleutian	50	7.5	57.9	50.0	2.00	1300	27
8 Kamchatka	50	9.3	36.9	25.9	3.86	673	27
9 Kurile	40	9.3	37.5	30.1	3.32	1386	47
10 (island)	44	9.3	34.5	25.9	3.86	1060	42
11 (land)	30	—	66.1	60.8	1.85	243	5
12 Honshu I*	30	9.7	50.8	30.4	3.29	820	28
13 Honshu II	30	9.7	53.3	34.0	2.94	1360	41
14 Izu	45	6.1	64.4	47.3	2.11	757	17
15 (island)	51	6.1	65.0	60.9	1.64	487	9
16 Bonin	65?	6.1	70.0	67.1	1.49	202	4
17 Mariana	60	4.0	81.1	77.5	1.29	621	9
18 Ryukyu	35	5.6	63.5	57.1	1.75	685	13
19 (island)	35	5.6	128.8	97.5	1.03	390	5
20 Java	48	7.1	37.4	28.2	3.54	1018	37
21 Sumatra	30*	6.6	68.8	67.1	1.49	1760	27
22 New Britain	50	—	52.0	47.4	2.11	948	21
23 Solomon	45	—	117.4	107.5	0.93	643	7
24 New Hebrides	68	2.7	143.6	150.3	0.67	1503	11
25 Tonga	43	8.9	87.2	73.0	1.37	732	11
26 Kermadec	55	6.4	66.2	58.5	1.71	175	4
27 New Zealand	49	5.5	75.8	62.5	1.60	370	6

*Northern part of Honshu.

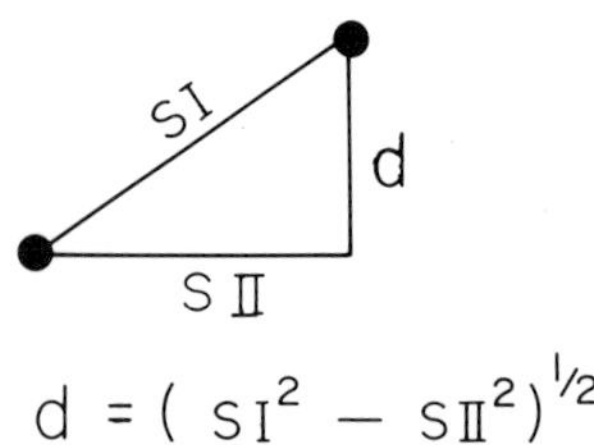

Fig. 3.   Schematic representation of the width of volcanic belt (d) which is calculated from Spacing I and II.

volcanic front of each arc. From the published figures of Wadati-Benioff zone (Ansell and Smith, 1975; Davies and House, 1979; Dewey and Algermissien, 1974; Engdahl *et al.*, 1977; Fitch, 1970; Isacks and Barazangi, 1977; Isacks and Molnar, 1971; Ishida, 1970; Katsumata and Sykes, 1969; Lahr, 1975; Sykes, 1966; Sykes and Ewing, 1965), we measured the dip angle as listed in Table 1. In Scotia and Sumatra, since we could not find the vertical cross section of Wadati-Benioff zone, the dip angle was taken from the papers by Uyeda and Kanamori (1979) and Fitch (1970).

Convergence rate of Table 1 was taken from Ruff and Kanamori (1980) and Uyeda and Kanamori (1979).

Relationship between dip and convergence rate of the oceanic plates is found to be in inverse relation as shown in Fig. 4. The similar evidence was found by Luyendyke (1970) who demonstrated the relationship of maximum dip angle and convergence rate. He interpreted that the vertical component of the sinking velocity of a plate is constant independent of the length of sinking plate and convergence rate provided that a plate sinks only by gravity. Therefore, the larger the convergence rate, the smaller the dip angle of a plate is. Furthermore, as can be seen in Fig. 4, trend of the relationship dip-convergence rate seems to be different for typical Type I and Type II arcs.

   4.   Volcano Spacing and Dip of Subducting Slab

Based on the parameters listed in Table 1, dip and spacing for 27 arcs are plotted as shown in Fig. 5, 6, and 7. In the figure, solid circles represent arc classified as Type II and open circles are arc of Type I. Double circles indicate the arcs where back-arc opening is recognized by several geophysical data (Uyeda and Kanamori, 1979). Among three figures, relationship between linear density and dip (Fig. 7) shows the most remarkable feature. According to this relation, the general trend is that linear density (volcano population) increases with increasing dip angle of the subducting slab. Uyeda and Kanamori (1979) classified trench-arc-back-arc system of island arc into two types, i.e., back-arc active and back-arc inactive. It was found that the arc of Type I of the present study corresponds to their arc of back-arc active and Type II corresponds to their type of back-arc inactive. Continental arcs have no back-arc basins. According

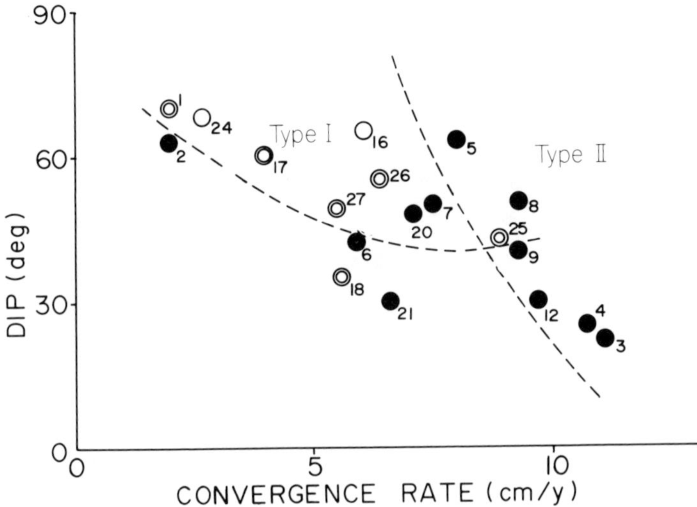

Fig. 4.   Relationship between dip angle of Wadati-Benioff zone and convergence rate. Solid circles represent the arcs of TYPE II and open circles are of Type I. Double circles are arcs of which back-arc spreading is confirmed. Numbers put aside of the circles correspond to the numbers of arc listed in Table 1.

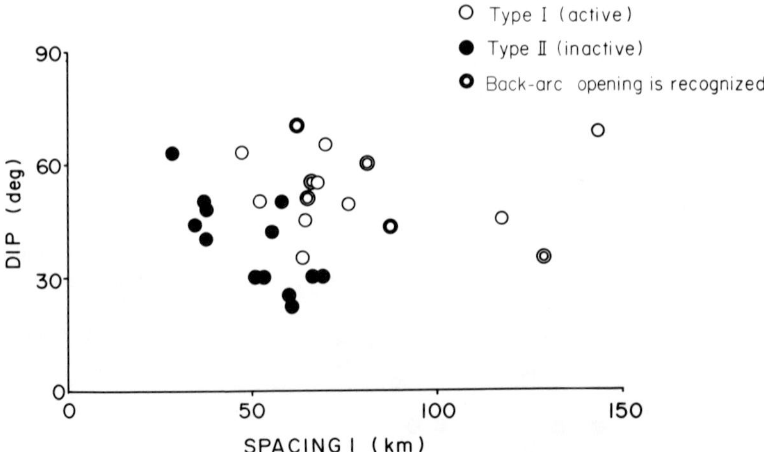

Fig. 5.   Relationship between dip angle of Wadati-Benioff zone and Spacing I.

to UYEDA and KANAMORI, the back-arc area of the Sumatra and Java arc is grouped with the continental arcs. Only exception is the case of central America. According to the present study, it should be classified as Type II. However, it has been grouped with the back-arc spreading type (back-arc active) because of the extensional volcanic grabens (UYEDA and KANAMORI, 1979, PLAFKER, 1976). Type II is considered

D. Shimozuru and N. Kubo

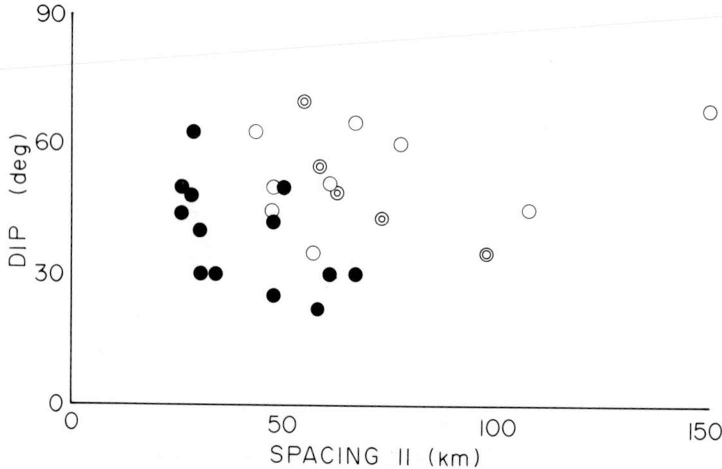

Fig. 6. Relationship between dip angle of Wadati-Benioff zone and Spacing II.

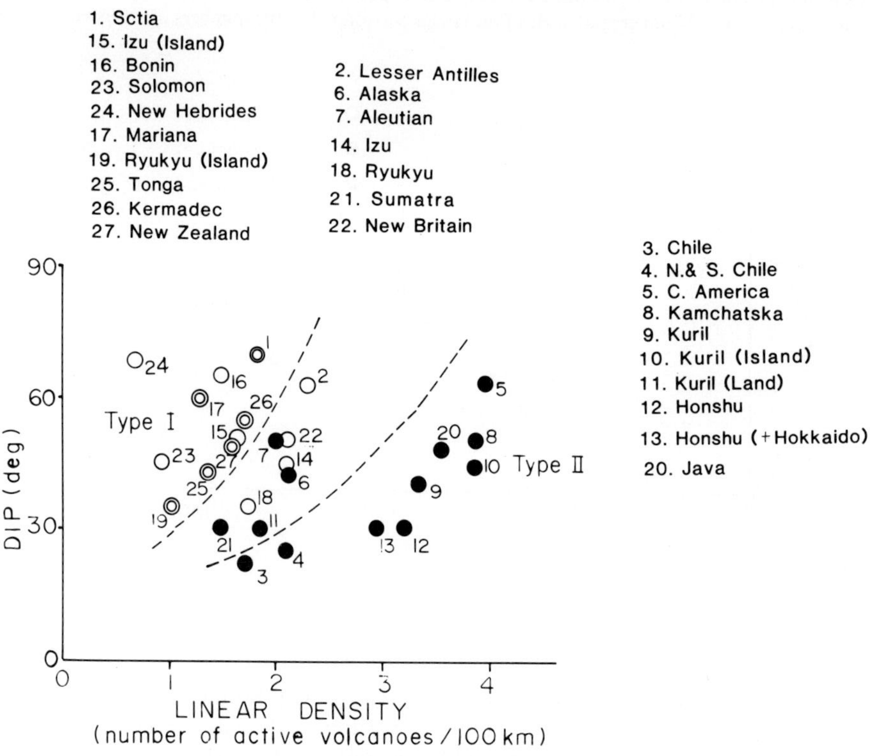

Fig. 7. Relationship between dip angle of Wadati-Benioff zone and linear density of volcanoes along volcanic front.

as well developed continental and island arc and Type I is undeveloped arc of oceanic type. Arcs located in the region between the two chain lines in Fig. 7 might be consumed as arcs under developing.

5. Discussion

As stated and shown in the previous section, population of volcanoes along volcanic front increases with increasing dip angle of subducting plates. Furthermore, with the same dip angle, linear density is larger for Type II arc than Type I arc. Based on the detailed mechanism studies and repeat time of great earthquakes, KANAMORI (1977) suggested that in Chile and Alaska the coupling and interaction between the oceanic and continental lithosphere are very strong. RUFF and KANAMORI (1980) introduced strength of coupling of the upper and lower plates and found that both Wadati-Benioff zone geometry and strength of coupling are significantly correlated to age of subducting oceanic lithosphere and plate convergence rate. The general trend is for low seismicity to correlate with the combination of older oceanic lithosphere and smaller convergence rates, while younger lithosphere and larger convergence rates are associated with great earthquakes—strongly coupled. UYEDA and KANAMORI (1979) classified the modes of subduction with two types. The extreme cases of each mode are represented by the Chilean arc and Mariana arc. At the former type subduction boundaries, the mechanical coupling between the upper and lower plates is strong and plate motion is seismic. At the Mariana type boundaries, the coupling is either much weaker or effectively diminishing and, hence, the plate motion is aseismic. They also examined the stress state in back-arc plate based on the available data of focal mechanism of intraplate earthquakes and concluded that back-arc of the Mariana type plate boundaries shows spreading feature and in tensional stress state, while at the Chilean type back-arc is in compressional state. The degree of coupling, convergence rate, dip angle and stress state of back-arc for two types of plate boundaries are listed in Table 2. An interpretation of the general trend of linear density and dip relationship is given below.

The large dip angle implies the smoothness of the downgoing movement of the subducting slab which yields the decrease of the degree of coupling of the upper and lower plates. Consequently the compressional tectonic stress decreases at the back-arc region. If the compressional stress decreases, new feeding channels may be formed by

Table 2. Comparison of degree of coupling of subducting slab, convergence rate of oceanic plates, dip angle of Wadati-Benioff zone and tectonic stress state of back-arc.

	Subd. slab	C. Rate	Dip	Back-arc
Type I (Mariana type)	Decoupled (Aseismic)	Small	Large	Tension
Type II (Chilean type)	Coupled (Seismic)	Large	Small	Compression

forceful intrusion of magma, and rate of upward transfer of magma due to the bouyancy of magmatic diapir may be accelerated resulting from the decrease of horizontal deviatoric stress. Contrary to the above, if the back-arc is dominated in compressional tectonic stress, rising magma unable to form new channels, but it may rise through the pre-existing conduit. This is the reason why the volcano spacing becomes large with the decreasing dip angle of the subducting plate and vise versa. It is worthwhile to point that, as can be found in Fig. 7, Alaska (6) and Aleutian (7) arc are deviated markedly from the group of Type II arc. Nakamura *et al.* (1977) investigated the orientation of the principal tectonic stresses at Aleutian and Alaskan volcanoes based on their surface features and faults. According to them, in the back-arc area, stress system is characterized by tensional deviatoric stress which suggests to assume an independent stress source. Their independent study seems to support the present evidence of deviation of two arcs from the group of Type II arc.

We benefited greatly from discussions with S. Uyeda and M. Sakuyama. We thank Mrs. Matsumoto and Mrs. Murakami for their preparation of manuscript.

REFERENCES

Acharya, H., Volcanism and aseismic slip in subduction zones, *J. Geophys. Res.*, **86**, 335–344, 1981.

Ansell, J. H. and E. G. C. Smith, Detailed structure of a mantle seismic zone using the homogeneous station method, *Nature*, **253**, 518–519, 1975.

Biot, M. A., Three-dimensional gravity instability derived from two-dimensional solutions, *Geophysics*, **31**, 153–166, 1966.

Daneš, Z. F., Mathematical formulation of salt-dome dynamics, *Geophysics*, **29**, 414–424, 1964.

Davies, J. N. and L. House, Aleutian subduction zone seismicity, volcano-trench separation, and their relation to great thrust type earthquakes, *J. Geophys. Res.*, **84**, 4583–4591, 1979.

Dewey, J. W. and S. T. Algermissien, Seismicity of the middle America arc-trench system near Managua, Nicaragua, *Bull. Seismol. Soc. Am.*, **64**, 1033–1048, 1974.

Engdahl, E. R., N. H. Sleep, and M. T. Lin, Plate effects in north Pacific subduction zones, *Tectonophysics* **37**, 95–116, 1977.

Fedotov, S. A., Mechanism of magma ascent and deep feeding channels of island arc volcanoes, *Bull. Volcanol.*, **39**, 241–254, 1975.

Fitch, T., Earthquake mechanisms and island arc tectonics in the Indonesian-Philippine region, *Bull. Seismol. Soc. Am.*, **60**, 565–591, 1970.

Isacks, B and P. Molner, Distribution of stress in the descending lithosphere from a global survey of focal-mechanism solutions of mantle earthquakes, *Rev. Geophys. Space Phys.*, **9**, 103–174, 1971.

Isacks, B. L. and M. Barazangi, Geometry of Benioff zones, lateral segmentation and downward bending of the subducted lithosphere, in *Island Arcs, Deep Sea Trenches and Back-Arc Basins*, edited by Talwani, M. and W. C. Pitman, pp. 99–114, 1977.

Ishida, M., Seismicity and travel-time anomaly in and around Japan, *Bull. Earthq. Res. Inst.*, **48**, 1032–1051, 1970.

Kaizuka, S., T. Matsuda, and K. Nakamura, Structure, earthquakes and volcanoes in the Japanese arcs, *Kagaku*, **46**, 196–210, 1976 (in Japanese).

Kanamori, H., Seismic and aseismic slip along subduction zones and their tectonic implications, in *Island Arcs, Deep Sea Trenches and Back-Arc Basins*, edited by Talwani, M. and W. C. Pitman, pp. 163–174, 1977.

Katsumata, M. and L. R. Sykes, Seismicity and tectonics of the western Pacific: Izu-Mariana-Caroline and

Ryukyu-Taiwan regions, *J. Geophys. Res.*, **7**, 5923–5948, 1969.

KATUI, Y. (ed), List of the World Active Volcanoes, with maps, in *Volcanological Society of Japan*, 1971.

LAHR, J. C., Detailed seismic investigation of Pacific-North American plate interaction in South Alaska, Ph.D. Thesis, Colombia Univ., New York, 1975.

LUYENDYKE, B. P., Dip of downgoing lithospheric plates beneath island arcs, *Geol. Soc. Am. Bull.*, **81**, 3411–3416, 1970.

MARSH, B. D. and Ian S. E. CARMICHAEL, Benioff zone magmatism, *J. Geophys. Res.*, **79**, 1196–1206, 1974.

MARSH, B. D., Island arc development: Some observations, experiments, and speculations, *J. Geology*, **87**, 687–713, 1979a.

MARSH, B. D., Island arc volcanism, *Am. Scient.*, **67**, 161–172, 1979b.

MOHR, P. A. and C. A. WOOD, Volcano spacing and lithospheric attenuation in the Eastern Rift of Africa, *Earth Planet. Sci. Lett.*, **33**, 126–144, 1976.

NAKAMURA, K., K. H. JACOB and J. N. DAVIES, Volcanoes as a possible indicators of tectonic stress orientation—Aleutian and Alaska, *Pageoph.*, **115**, 87–112, 1977.

PLAFKER, G., Tectonic aspects of the Guatemala earthquake of 4 Feb. 1976, *Science*, **193**, 1201–1208, 1976.

SELIG, F., A theoretical prediction of salt dome patterns, *Geophysics*, **30**, 633–643, 1965.

SYKES, L. R., Seismicity and deep structure of island arcs, *J. Geophys. Res.*, **71**, 2981–3006, 1966.

SYKES, L. R. and M. EWING, The seismicity of the Caribbean region, *J. Geophys. Res.*, **70**, 5056–5074, 1965.

UYEDA, S. and H. KANAMORI, Back-arc opening and mode of subduction, *J. Geophys. Res.*, **84**, 1049–1061, 1979.

VOGT, P. R., Volcano spacing, fractures, and thickness of the lithosphere, *Earth Planet. Sci. Lett.*, **21**, 253–252, 1974.

Arc Volcanism: Physics and Tectonics, edited by D. Shimozuru and I. Yokoyama, 153–163.
Copyright © 1983 by Terra Scientific Publishing Company (TERRAPUB), Tokyo.

Fore-Arc Volcanism and Cycles of Subduction

Kazuo Kobayashi

Ocean Research Institute, University of Tokyo
1–15–1 Minamidai, Nakano-ku, Tokyo 164, Japan

Volcanic rocks including boninites have been collected in the fore-arc regions of Izu-Bonin and Mariana trenches. In the southwestern and northeastern Japan the fore-arc volcanism appears to be more felsic. These short-lived fore-arc volcanisms occurring at positions much closer to the trench axis than usual volcanic front are correlatable to the initial stage of subduction which introduces H_2O and CO_2 to the asthenosphere beneath the fore-arc but does not cool it down. Coexistence of harzburgites and arc basalts with boninites in the fore-arc region found by dredge hauls at the Bonin trench in our recent cruise as well as at the Mariana trench by Mendeleev seems to imply possibility of obduction of the fore-arc region together with the oceanic lithosphere to form the ophiolite.

1. Fore-Arc Volcanism in the Western Pacific

Active volcanoes in island arc are so systematically aligned that a volcanic front is defined as a line seaward of which no volcanoes exist (Sugimura, 1960). Density of active volcanoes is the largest immediately landward of the front and decreases landward as distance from the front increases.

The volcanic front is nearly parallel to the axis of deep-sea trench in any arc-trench systems in the world. In northeastern Honshu, Japan the volcanic front is situated at a distance of about 300 km from the trench axis. Distance of the front from the trench axis slightly varies at different locations but is no less than 200 km.

Location of the volcanic front seems to be better correlatable to depth of deep-focus earthquakes. Depth of deep-focus earthquakes beneath the volcanic fronts is 90 to 150 km, and in most arcs 100 to 130 km. If angle of deep-focus earthquake plane is larger, distance between the trench axis and volcanic front is shorter (Fig. 1). It implies that magma is generated in the mantle beneath the arc where the oceanic lithosphere sinks to depths greater than about 100 km.

Volcanic front can also be defined with volcanoes of older geological ages. Fronts of Quaternary volcanoes are generally concordant with those of the present-day active volcanoes. In northeastern Honshu, Japan the front of early Miocene volcanoes is situated about 20 km east of the present volcanic front. This discrepancy may possibly be explained either by a relative shift of northeastern Honshu of about 20 km toward the axis of the Japan Trench or by a decrease in angle of deep-focus earthquake plane of about 10 degrees since early Miocene.

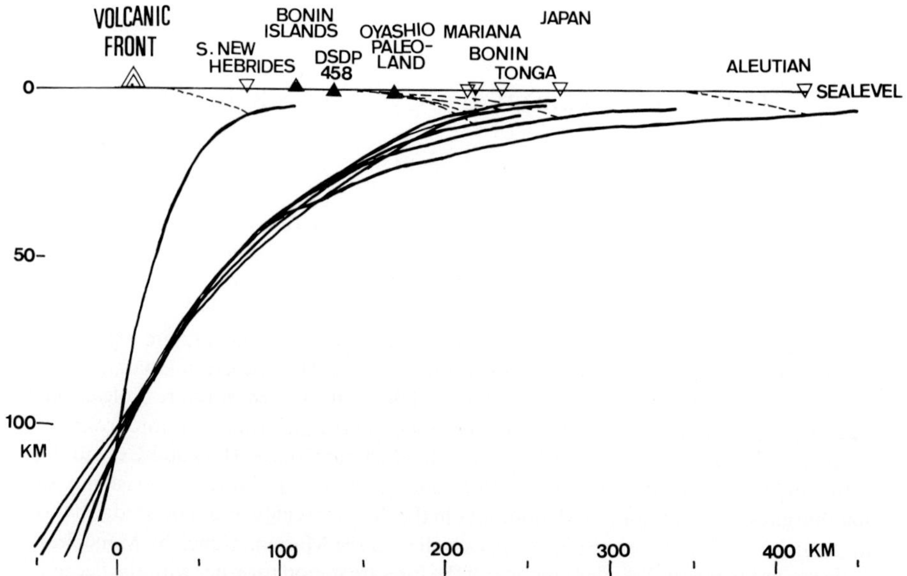

Fig. 1. Positions of trench and subduction zone relative to the volcanic front in several island arcs (modified from KARIG and SHARMAN, 1975).

In the sharp contrast to such small shifts of volcanic fronts in geological time, volcanic materials have been collected from fore-arc regions of some arcs much closer to the trench axis. The most well-known example of the fore-arc outcrops of volcanic masses is the Bonin Islands (Fig. 2). The Bonin Islands are situated at a distance of only 100 km or less from the axis of the Bonin Trench, while a chain of active volcanic islands including Nishinoshima and Iwojima is located at a distance of about 250 km from the trench axis.

The Bonin Islands are characterized by occurrence of boninite, which is a type of high MgO ($> 9\%$), high SiO_2 ($> 57\%$), low TiO_2 ($< 0.4\%$) andesites (KURODA and SHIRAKI, 1975; CAMERON et al., 1979). Boninite occurs as pillow lavas, dykes and breccias. Pillow structures with high vesicularity of some boninites seem to indicate that they erupted in a relatively shallow ocean bottom. Age of eruption was determined with the K-Ar method to be about 40 Ma (TSUNAKAWA, 1980). Similar rocks were recovered from fore-arc toe of the Mariana arc-trench system (about 85 km landward of the trench axis) at site 458 of DSDP leg 60 (HUSSONG et al., 1982). KAY (1978) described magnesian andesites in Aleutian Islands and attributed them to remelts of subducted oceanic crust.

KUSHIRO (1972, 1982) has demonstrated by his laboratory experiments that boninite is possibly formed under a very hydrous condition. It has also been shown that boninites were formed from a very primitive magma quickly brought up to the surface in an extensional stress field.

Volcanic activities in the fore-arc regions within 100 km from the present trench

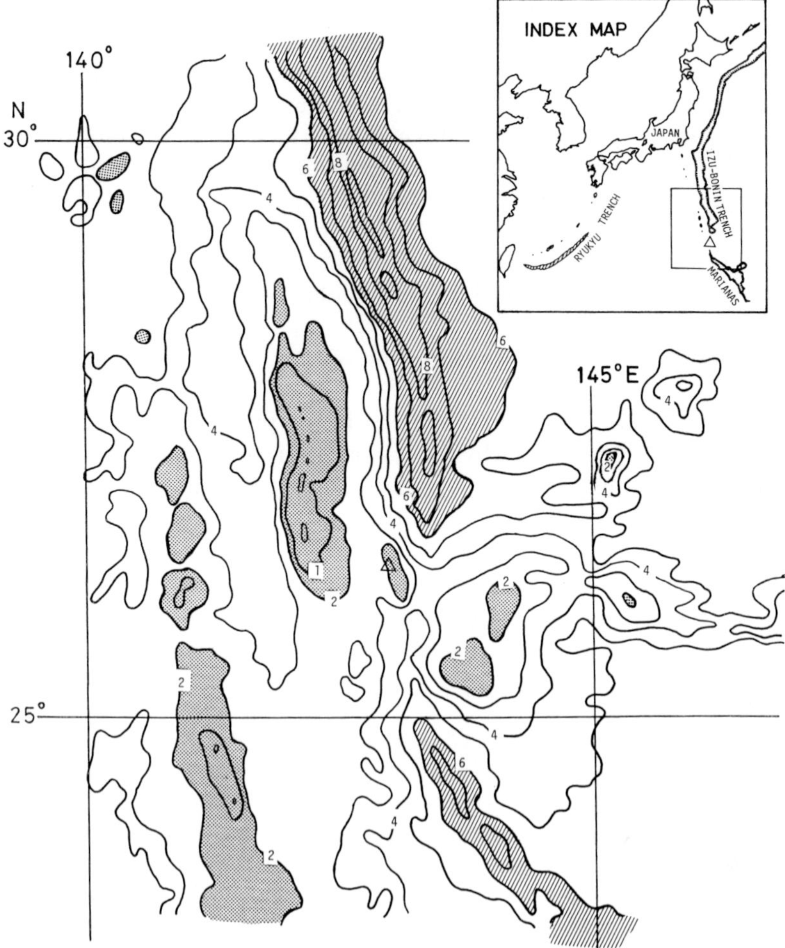

Fig. 2.   Bathymetry around the Bonin Islands (from Tomoda and Fujimoto, 1982). Contours in km. Δ indicates submarine topography from which harzburgites and boninites were dredged.

axis are also known in other arc-trench systems. Along the Japan trench a subaerial eruption of dacite dated to be 23 Ma at the Oyashio paleoland was revealed by site 437 of DSDP leg 57 (Von Huene *et al.*, 1980). A chain of felsic volcanic and plutonic rocks has been reported along the southern coast of southwestern Japan and dated to be 13 to 15 MaBP (Shibata, 1978). No boninite has been found in the fore-arc regions of Honshu and Shikoku, Japan. Boninite may be covered by more differentiated felsic rocks, as boninite at the Bonin Islands is overlain by andesite and dacite.

High magnesian andesites including Sanukite and Sanukitoid with slightly different chemical composition and mineral assemblage occur in Setouchi (inland sea region) of the southwestern Honshu (Nakada and Takahashi, 1979; Tatsumi and

ISHIZAKA, 1981; TATSUMI, 1982). Their age of eruption is nearly the same as or slightly younger than fore-arc felsic rocks in southwestern Japan. However, their location is at a greater distance (approximately 200 km) from the trench axis and their petrogenesis seems to be somehow different from that of boninite.

In the present paper the author will not deal with detailed petrological properties of these rocks but attempt to consider their geotectonic settings in a framework of plate tectonic model of subduction cycles (KOBAYASHI and ISEZAKI, 1976; KOBAYASHI, 1983). It will be suggested here that occurrence of the fore-arc magmatism is correlatable to the initial stage of subduction. Relevance of the present conclusion to the origin of some other boninites accompanied by the ophiolite complexes will then be discussed.

2.   Cycles of Subduction and Generation of Boninite Magma

A model upon which the author relies in the present consideration is that of subduction cycles first postulated by KANAMORI (1971) to explain various types of focal mechanisms of earthquakes occurring in the lithosphere of the subduction zones. The author (KOBAYASHI and ISEZAKI, 1976; KOBAYASHI, 1983) has modified his seismic model to better suit other phenomena such as volcanic and tectonic events in the arc-trench systems. A cycle consists of four distinctive stages illustrated in Fig. 3. Stages 2 and 3 are those of normal arc volcanism and back-arc opening, which are not

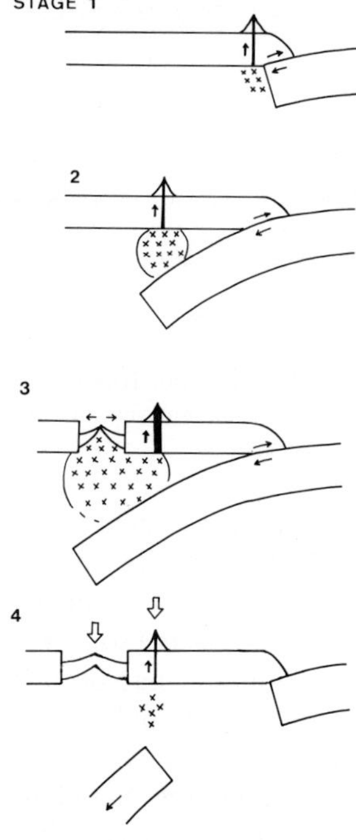

Fig. 3.   A model of a subduction cycle. Arrow indicates direction of motion.

considered in this paper. Stage 4 is generally followed by Stage 1 of the next cycle so that the cycles are endlessly continued except for the case in which the converging plates drastically change direction of their relative motion.

In Stage 1 the downgoing slab of the oceanic lithosphere is still too short (less than 100 km) and its bottom end is too shallow (20–40 km) to be melted by the heat supplied from the surrounding asthenosphere. Although temperature at the asthenosphere beneath the continental or island-arc lithosphere may be higher than the melting temperature of some minerals, it takes much time to warm up the cold upper surface of the subducted oceanic slab. On the contrary, materials in the asthenosphere beneath the continental or island-arc lithosphere melt to form a large volume of magma, because the melting temperature is much lowered with H_2O and CO_2 brought by the downgoing oceanic slab containing them as hydrated rocks and carbonates as well as pore water (SUN and NESBITT, 1978; DUNCAN and GREEN, 1980).

Magma thus formed is very likely to be Mg-rich, SiO_2-rich and Ti-poor andesite. If the overriding continental or island-arc lithosphere is apt to be extensional, it will easily be rifted by the effect of swell in volume due to generation of magma. The high magnesian andesite magma will then rapidly ascend to the surface of the earth to form boninite lava.

The best examples of this process of boninite genesis may be seen in the Bonin Islands and DSDP site 458 at the Mariana fore-arc toe. A line comprising the present Izu-Bonin and Mariana trenches acted as a transform fault separating the old Pacific and Philippine-Sea plates (UYEDA and BEN-AVRAHAM, 1972; UYEDA and MIYASHIRO, 1974). Shikoku Basin (WATTS and WEISSEL, 1975; KOBAYASHI and NAKADA, 1978), Parece Vela Basin (MROZOWSKI and HAYES, 1979) and the Mariana Trough (KARIG *et al.*, 1978) had not opened so that the Izu-Bonin and Mariana adjoined the Kyushu-Palau Ridge (KLEIN and KOBAYASHI, 1980; 1981). It is quite plausible that the Philippine Sea side of the transform fault is younger than the Pacific side at the positions of the present Bonin and Marianas Islands, since magnetic lineations and DSDP results have indicated that the age of the Philippine Basin ranges from 40 to 60 Ma, while those of the western Pacific are older than 100 Ma.

Traces of the Hawaiian hot spot along the Emperor-Hawaii chain have shown that the Pacific plate changed direction of its motion at about 42 MaBP from NWN to WNW (MORGAN, 1972; DALRYYMPLE *et al.*, 1980). The motion given from the hot spot traces is a roughly absolute motion, unless the mantle plume moved appreciably. The absolute motion of the Philippine-Sea plate is unknown. If we assume that the Philippine-Sea plate was fixed relative to the earth's rotational axis at 42 MaBP, the Pacific and Philippine-Sea plates started to converge after the change in the direction of the Pacific plate motion. It is natural that the older Pacific lithosphere started to subduct beneath the younger Philippine-Sea lithosphere. As the Philippine-Sea was only about 10 Ma old at the time of the initial subduction of the Pacific lithosphere, thickness of the overriding lithosphere is only 20 to 25 km (YOSHII, 1975). When the bottom end of downgoing slab of the Pacific reached this depth, a large amount of materials beneath the eastern edge of the Philippine-Sea lithosphere melted at a temperature of 1,100°C to 1,200°C with a supply of H_2O and CO_2 to form high magnesian andesite magma. KUSHIRO (1982) has concluded that the original magma of high magnesian andesite collected from DSDP site 458 at the Mariana fore-arc toe must have been formed at temperatures higher than 1,200°C and under pressures lower than 8 kbar (i.e. at depth shallower than 25 km). The circumstances in the initial stage of subduction of the Pacific plate illustrated here are quite consistent with Kushiro's experimental and petrological conclusion.

Existence of Ryukyu and Philippine trenches at around 42 MaBP has not been proved but is very likely. If these trenches existed, the Philippine-Sea plate also moved westward when the subduction of the Pacific beneath the Philippine-Sea started. So the stress field in the eastern margin of the overriding Philippine-Sea lithosphere was extensional with the principal axis roughly perpendicular to the line of convergence. These situations are quite favourable to the extrusion of boninite in the Bonin Islands and the Mariana fore-arc toe. As the temperature immediately below the bottom of lithosphere is sufficiently high to produce tholeiitic magma also, it is plausible that a small amount of tholeiite erupted before the extrusion of boninite, as found in the DSDP hole 458 (HUSSONG, *et al.*, 1982) and in the Bonin Islands (SHIRAKI *et al.*, 1978, 1980).

During the boninite eruption the fore-arc region was uplifted to be emerged above sea level. A megafossil of giant foraminifera *nummulites boninensis* in Hahajima, Bonin Islands were deposited in this period. The fore-arc region may have subsided soon after the release of boninite magma. As the subduction of the Pacific plate further proceeded the asthenosphere beneath the eastern edge of the Philippine-Sea was cooled by the cold oceanic lithosphere and no more high magnesian andesite magma was generated at that depth. Magma of the normal island-arc type was formed after the downgoing slab reached a depth of about 100 km. As the rate of plate motion was about 6 cm/yr, the above-mentioned process of fore-arc magma generation was active for only a short time, perhaps for less than 1 Ma. It is thus understandable that the boninite eruption was a very short-lived event.

At present an oceanic plateau in the Pacific plate is colliding with the Bonin Islands region. Deep topographic features of the Izu-Bonin Trench are lost at the collision zone, as seen in Fig. 2. It is quite likely that the Bonin Islands region is being heaved up by the effect of the compressive force caused by the collision. Figure 4 shows a simplified map of free-air gravity anomaly in the same area as given in Fig. 2. It is clearly seen in Fig. 4 that the free-air gravity anomaly is very high (amounting to 380 mgal) on the fore-arc ridge including the Bonin Islands. This anomaly indicates that the topography of the fore-arc ridge is nonisostatically maintained by a force due to the collision of the Ogasawara Plateau.

From the crest of a topographic high about 100 km ESE of Hahajima, many boulders of serpentinized harzburgite and dunite were collected by dredge hauls (ISHII, 1981) in the cruise KH80-4 of R.V. Hakuho-maru directed by the present author. A block of boninite was collected together with many pieces of basaltic lavas from the western slope of the same topographic high. Boninite appears to coexist with harzburgites and basalts in the Bonin ridge, although no ultramafic rocks have been found on the Bonin Islands. It seems likely that the harzburgites were brought up to the sea bottom when the fore-arc region was rifted and they have been further upheaved by the ongoing collision.

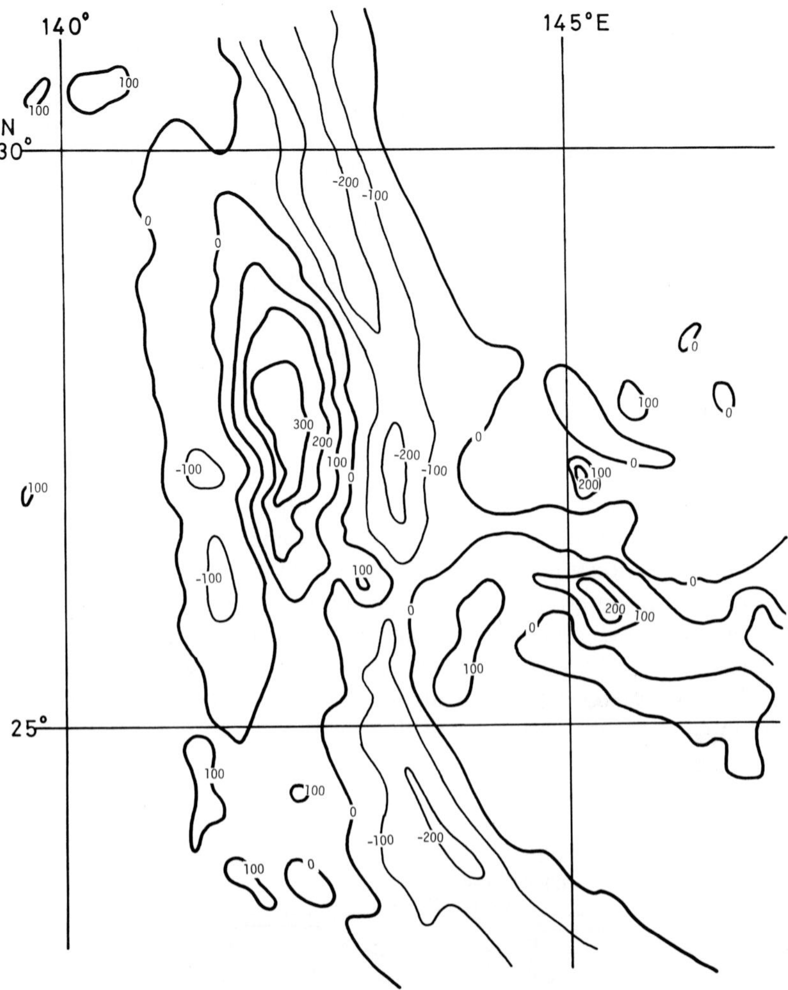

Fig. 4.   Free-air gravity anomaly around the Bonin Islands (from TOMODA and FUJIMOTO, 1982), Contours in mgal.

Harzburgites were collected by dredge hauls from the inner wall of the Mariana Trench together with gabbros and basalts with chemical composition of island arc type (BLOOMER, 1981). Boninites were also dredged by R.V. Dmitry Mendeleev in her 1976 cruise for IGCP Project "Ophiolites" from the inner slope of Mariana trench at 12°N (BOGDANOV, 1977; DIETRICH *et al.*, 1978). Occurrence of boninites together with suites of ultramafic rocks, gabbros, basalts apparently similar to the sections of an ophiolite in the inner wall or fore-arc toe of the Mariana trench seems to imply that these rocks were emplaced close to the bottom surface when the fore-arc region was rifted and boninite magma erupted during Stage 1 of the subduction cycle illustrated in Fig. 3.

3. Origin of Boninites in the Ophiolites

High magnesian andesites conforming to the definition of boninite have been found in some ophiolite complexes. In the Setogawa ophiolite belt at Shizuoka Prefecture, south-central Japan, boninites occur as discrete blocks of lava sheet or volcanic breccia adjacent to serpentinized ultramafic bodies (OHASHI and SHIRAKI, 1981). Samples with similar compositions and textures have been reported from the uppermost pillow lava of Troodos massif and adjacent fault belt in Cyprus (SMEWING et al., 1975; SIMONIAN and GASS, 1978), Cape Vogel Peninsula, eastern Papua (DALLWITZ et al., 1966) and several other places.

Occurrence of boninites both in the fore-arc regions of trench slope and in some ophiolites seems to provide a cruicial information on the sources of materials constituting the ophiolites. When an oceanic plateau or a microcontinental block approaches a trench and eventually collides with the continental margin or an island arc, parts of the subducted lithosphere will be accreted and obducted on the continental margin together with the fore-arc blocks including boninites, arc basalts and harzburgites, as seen in the southern part of the Bonin trench (Fig. 5).

If this consideration is correct, the isotope age of boninites included in ophiolites should be that of the initiation of the relevant subduction but not the age of a tectonic process generating the ophiolite, i.e. accretion. Although age determination of boninites and other rocks included in ophiolites is often difficult due to alteration and metamorphism, it would yield a key to identify the process of ophiolite emplacement as well as an initial stage of subduction at that district.

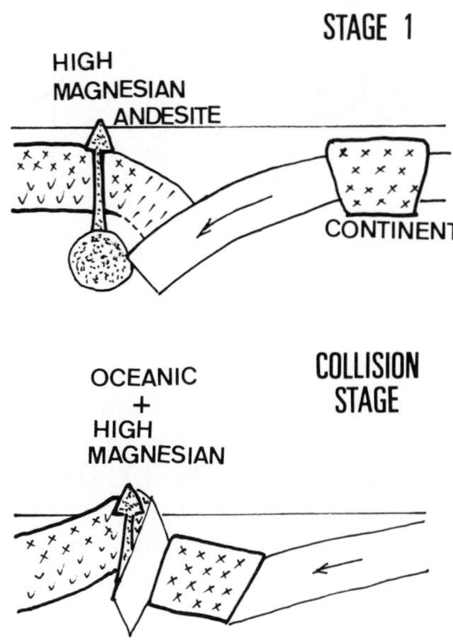

Fig. 5. A model illustrating generation of ophiolite together with boninites and arc basalts.

4. Generation of High-Magnesian Andesites and Felsic Rocks in Some Modified Situations

The case of the Izu-Bonin-Mariana arc was mainly considered in Chapter 2 of this paper as a typical example of boninite genesis. Change from transform fault to subduction at 42 MaBP and the thin overriding lithosphere under extensional stress are rather special circumstances which may seldom be brought about in other localities.

Change from transform to subduction seems to have occurred in the Nankai Trough about 15 MaBP. This happened during the opening of the Shikoku Basin in a direction roughly parallel to the Nankai Trough, which was primarily a kind of transform fault. Much geological evidence shows that the Nakai Trough became subduction immediately after the opening ceased (KOBAYASHI, 1983).

However, the thickness of overriding lithosphere in the continental margin (Shikoku and Kii Peninsula) exceeds 40 km (ASADA and ASANO, 1972) and temperature gradient in the upper mantle is, therefore, much smaller than that at the Bonin and Marianas at 42 MaBP. Shikoku and Kii Peninsula are parts of the Eurasia plate which tends to converge toward the Philippine-Sea plate so that compressional stress usually prevails in the overriding lithosphere.

After subduction proceeds to Stage 1 of the cycle (Fig. 2), magma is generated at a depth of the fore-arc region by introduction of H_2O and CO_2 from the subducted slab but the magma can not rapidly ascend, because the overriding lithosphere is thick and in a compressional state. The magma is, therefore, differentiated and reacts with the overriding continental crust to form felsic rocks when it eventually intrudes and extrudes to the fore-arc region of Shikoku and Kii Peninsula. Felsic volcanic and plutonic rocks in the outer zone of southwestern Japan dated to be 13 to 15 Ma may probably have been formed by this process.

The subducted oceanic lithosphere from the Shikoku Basin was very young when it started to sink. In addition extensive off-ridge volcanism in the Shikoku Basin at about 15 MaBP was revealed by the results of DSDP leg 58 (KLEIN et al., 1978). Therefore, the subducted lithosphere was very thin and could not appreciably cool down the surrounding asthenosphere beneath the overriding southwestern Japan lithosphere. Angle of subduction was probably very gentle due to its small negative buoyancy. High magnesian andesite magma was generated even after the subducted slab reached beneath the Setouchi (Inland Sea) region.

Since the Setouchi region is more than 200 km away from the axis of the Nankai Trough, the compressional stress field caused by convergence of the two plates at the trough would not propagate much toward the Setouchi region. The region was thus so affected by swell of volume in underlying Mg-rich magma as to be in an extensional state. The Setouchi region then rifted and sanukitoids were erupted (NAKADA and TAKAHASHI, 1979; TATSUMI, 1983).

The present consideration has implied that magmatic events in the fore-arc regions of island arcs mark the initiation of one cycle of subduction. By identifying extent and age of the fore-arc events the evolutionary history of island arcs will be better interpreted in a framework of the plate tectonics.

REFERENCES

ASADA, T. and S. ASANO, Crustal structures of Honshu, Japan, in *The Crust and Upper Mantle of the Japanese Area, Part I, Geophysics*, edited by Jap. Nat. Comm. UMP, pp. 45–55, 1972.

BLOOMER, S., Marianas forearc ophiolite-structure and petrology *EOS, Abstr. AGU*, **62**, 1086, 1981.

BOGDANOV, N. (ed.) Initial Report of the Geological Study of the Oceanic Crust of the Philippine Sea Foor, *Ofioliti* **2**, 137–168, 1977.

CAMERON, W. E., E. G. NISBET, and V. J. DIETRICH, Boninites, Komatiites, and ophiolitic basalts, *Nature*, **280**, 550–553, 1979.

DALLWITZ, W. B., D. H. GREEN, and J. E. THOMPSON, Clinoenstatite in a volcanic rock from the Cape Vogel area, Papua, *J. Petrol.*, **7**, 375–403, 1966.

DARLYMPLE, G. B., M. A. LANPHERE, and D. A. CLAGUE, Conventional and $^{40}Ar/^{39}Ar$, K-Ar ages of volcanic rocks from Ojin (site 430), Nintoku (site 432) and Suiko (site 433) seamounts and chronology of volcanic propagation along the Hawaiian-Emperor chain, in *Initial Reports of Deep Sea Drilling Project, V. 55*, edited by E. D. Jackson, I. Koizumi *et al.*, pp. 659–676, U.S. Govern. Print. Office, Washington, D.C., 1980.

DIETRICH, V., R. EMMERMANN, R. OBERHANSLI, and H. PUCHELT, Geochemistry of basaltic and gabbroic rocks from the west Mariana basin and the Mariana Trench, *Earth Planet. Sci. Lett.*, **39**, 127–144, 1978.

DUNCAN, R. A. and D. H. GREEN, Role of multistage melting in the formation of oceanic crust, *Geology*, **8**, 22–26, 1980.

HUSSONG, D., S. UYEDA *et al.*, *Initial Reports of Deep Sea Drilling Project, V. 60*, U.S. Govern. Print. Office, Washington, D.C., 1982.

ISHII, T., Descriptions of dredge samples, in *Preliminary Report of R. V. Hakuho Maru Cruise KH80-3*, edited by K. Kobayashi, pp. 105–163, 1980.

KANAMORI, H., Great earthquakes at island arcs and the lithosphere, *Tectonophysics*, **12**, 187–198, 1971.

KARIG, D. E. and G. SHARMAN, Subduction and accretion in trenches, *Geol. Soc. Am. Bull.*, **86**, 377–389, 1975.

KARIG, D. E., R. N. ANDERSON, and L. D. BIBEE, Characteristics of back-arc spreading in the Mariana Trough, *J. Geophys. Res.*, **83**, 1213–1226, 1978.

KAY, R. W., Aleutian magnesian andesites: melts from subducted Pacific ocean crust, *J. Volcanol. Geotherm. Res.*, **4**, 117–132, 1978.

KLEIN, G. and K. KOBAYASHI, Geological summary of the north Philippine Sea, based on Deep Sea Drilling Project Leg 58 results, in *Initial Reports of Deep Sea Drilling Project, V. 58*, G. Klein, K. Kobayashi *et al.*, pp. 951–962, 1980.

KLEIN, G. and K. KOBAYASHI, Geological summary of the Shikoku Basin and northwestern Philippine Sea, Leg 58, DSDP/IPOD drilling results, in *Oceanologia Acta, Proc. 26th ICG*, pp. 181–192, 1981.

KLEIN, G., K. KOBAYASHI, H, CHAMLEY, D. M. CURTIS, H. J. B. DICK, D. J. ECHOLS, D. M. FOUNTAIN, H. KINOSHITA, N. G. MARSH, A. MIZUNO, G. V. NISTERENKO, H. OKADA, J. R. SLOAN, D. M. WAPLES, and S. M. WHITE, Off-ridge volcanism and sea-floor spreading in the Shikoku Basin, *Nature*, **273**, 746–748, 1978.

KOBAYASHI, K., Cycles of subduction and Cenozoic arc activity in the northwestern Pacific margin, in *Geodynamics of the Western Pacific*, T.W.C. Hilde (ed.), Geodynamics Ser. No. 9, AGU/GSA, 1983.

KOBAYASHI, K. and N. ISEZAKI, Magnetic anomalies in the Sea of Japan and the Shikoku Basin: Possible tectonic implications, in *The Geophysics of Pacific Ocean Basin and Its Margins, Geophys. Mongr.* **19**, pp. 235–251, Am. Geophys. Union, Washington, D.C., 1976.

KOBAYASHI, K. and M. NAKADA, Magnetic anomalies and tectonic evolution of the Shikoku Inter-Arc Basin, *J. Phys. Earth*, **26**, Suppl. 391–402, 1978.

KURODA, N. and K. SHIRAKI, Boninite and related rocks of Chichijima, Bonin Islands, Japan, *Rep. Fac. Sci. Shizuoka Univ.*, **10**, 145–155, 1975.

KUSHIRO, I., Melting of hydrous upper mantle and possible generation of andesitic magma: An approach from synthetic systems, *Earth Planet. Sci. Lett.*, **22**, 294–299, 1974.

KUSHIRO, I., Petrology of high-MgO bronzite andesite resembling boninite from site 458 near the Mariana trench, in *Initial Reports of Deep Sea Drilling Project, V. 60*, D. Hussong, S. Uyeda *et al.*, pp. 731–734, U.S. Govern. Print. Office, Washington, D.C., 1982.

MORGAN, W. J., Deep mantle convection plumes and plate motions, *Am. Assoc. Petrol. Geol.* **56**, 203–213, 1972.

MROZOWSKI, C. L. and D. E. HAYES, The evolution of the Parece Vela basin, eastern Philippine Sea, *Earth Planet. Sci. Lett.*, **46**, 49–67, 1979.

NAKADA, S. and M. TAKAHASHI, Regional variation in chemistry of the Miocene intermediate to felsic magmas in the outer zone and the Setouchi province of southwest Japan, *J. Geol. Soc. Japan*, **85**, 571–582, 1979.

OHASHI, F. and K. SHIRAKI, High-magnesia and high-silica volcanic rock in the Setogawa ophiolite, *J. Jpn. Assoc. Min. Petrol. Econ. Geol.*, **76**, 69–79, 1981 (in Japanese).

SHIBATA, K., Contemporanity of Tertiary granites in the outer zone of southwest Japan, *J. Geol. Surv. Japan*, **29**, 551–554, 1978.

SHIRAKI, K., H. URANO, and S. MARUYAMA, Clinoenstatite in boninites from the Bonin Islands, Japan, *Nature*, **285**, 31–32, 1980.

SHIRAKI, K. N. KURODA, S. MARUYAMA, and H. URANO, Evolution of the Tertiary volcanic rocks in the Izu-Mariana arc, *Bull. Volc.*, **41**, 548–562, 1978.

SIMONIAN, K. O. and I. G. GASS, Arakapas fault belt, Cyprus: a fossil transform fault, *Geol. Soc. Am. Bull.*, **89**, 1220–1230, 1978.

SMEWING, J. D., K. O. SIMONIAN, and I. G. GASS, Metabasalts from the Troodos massif, Cyprus: genetic implications deduced from petrography and trace element geochemistry, *Contr. Mineral. Petrol.*, **57**, 49–64, 1975.

SUGIMURA, A., Zonal arrangement of some geophysical and petrological features in Japan and its environs, *J. Fac. Sci., Univ. Tokyo, Sec. 2*, **12**, 133–153, 1960.

SUN, S. -S. and R. W. NESBITT, Geochemical regularities and genetic significance of ophiolitic basalts, *Geology*, **6**, 689–693, 1978.

TATSUMI, Y., Miocene Setouchi volcanic belt, southwest Japan and origin of the Shikoku Inter-arc basin, in *Geodynamics of the Western Pacific*, T.W.C. Hilde (ed.), Geodynamics Ser. No. 9, AGU/GSA, 1983.

TATSUMI, Y. and K. ISHIZAKA, Existence of andesitic primary magma: an example from southwest Japan, *Earth Planet. Sci. Lett.*, **53**, 124–130, 1981.

TOMODA, Y. and H. FUJIMOTO, Maps of gravity anomalies and bottom topography in the Western Pacific and Reference book for gravity and bathymetric data, with 6 charts, *Bull. Ocean Res. Inst., Univ, of Tokyo*, **14**, 158, 1982.

TSUNAKAWA, H., K-Ar dating of volcanic rocks in the Bonin islands (abstract), *Bull. Vol. Soc. Japan*, **25**, 307, 1980.

UYEDA, S., Subduction zones: An introduction to comparative subductology, *Tectonophysics*, **81**, 133–159, 1982.

UYEDA, S. and Z. BEN-AVRAHAM, Origin and development of the Philippine Sea, *Nature, Phys. Sci.*, **240**, 176–178, 1972.

UYEDA, S. and A. MIYASHIRO, Plate tectonics and the Japanese islands: a synthesis, *Geol. Soc. Am. Bull.*, **85**, 1159–1170, 1974.

VON HUENE, R., M. LANGSETH, N. NASU, and H. OKADA, Summary, Japan trench transect, in Scientific Party, *Initial Reports of Deep Sea Drilling Project, V. 56–57*, Part 1, pp. 473–488, 1980.

WATTS, A. B. and J. K. WEISSEL, Tectonic history of the Shikoku marginal basin, *Earth Planet. Sci. Lett.*, **25**, 239–250, 1975.

YOSHII, T., Regionality of group velocities of Rayleigh waves in the Pacific and thickening of the plate, *Earth Plenet. Sci. Lett.*, **25**, 305–312, 1975.

Arc Volcanism: Physics and Tectonics, edited by D. Shimozuru and I. Yokoyama, 165–175.
Copyright © 1983 by Terra Scientific Publishing Company (TERRAPUB), Tokyo.

Thermal and Mechanical Processes Producing Arc Volcanism and Back-Arc Spreading

Yoshiaki IDA

Ocean Research Institute, University of Tokyo,
Nakano-ku, Tokyo 164, Japan

Some possible mechanisms that may govern thermal and mechanical processes associated with the subduction of the lithosphere are examined to explain the heat and material sources of arc volcanism and back-arc spreading. It is proposed that the evolutional processes in the subduction zone comprise the following three stages. In the first stage, circulation of mantle material is induced in the asthenosphere above the slab by upward migration of the magma that has been produced by partial melting of subducted oceanic crust. Crustal deformation and stress in the island arcs result from the process of magma migration into the crust of the arcs. Abundance of incompatible elements in the volcanic rocks reflects the circulating flow of mantle material that emits the volcanic magmas. In the second stage, the horizontal temperature difference associated with cool slab causes a thermal convection. Subducted oceanic crust is subject to substantial melting and accumulates in the asthenosphere. In the third stage, the asthenosphere containing gradually accumulated melt expands and pushes the slab aside. Sharp dip angles of deep seismic zones are produced by the bending of the slab that is weakened by inelastic deformation. The retreat of slab tears the arc on the one hand, and allows hot material to upwell from deeper mantle through the mesosphere on the other hand. Back-arc spreading begins under these circumstance. Marginal basins have different types of crustal rock from island arcs, because of such difference of magma source and upwelling process.

1.   Introduction

Two distinctive types of volcanic activity occur in the subduction zones, one with calc-alkaline magma in the island arcs, and the other with basaltic magma in the marginal basins. Those two volcanisms produce continental and oceanic crusts, respectively. It is quite interesting that both of these activities are commonly subject to the effect of the descending lithosphere.

Some hypotheses have been proposed to describe the process producing arc volcanism and back-arc spreading. For instance, a hot diapir that ascends through the mantle was assumed as the origin of either arc volcanism (SAKUYAMA and KUSHIRO, 1981) or back-arc spreading (KARIG, 1971). As another mechanism to carry heat and material for the volcanic eruptions, TOKSÖZ and BIRD (1977) proposed a convection that is mechanically induced by the down-going motion of slab. On the other hand, UYEDA and KANAMORI (1979) attributed the two types of volcanism to the different modes of the plate motion in the deep-sea trench. The present author (Ida, 1978, 1981) gave another model of thermal and mechanical processes, assuming that the subducted oceanic crust would be remelted and standing in the wedge shaped mantle above the slab.

Although these theories succeeded in explaining some aspects of the trench-arc-back arc system, the physical mechanism to cause the process has not yet been clarified completely by them. We cannot tell definitely why the rock types are so different between the arc volcanism and back-arc spreading as actually observed. What factor controls the switching of the two types of volcanic activity has neither been answered satisfactorily. A more fundamental but unsolved question is what is the energy, or heat, source to maintain these volcanic eruptions.

In this paper, we try to construct a possible model to answer these questions. In some meaning, the questions are concerned with almost all fields of solid earth sciences. Here we pick up some pieces of evidence that seem to characterize most basic aspects of the subduction zones.

There has been a big debate on whether the magma of arc volcanism comes from the wedge mantle or subducted oceanic crust (Green, 1980; Sakuyama and Kushiro, 1981). Although this is one of the most fundamental questions in petrology, a definite answer has not been given by any petrological constraint alone.

In this paper, we assume that the subducted oceanic crust is important material source of the volcanisms in the subduction zones. This assumption seems to be inevitable to assure material source for the volcanisms. Namely it is difficult to derive reasonably thick silicic crust in the island arc from the restricted volume occupying the wedge mantle. On the other hand, subducted oceanic crust is carried unlimitedly by the moving lithosphere. Only a minor part of the sinking crustal material is necessary to explain the rate of volcanic extrusion in the island arcs (Ida, 1978). The melting of subducted oceanic crust also gives a simple explanation to some other phenomena, as is shown below.

2. Heat Available to Magma Generation

A most serious paradox in geothermal problems of the subduction zones is what mechanism could generate heat to allow partial melting for volcanisms. It is generally accepted that a down-going slab should cool its environment so exhaustively that neither frictional heating nor other heat sources associated with radiogenic or chemical reactions could raise the temperature of the mantle enough, competing with the cooling effect (e.g., Anderson *et al.*, 1978).

It is sometimes told that arc volcanism is made possible by the effect of water or other volatile components. Indeed some volatile components, which are supposed to be introduced into the mantle by the descending lithosphere, could substantially reduce solidus temperature. Therefore partial melting will occur in the presence of the volatile components even at very low temperatures.

However the idea that the occurence of partial melting is mainly due to the effect of volatile components, in particular water, is not consistent with petrological evidence that erupted magmas are not saturated with those components (Sakuyama and

KUSHIRO, 1981). Furthermore some magmas show higher temperatures even on the earth's surface than is expected from the mantle with cool slab. More direct geophysical observations, including high surface heat flow and well-developed low-velocity layer in the island arcs and marginal basins, suggest that the subduction zones must have really higher temperature, compared with other regions, in spite of the strong cooling effect of descending slab.

It is also noted that there exists remarkable low-velocity, low-Q zone just adjacent to the slab. The boundary between the hot low-velocity region and cool slab is so sharp (probably less than 10 km) that seismic waves are refracted or reflected there (SNOKE *et al.*, 1977; FUKAO *et al.*, 1978).

The high temperature state in the wedge mantle above the slab means that effective heat transfer takes place there. As a candidate of the desired heat transport, however, thermal conduction is excluded, because conductive heat transport takes too long. Table 1 gives the characteristic time of thermal conduction for some typical distances. For instance, conductive heat is able to diffuse only up to the distance of 10 km, while the subducted plate sinks over 100 km. It takes more than several hundred million years for conductive heat to influence the entire volume of the wedge mantle. Various important events that occured in the subduction zones did not wait for such a long period.

Therefore signifcant heat transport can be made only by some non-conductive heat flux involving material flow. A convection induced by heating from below is hindered, however, by deep-seated cool slab. Instead TOKSÖZ and BIRD (1977) proposed a convection that is mechanically driven by the down-going motion of slab. However the evidence of hot material just adjacent to the slab suggests that the mechanical coupling between the slab and neighboring mantle may not be strong enough to drive the mantle flow. As other possible mechanisms of heat transport, we here examine the following two probable processes that involve material circulation in the wedge mantle.

First, we propose a process that might produce arc volcanism, as is schematically described in Fig. 1. Suppose that the subducted oceanic crust is remelted at a certain

Table 1.   Characteristic distance Δx_c of thermal conduction corresponding to given time interval Δt.

Δt	Δx_c	Δx_m
1 year	10 m	10 cm
10^4 years	1 km	1 km
1 m.y.	10 km	100 km
10 m.y.	30 km	10^3 km
100 m.y.	100 km	10^4 km
10^3 m.y.	300 km	10^5 km

The table is based on the relation $\Delta t = (\Delta x_c)^2/\kappa$, where the thermal diffusivity κ is assumed to be 10^{-2} cm^2/s. For a comparison, the distnace Δx_m over which the plate moves at the velocity of 10 cm/year in the same time interval is also given.

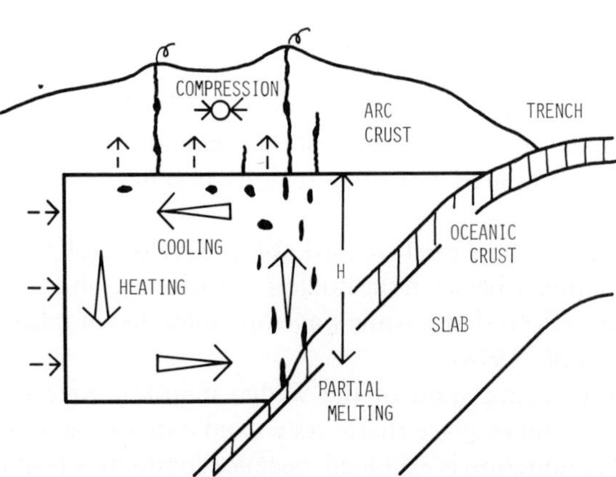

Fig. 1.   Stage of arc formation. The circulation in the asthenosphere is induced by upward migration of magma that has been produced by the partial melting of the subducted oceanic crust. Thickening and thinning arrows represent the flows with rising and falling temperature, respectively. Dashed arrows show the direction of heat flux. The melt is separated at the bottom of arc crust, and further ascends as magmas for arc volcanism. Crustal stress tends to be compressive as a whole, even if short-term irregurarities disturb the field, corresponding to elastic deformation associated with irregular migration of volcanic magma. Incompatible elements in the magma decreases with the distance from the trench axis, because temperature is lowered as material flows horizontally in the upper level of the asthenosphere.

depth of the mantle. The mantle block that has absorbed resultant low-density melt tends to ascend, being subject to buoyancy force. When the ascending block reaches the bottom of the crust, however, it can no longer go up as a whole. Therefore only the melt that has been separated from the mantle block is able to upwell through the crust. The block that has lost melt becomes heavier and again sinks into deeper parts.

In this manner, a circulation is induced by upward motion of melt in the asthenosphere above the slab. In the way of the circulation, heat would be absorbed from hotter part of the environment, and brought to the cool slab. The circulation itself could also produce some amount of heat as a result of viscous dissipation of motion. The gravitational energy to drive the circulation is written as $fgH\Delta\rho$ per volume, where f is volume fraction of melt, g is gravitational acceleration, H is the vertical distance of ascending pass, and $\Delta\rho$ is the density difference between solid rock and melt. If all of the gravitational energy is relaxed by the viscous heating, the temperature of ascending material is elevated by about $100°C$ for $H = 100\,\mathrm{km}$ and $f\Delta\rho/\rho = 0.1$.

The relaxed gravitational energy is usable to produce melt in the next step, when the resultant heat is brought back to the slab by the circulation. The process would be maintained by its own energy source, if the relaxed energy could cover the latent heat to consume. Unfortunately, accurate estimate of latent heat is not available for partial melting. If the latent heat is 20 cal/g, the self-contained process works for $\Delta\rho/\rho = 0.1$. We must assume $\Delta\rho/\rho = 0.5$, however, if the latent heat is as much as 100 cal/g. The estimate of 110 to 160 cal/g made by FUKUYAMA (1982) is rather close to the latter case, and it suggests that external heat supply from the environment is necessary for the partial melting.

Another type of convection is also possible. Generally thermal convection could be induced more easily by a horizontal temperature difference than by a vertical difference. There is no critical value, similar to the Rayleigh number, for horizontal temperature gradient to exceed in inducing convection (e.g., ELDER, 1966). A strong horizontal gradient actually exists in the subduction zones, owing to the sinking cool slab. It is quite likely that a convection of mantle material is induced by this horizontal gradient (RABINOWICZ *et al.*, 1980), as is shown in Fig. 2.

Convection driven by a horizontal temperature difference always carries heat upward, even against a reversed geothermal gradient with lower temperatures deeper (IDA, 1981). This conveniently explains the peculiar coexistence of well-developed warm asthenosphere with underlying cool slab. Namely, the convection could continuously concentrate heat into upper level of the asthenosphere, and maintain the temperature contrast, dominating the effect of thermal conduction.

3.   Slab Geometry and Crustal Stress

If some melt is stripped from the descending lithosphere and remains in the asthenosphere, the standing melt has various important effects on the mechanical processes in the subduction zones. The presence of such melt should cause positive gravity anomaly, if it is not isostatically adjusted. The distribution of positive gravity anomaly is actually observed in both island arcs and marginal basins (WATTS and TALWANI, 1974). The magnitude of observed anomaly is explained by subducted oceanic crust that stands temporarily for a few million years, prior to final isostatic adjustment (IDA, 1978).

It gives an important constraint to the model that arc volcanism is associated with relatively low dip angles of the subducted slab, whereas back-arc spreading occurs with high dip angles, sometimes close to the vertical (UYEDA and KANAMORI, 1979). An analysis of the lithospheric bending (IDA, 1982) shows that such sharp bending as

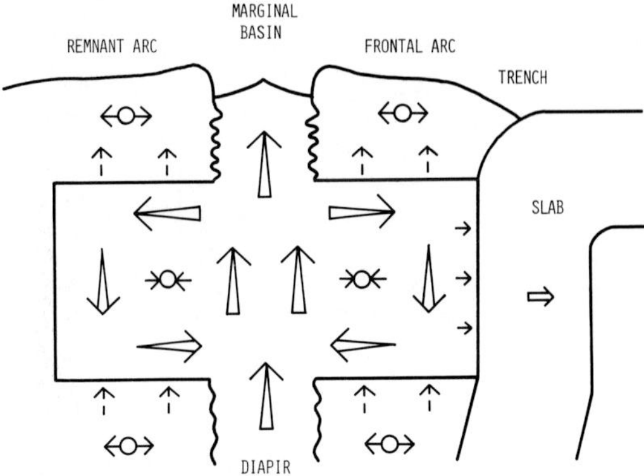

Fig. 2.   Stage of material accumulation. A thermal convection is induced by horizontal temperature difference associated with the descending cool slab. The convection concentrates heat into the upper level of the asthenosphere. Most of melted oceanic crust accumulates in the asthenosphere.

                              Y. IDA

characterizes back-arc spreading is realized by the following combination of forces: (1) pull in the plate boundary, (2) compressive stress sustained in the lithosphere, and (3) some force causing downward bending.

All the three forces can be simultaneously produced, if some excess material is put into the asthenosphere above the slab. Namely such excess mass pushes the slab so as to bend it downward. Corresponding to the retreat of the slab, the arc overlying the expanded asthenosphere is stretched, so that the two plates pull each other in the plate boundary. To react the pressure, a compressive stress arises in the oceanic lithosphere.

It is thus interpreted that high dip angles reflect the accumulation of subducted oceanic crust in the asthenosphere. This understanding further leads us to the idea that back-arc spreading itself might be caused by an excessive accumulation of material (Fig. 3). As has been pointed out above, the volume increase of the asthenosphere due to the stored material must stretch the overlying island arc. The continued accumulation of material would make the effect so strong that the island arc might be finally torn away. Such fracture of the arc would be identified as the start of back-arc spreading.

It is known that arc volcanism tends to be accompanied by compressive crustal stress, whereas back-arc spreading begins in the state of extensional stress (UYEDA and KANAMORI, 1979). Such fundamental feature of stress state is interpreted by the above-mentioned scheme, as follows. First of all, extensional stress in the stage of back-arc spreading clearly corresponds to the stretching of the arc, due to excessive accumulation of material in the asthenosphere. On the other hand, the state of stress associated with arc volcanism reflects the process of growing continental crust. New material

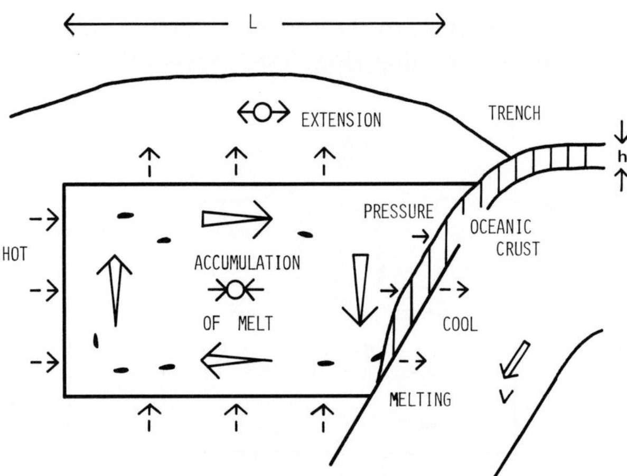

Fig. 3.   Stage of back-arc spreading. The asthenosphere expands due to accumulated melt and pushes the slab aside. Bending and seaward retreat of slab are accelerated by inelastic weakening of slab caused by the deformation. The retreat of slab tears the arc on the one hand, and allows hot material to come up from deeper mantle through the mesosphere on the other hand. Back-arc spreading begins in the fractured arc with the heat and material supply made by upwelling mantle flow.

added to the arc as volcanic magmas increases the crustal volume of arc, which causes compressive stress in the crust as a whole.

Actual stress field is not so simple, however, as has been described above. A systematic stress gradient is usually observed in the crust of the subduction zone (NAKAMURA and UYEDA, 1980). Such non-uniform stress distribution cannot be explained by a simple pull or push against the entire plate. It is more natural to consider that the stress gradient results from more or less systematic distribution of excessive material in the earth's interior.

The deformation of an island arc has been revealed by detailed surveys in Japan (For the review of the data, see KASAHARA, 1978; SUGIMURA, 1978). The geodetic measurement for these 60 years has given the data of recent crustal movements. The result of vertical deformation shows that both rising and sinking are distributed over the islands of Japan, being mixed with each other. The rate of the vertical motion varies within a certain range, and its mean amplitude is about 0.3 cm/year. Horizontal strain rate also has non-uniform distribution and its representative magnitude is roughly 3 $\times 10^{-7}$ /year. On the other hand, the vertical deformation during last 2 m.y. was determined from geographical and geological method, and it displays the pattern of more uniform rising with the mean rate of about 0.05 cm/year.

Those features of observed deformation can be interpreted as a result of the continuous intrusion of about one fifth of subducted crustal material into the arc crust (IDA, 1980). The magnitudes of recent crustal movement is considered to represent elastic straining. Non-uniform distribution of deformation would reflect the irregularity of upwelling migration of magma. In particular, crustal sinking occurs around volcanic mountains at which magma has extruded out of the crust. However the elastic stress attains the crustal strength after the intrusion of magma goes on for about several thousand years. Therefore elastic strain must be relaxed every several thousand years. In much longer time scale, more gentle and more slow deformation should be observed, because irregular short-term change is smoothed out. This is considered to correspond to the observed deformation in much longer period of the Quaternary.

Next we consider the process to start back-arc spreading in more detail. We have already pointed out that the accumulation of material in the asthenosphere causes the pressure that pushes the slab aside. If the entire oceanic crust of thickness h is continuously added to the volume of mantle that extends over the distance L starting from the trench axis, the rate of pressure increase equals $gV\rho'h/L$, where g is the gravitational acceleration, V is the velocity of plate subduction, and ρ' is the density of accumulated material. We have 400 bar/m.y. as a rough estimate of the rate of pressure increase, assuming that $V = 10$ cm/year, $\rho' = 3$ g/cm^3, $h = 6$ km, and $L = 500$ km.

On the other hand, the pressure inhomogeneity that is maintained in the earth's interior is expected to be at most of several kilobars, because larger inhomogeneity should produce vertical crustal movement over several kilometers. Therefore the increased pressure due to the material accumulation should become sufficiently effective within the period of one or two million years since substantial accumulation of material starts.

An estimate based on elastic bending of the lithosphere (IDA, 1982) shows, however, that the dip angle of slab cannot be changed largely by above-mentioned magnitude of pressure. Probably the process of bending is promoted by the inelastic weakening of the oceanic lithosphere due to the deformation. Namely the bending and retreat of the slab are rapidly accelerated as the pressure increases beyond a certain critical value. Such accelerated retreat of the slab would be accompanied by sudden stretching of the overlying island arc, which could start the back-arc spreading.

Same effect is also expected to work on the underlying mesosphere. Namely, space that some material could enter would be made in the mesosphere, as the slab retreats seaward (See Fig. 3). This allows hot material to ascend from deeper mantle. Such upwelling of hot material is able to provide with heat and material to maintain the activity of newly formed small ridge system in the marginal basin. The resultant volcanic activity should have a quite different character from arc volcanism, because of different source of heat and material.

4.   Tectonic Cycle in the Subduction Zone

In the previous sections, we have proposed three possible mechanisms, which could sustain heat and material for volcanic activities. These mechanisms are summarized in Figs.1, 2, and 3, respectively.

The present author imagines that these three mechanisms occur sequentially in the evolutional process of the subduction zone and constitute a tectonic cycle, as follows.

1)   Magma caused by partial melting of subducted oceanic crust ascends and increases the crust of the island arc. The upward migration of magma induces a circulation of mantle material, which works as a carrier of material and heat (Fig. 1).

2)   Horizontal temperature gradient caused by sinking cool slab drives a convection, which effectively concentrates heat into the upper level of the asthenosphere. Material derived by partial melting of subducted oceanic crust is mainly stored in the asthenosphere (Fig. 2).

3)   Expansion of the asthenosphere due to the material accumulation pushes the slab aside and enforces it to retreat seaward. The retreat of slab allows hot mantle material to upwell from deeper mantle through the mesosphere. The lighosphere is also torn away by the same effect, and back-arc spreading begins (Fig. 3).

Here we would like to consider the relations of these three stages. In particular, the process of transition from one stage to another contains some problems to be examined in more detail.

It has been pointed out that the first stage of arc volcanism consumes only a minor part of the entire subducted crust. In other words, the magma is produced by further partial melting of the oceanic crust that itself comprises the chemical components quite feasible to partial melting. The partial melting in the first stage could start, utilizing such material with low melting point, even where temperature is not always favorably high. Partial melting of subducted crust could produce silicic magma of low density (GREEN, 1980). It is understood that the resultant low-density magma sustains the circulation proposed in Fig. 1, and finally constitutes continental crust that always floats above the mantle.

According to this interpretation, the difference of rock types between island arc and ocean basin is attributed to the difference of original material to melt. Namely, the volcanic material erupted in the island arc is derived from the partial melting of oceanic crust, wheareas the mid-ocean ridge in the marginal basin involves partial melting of upwelling mantle material. It is also noted that the process leading to volcanic eruption is quite different between the two. The arc magma must move upward along a long pass in the crust after it has been separated from the mantle material. In contrast to this, the magma for the oceanic crust ascends with hot mantle material up to near the earth's surface. The variety of rock types contained in the arc crust, compared with relative uniformity of ocean crust, is obviously related to such difference of ascending process of magma.

As is well known, such incompatible elements as potassium becomes more abundant in the rocks with similar composition of major elements, as the location of extrusion is apart from the trench axis. Such enrichment of incompatible elements is interpreted most simply by the decrease of the degree of partial melting in the original material providing the magma. The first stage displayed in Fig. 1 contains the flow in the upper layer that moves in a contrary direction to the trench, being cooled by the loss of heat to the surface. Therefore the flow should emit the magma of decreasing degree of partial melting, and thus increasing content of incompatible elements. It is supposed in this picture that the upwelling current occurs just beneath the so-called volcanic front that has most abundant volcanic material. The volume of volcanic material decreases rapidly with increasing distance from the volcanic front.

The most important role given to the second stage of the tectonic cycle is the accumulation of material, which leads to the third stage. This exibits the contrast to the first stage that is accompanied by the release of material into the arc crust. It should be noted here that the direction of circulating flow is opposite between the two stages, corresponding to the difference of driving mechanism. Therefore the two processes compete with each other without coexisting cooperatively.

The second stage takes the place of the first, when the horizontal temperature gradient dominates the gravitational effect due to the inclusion of low-density melt. The transition would occur when the cooling effect by the slab has more dominant influence, or when the density of melt is no longer sufficiently low. It requires more quantitative analysis, however, to determine which is the more important factor.

In the islands of Japan, compressive and extensional stress field appeared in turn every few million years. The transition of stress field occurred simultaneously over a wide area, sometimes over the whole Honshu island. The fact that the transitions between compressive and extensional stresses were accompanied by violent volcanic activities (TSUNAKAWA, 1982) supports the present idea that the stress field is governed by the migration of volcanic magma. The repeated transitions may suggest that the first and second stages continue to compete for a long time, alternating with each other several times.

The mechanism to shift the second stage to the third must give an explanation to the question why the deep-sea trenches in the Peru-Chile region do not have any marginal basins. It seems, to the present author, that the answer to this question is related to the presence of the Mid-Atlantic ridge. In the third stage, abundant hot material must upwell from deeper parts of the mantle. Such process will be hindered when other mid-ocean ridge systems landward of the trench are actively absorbing substantial material from below. According to this understanding, a back-arc spreading occurs only in the subduction zone that is well separated from any great mid-ocean ridge system. This idea is consistent with extremely biased distribution of marginal basins in the Western Pacific. In connection with this, it is interesting that the slab in the Peru has such a peculiar form as if it were pulled toward the direction of the Mid-Atrantic ridge (HASEGAWA and SACKS, 1981).

5. Concluding Remarks

Generally the energy source to maintain tectonic activities of subduction zones may be classified into the following three categories.
1)   The energy associated with the descending motion of the slab.
2)   Heat collected from hot parts of the environment.
3)   Heat carried by hot material entering the system.

A most typical example included in the first category is frictional heat that is produced on the boundary surface of slab. Gravitational energy that has been considered in Fig. 1 also belongs to this category, because the energy is stored by such downgoing motion of slab as holds light material against the gravitational effect. The circulation induced by horizontal temperature difference in Fig. 2 could achieve the second process effectively. The third category usually involves upward migration of material in the earth's interior, as is proposed in Fig. 3. In our model, all the three energy sources thus play their own role in the respective stages of the tectonic cycle.

It has been shown that the model could consistently arrange various features of the subduction zones. More quantitative examination in future will reveal the mechanism more clearly.

REFERENCES

ANDERSON, R. N., S. E. DELONG, and W. M. SCHWARZ, Thermal model for subduction with dehydration in the downgoing slab, *J. Geol.*, **86**, 731–739, 1978.

ELDER, J. W., Numerical experiments with free convection in a vertical slot, *J. Fluid Mech.*, **24**, 823–843, 1966.

FUKAO, Y., K. KANJO, and I. NAKAMURA, Deep seismic zone as an upper mantle reflector of body waves, *Nature*, **272**, 606–608, 1978.

FUKUYAMA, H., Heat of formation of basaltic magma (priprint), 1982.

GREEN, T. H., Island arc and continent-building magmatism—A review of petrogenic models based on experimental petrology and geochemistry, *Tectonophysics*, **63**, 367–385, 1980.

HASEGAWA, A. and I. S. SACKS, Subduction of the Nazca plate beneath Peru as determined from seismic observations, *J. Geophys. Res.*, **86**, 4971–4980, 1981.

IDA, Y., Oceanic crust in the dynamics of plate motion and back-arc spreading, *J. Phys. Earth*, **26**, S55–S67, 1978.

IDA, Y., Crustal stress due to the intrusion and migration of magma, *Chikyu*, **2**, 615–619, 1980 (in Japanese).

IDA, Y., Thermal circulation of partial melt and volcanism behind trenches, *Oceanol. Acta*, N°SP, 241–244, 1981.

IDA, Y., Forces acting on the subducted lithosphere revealed by the geometry of deep-sesmic zones, in *Accretion Tectonics in the Circum-Pacific Regions*, edited by H. Hashimoto and S. Uyeda, pp. 335–344, Terrapub, Tokyo/Reidel, Dordrecht, Holland, 1983.

KARIG, D. E., Origin and development of marginal basins in the western Pacific, *J. Geophys. Res.*, **76**,

2542–2561, 1971.

KASAHARA, K., Earthquakes and tectonics, in *Iwanami Earth Science Series, Vol. 10*, pp. 33–88, Iwanami, Tokyo, 1978 (in Japanese).

NAKAMURA, K. and S. UYEDA, Stress gradient in arc-back arc regions and plate sudbuction, *J. Geophys. Res.*, **85**, 6419–6428, 1980.

RABINOWICZ, M., B. LAGO, and C. FROIDEVAUX, Thermal transfer between the continental asthenosphere and the oceanic subducting lithosphere: Its effect on subcontinental convection, *J. Geophys. Res.*, **85**, 1839–1853, 1980.

SAKUYAMA, M. and I. KUSHIRO, Subduction and volcanic zones, *Kagaku*, **51**, 499–507, 1981 (in Japanese).

SNOKE, J. A., I. S. SACKS, and H. OKADA, Determination of the subducting lithosphere boundary by use of converted phases, *Bull. Seismol. Soc. Am.*, **67**, 1051–1060, 1977.

SUGIMURA, A., Major relief, volcanoes and earthquakes in island arcs, in *Iwanami Earth Science Series, Vol. 10*, pp. 159–181, Iwanami, Tokyo, 1978 (in Japanese).

TOKSÖZ, M. N., and P. BIRD, Formation and evolution of marginal basins and continental plateaus, in *Island Arcs, Deep Sea Trenches and Back-Arc Basins*, edited by M. Talwani and W. C. Pitman, III, pp. 379–393, Am. Geophys. Union, Washington, D. C., 1977.

TSUNAKAWA, H., Ancient stress field in Japan based on K-Ar ages of Neogene dikes and boninite, Ph. D. thesis, University of Tokyo, 1982.

UYEDA, S. and H. KANAMORI, Back-arc opening and the mode of subduction, *J. Geophys. Res.*, **84**, 1049–1061, 1979.

WATTS, A. B. and M. TALWANI, Gravity anomalies seaward of deep-sea trenches and their tectonic implications, *Geophys. J. R. Astron. Soc.*, **36**, 57–90, 1974.

Arc Volcanism: Physics and Tectonics, edited by D. Shimozuru and I. Yokoyama, 177–189.
Copyright © 1983 by Terra Scientific Publishing Company (TERRAPUB), Tokyo.

Evolution of Arc Volcanism Related to Marginal Sea Spreading and Subduction at Trench

Eiichi Honza

*Geological Survey of Japan, Higashi, Yatabe,
Tsukuba, Ibaraki 305, Japan*

There may be a few fundamental elements which constitute arcs in the circum Pacific Rim where several geomorphologic variations may occur in them.

A possible mechanism for arc volcanism is suggested as an alternative activity with marginal sea spreading and also a related activity to the subduction complex in the trench on the basis of convection current under the arc formed by the frictional heating along the boundary between subducted oceanic slab and asthenosphere under the arc.

Intense volcanism may occur while marginal sea spreading is slightly active. It is also inferred that the period of formation of subduction complex in the trench may be correlated to that of volcanism on the arc. Low level volcanism may occur when marginal sea is spreading. The period of consumption in the trench is also correlated to that of low level volcanism.

1. Introduction

Arc volcanism occurs in a chain or in a zone along a part of an arc, some hundreds of kilometers landward of trench. This feature of arc volcanism is noticed as a suggestion in possible control of subduction of an oceanic plate under an arc. However, some problems remain on heat mechanism of arc volcanism to drive to the surface from deeper part under the arc, whenever some possible mechanisms are suggested as related occurrence to the subducted oceanic plate since the end of the 1960's.

A chain distribution of arc volcanoes is commonly observed in most of the arcs of the circum-Pacific Rim, except for these in which a zonal distribution is observed. A zonal distribution of modern arc magma is well known in the Tohoku and the northern Ogasawara Arcs in which three main volcanic rock series of tholeiitic, calc-alkali and alkalic series are recognized (Kuno, 1968; Miyashiro, 1974). The zonal distribution is also observed in the Kamchatska, parts of the Philippine, the Mexico and a part of the Chilean Arcs.

The difference between a chain and a zonal distribution in modern arc volcanism seems to be related to the difference in geometric scale of arcs. At least, as a surficial phenomenon, relatively wide arcs have a zonal distribution and narrow arcs have a chain volcanism. This fact may imply that arc volcanism essentially takes place in a chain eruption and is variable to zonal distribution by enlargement of geometric scale of arc body. Some exceptions may appear in the junction of arcs where some complex tectonics may develop as suggested in the Mexican Arc.

In this paper, a possible mechanism for arc volcanism and arc tectonics which may be related to the volcanism as a part of the fundamental framework constituting arcs are discussed on the basis of data obtained mainly in the NW Pacific Rim (Fig. 1).

2. Geomorphology of Arc

Many types of morphology have been distinguished in the circum Pacific arcs by the recent detailed works from which many morphologic terms are proposed for arcs or arc structures (Matsuda and Uyeda, 1970; Dickinson, 1973, Karig and Sharman, 1975; Honza *et al*, 1977; Seely, 1979; Honza, 1981a). Here, the term arc is applied to the area between the marginal sea and the outer edge of the oceanic trench (Seely, 1979; Honza 1981a).

There are some common features which are associated with arcs. They are an active volcanic chain, a forearc basement high and a trench. Even if, there may be

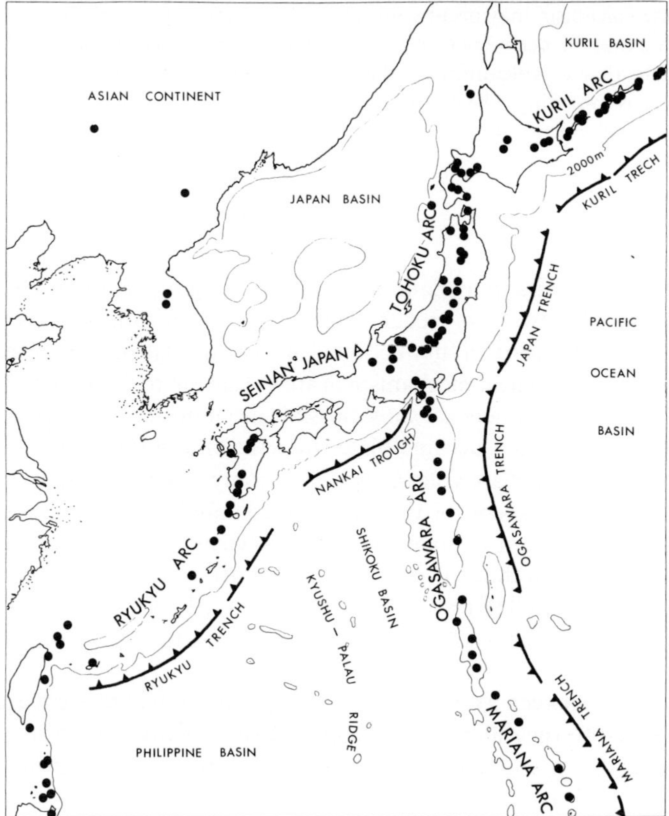

Fig. 1.   Island arcs in the NW Pacific Rim. Holocene volcanoes are shown in closed circles (Ono *et al.*, 1981; Simkin and Siebert, 1981).

discontinuities in some of the features in them (Fig. 2). A marginal sea is also one of the common features in the NW Pacific Rim. These features are noticed to be fundamental geomorphogic elements to constitute arc (HONZA, 1981a).

A few morphologic patterns of arcs may be categorized on the basis of the fundamental elements taking place in the arc, especially in the forearc areas (HONZA, 1981). The first category is arcs consisting of landward slopes underlain by a continental crust and the second consists of parts of landward slopes underlain by a accretionary prism. Arcs of the former types can be categorized as consumption. The latter is also divided in two types, the one dominantly underlain by a continental crust being categorized as a continental arc and the other dominantly underlain by an accretionary prism categorizing those as accretionary arc in which the subduction complex extends upward over the trench slope break, nevertheless some problems remain on the formation of accretionary material in the subduction complex (HONZA, 1981a).

3. Historical Evolution of Arc Volcanism

Activity of arc volcanism is not steady during its entire history which is well suggested in modern arcs in the NW Pacific Rim. Radiometric age determinations on

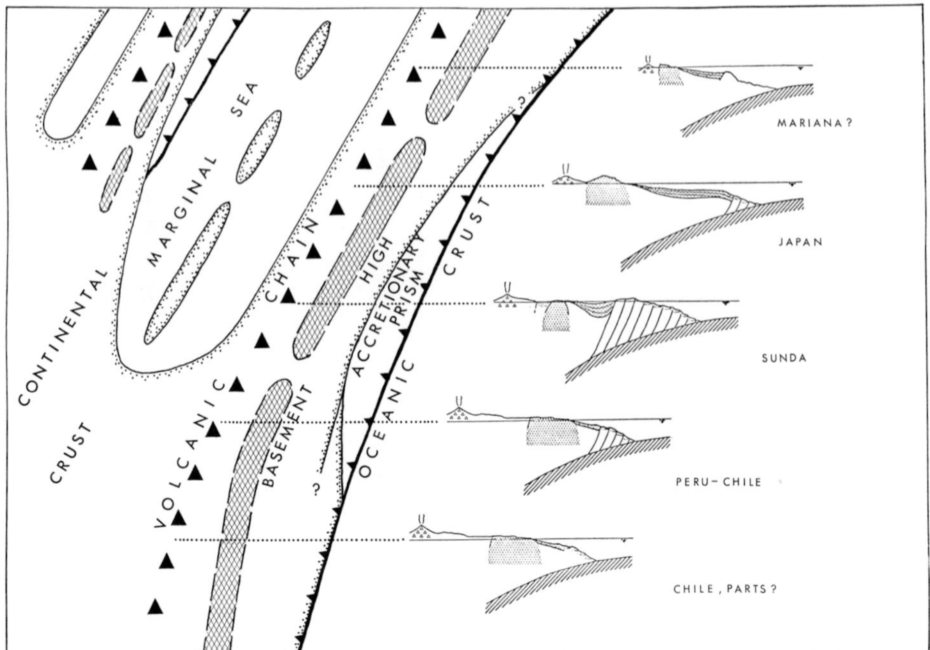

Fig. 2.   Arc geomorphology based on the fundamental elements which constitute arcs. Some of these elements are discontinuous and variable.

the Tohoku, Seinan Japan and the Ogasawara Arcs (Kaneoka *et al.*, 1970; Tanaka and Nozawa, 1977; Tsuchi, 1979; Konda and Uyeda., 1980) show histories of arc volcanism and its discontinuities (Fig. 3).

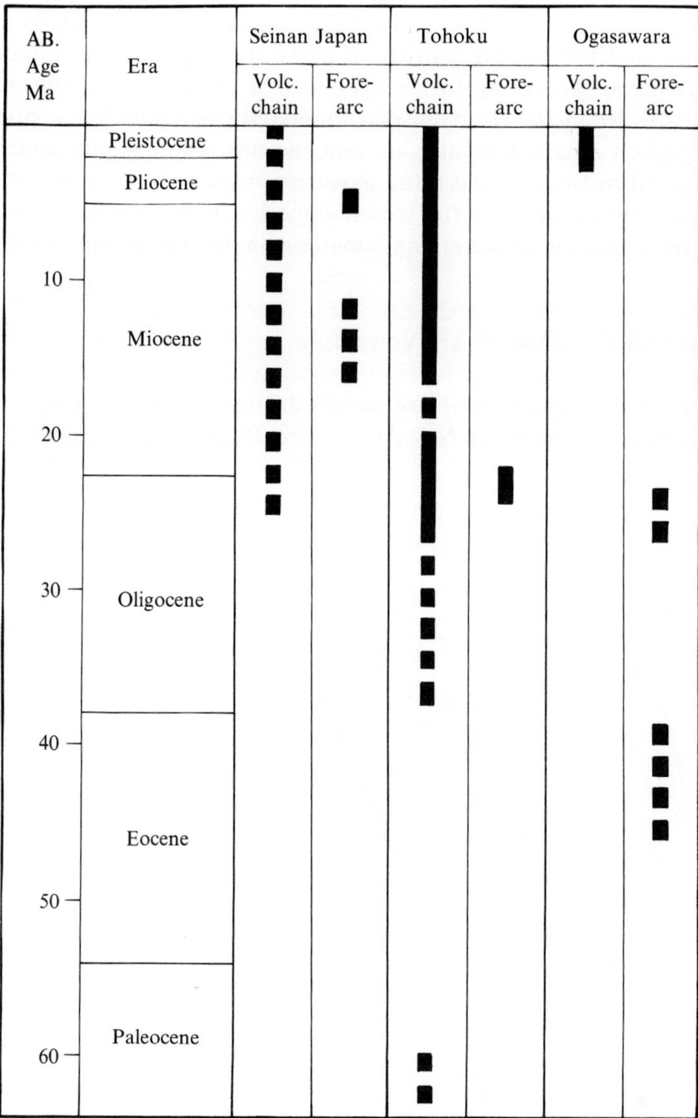

Fig. 3.   Historical evolution of volcanism in the Seinan Japan, the Tohoku and the Ogasawara Arcs based on the radio-metric age determinations without any consideration on intensity of volcanism. The Seinan Japan Arc from Tsuchi. (1979), the Tohoku Arc from Tsuchi ed. (1979), Konda and Uyeda (1980), and the Ogasawara Arc from Kaneoka *et al.* (1970).

Some uncertainties remain on the beginning of volcanism in the Tohoku Arc in which stratigraphic and biological studies support the beginning in late to latest Oligocene, nevertheless earlier ages back to the early Oligocene were obtained by absolute age determinations for several samples. The beginning of the activity of the Tohoku Arc is also suggested in the forearc area where some modern sedimentary basins were formed since late Oligocene (HONZA, 1981a; SCIENTIFIC PARTY, 1980).

Apparently, continuous eruptions have occurred in the Tohoku Arc since late Oligocene, although the intensity of volcanism were variable. Intense volcanism is suggested during early Miocene, gradually decreasing its activity until later early Miocene; less volcanism during middle Miocene and gradually increasing again from late Miocene (MATSUDA *et al.*, 1967; ISSHIKI, 1977).

Volcanic history is not clear in the Seinan Japan Arc, because of sparse age data. However, apparent continuous eruption has been taking place during its development (TSUCHI, 1979). A unique volcanism occurred in the Setouchi forearc area during the middle Miocene (TSUCHI, 1979).

Volcanism of the Ogasawara and the Mariana Arcs has been intermittent since middle Eocene, although the data obtained is only from the islands which occupy small areas on the ridge (Table 1). There are probably three stages of volcanism in the Ogasawara and the Mariana Arcs, the first from middle to late Eocene, the second during late Oligocene and the last since late Pliocene. The latest Miocene (7Ma) was suggested by recent age determination from the northern margin of the Mariana Ridge, while in the Ogasawara Arc, no latest Miocene has been obtained from both of the Ogasawara and the Shichito Ridges (HONZA *et al.*, 1981).

4.   Morphologic Variation of Arc Volcanism

The last volcanism which constructed a volcanic chain of the modern Ogasawara Arc occurred on a different line as compared with that of the earlier two stages which occurred on the Ogasawara Ridge. The Ogasawara Ridge forms a forearc basement high in the modern Ogasawara Arc. The volcanic chain (Shichito Ridge) is approximately 100 km west of the earlier volcanism (Ogasawara Ridge) intervened by the Ogasawara Trough between them.

The Ogasawara Ridge does not extend to the northern Ogasawara Arc where the Shichito Ridge consists of an only high in the Arc with modern volcanic islands on it. The northern continuation of the Ogasawara Ridge may consist a high of trench slope break bordering continental slope and inner trench slope.

The Mariana Arc located on the same morphologic line as the Ogasawara Arc shows slightly different features. There are, from west to east, the West Mariana Ridge, Mariana Trough which is noticed as a possible active spreading backarc basin, and Mariana Ridge consisted of both volcanic chain and forearc basement high. Volcanic chain and forearc basement high seem to seperate gradually southwards towards the center of the arc (HESS, 1948). No modern volcanism has formed in the area south of Saipan Island.

Geomorphologic variation of arc volcanism is also observed in the Ryukyu Arc where quite different features are distinguishable between the northern and the

Table 1. Stratigraphy of the Ogasawara and the Mariana Arcs.

	Shikoku Basin	Shichito Ridge	Ogasawara Ridge	Mariana Ridge
Quaternary	hemipelagic sediments and turbidites	volcanic rocks		Mariana Limestone
Pliocene				
Miocene	?			
	volcano-clastics			Alifan Ls.
				Bonya Ls.
Oligocene			Minamizaki F.	Umatac Formation
			volcanoclastics	
			Sekimon F.	Alutom Formation
Eocene			Okimura F.	
			Yusan F.	
Paleocene				

Shikoku Basin from KARIG *et al.* (1975), deVRIES KLEIN *et al.* (1980), and WATTS and WEISSEL (1974). Ogasawara Arc from KUNO (1968), MIYASHIRO (1974), KANEOKA *et al.* (1970), HANZAWA (1947), IWASAKI and AOSHIMA (1970), MATSUMARU (1974), and UJIIÉ and MATSUMARU (1977). Mariana Arc from LADD (1966).

southern parts of the arc. Common features which are observed for both northern and southern parts are a trench (Ryukyu Trench) and a uplifted forearc basement high (Ryukyu Ridge) dotted with islands. The northern arc has a volcanic chain (Tokara Volcanic Chain) while the backarc basin is not observed clearly, only showing gradual deepening toward the south. The southern arc has no volcanic chain. A few historic eruptions were reported in a relatively deep backarc basin (Okinawa Trough). There are a few volcanics in the narrow faulted depressions along the center of the Okinawa Trough (HONZA, 1976, 1977; HERMAN *et al.*, 1978). However, it is not certain whether the volcanics are identified as a part of a modern volcanic chain or not.

A slight shift of the volcanic chain is also noticed for the Tohoku Arc during the Neogene in which an older volcanic chain might have located a little forearc side as compared with that in recent years. (MATSUDA *et al.*, 1967; NAKAMURA, 1969).

These facts suggest that arc geomorphology is variable both in space and in geological time. There may be variable features even in the different parts within an arc, variable from intense to no volcanism, variable from uplifted forearc basement high having a range of islands to no uplifted high only showing a break bordering continental slope and inner trench slope, and variable from deep backarc basin to gradual deepening without any basin feature.

There may be a similar temporal change, as mentioned above, even in the same place within an arc, as suggested by periods of intense to no volcanism, periods of uplift and subsidence of forearc basement high and alternating periods of spreading to no spreading of backarc basins.

Certain features are suggested to be alternative during their evolutions. A volcanic ridge during earlier geological age might have been converted to have a different roll in the later stage of the arc as to show a nonvolcanic forearc basement high on the forearc area resultant in the formation of a new volcanic chain on the inner side of the arc.

5.   Alternative Occurrence of Arc Volcanism and Marginal Sea Spreading

There may be a certain corelation between arc volcanism and marginal sea spreading.

Studies on magnetic anomaly in the Shikoku Basin revealed lineations striking northwest, parallel to the volcanic chain of the Ogasawara Arc. The lineation pattern is identified as a sequence of anomalies 7 (27 Ma) through 5 (10 Ma) between the base of the Kyushu-Palau Ridge and east margin of the Basin with tentative identification on the eastern side on account of rough and complex morphology of the basement (WATTS and WEISSEL, 1975). The other identification is based on a symmetrical anomaly profile with a spreading center along the center of the basin (KOBAYASHI and NAKADA, 1979).

Spreading of the Shikoku Basin, thus appears to have begun in late Oligocene and continued until the middle to late Miocene. On the other hand, no volcanism is known on the Ogasawara Arc from Miocene to early Pliocene.

Morphologic variation in the Ryukyu Arc also suggests that there may be a relative deeper backarc basin when there is no volcanic chain and there may be no deeper backarc basin when modern volcanic chain is active.

These facts from modern geomorphology and historical evolution may imply that arc volcanism is in direct opposition to spreading of marginal sea. In other words, alternative activity may be taking place for arc volcanism and spreading of marginal sea. Low level volcanism may occur on the volcanic chain while marginal sea is opening and slight spreading may occur on the marginal sea when volcanism is active on the volcanic chain.

6.   Subduction Complex in the Trench

Subduction complex in the inner trench slope consists of melange and terrigenous sediments overlying on the melange supplied from a shallower slope. Melange is interpreted to consist of both terrigenous sediments and pelagic sediments scraped off onto inner trench slope (DICKINSON, 1973; KARIG and SHERMAN, 1975; SEELY, 1979; HONZA, 1981a). Thrusts are dominant features to form the complex, which may be formed by the horizontal compressional force exerted by the subduction of oceanic plate. Accreted material may be added only to the outermost slab bounded by thrusts in the subduction complex. Amount of terrignuous sediments may increase toward inner slab of the complex when terrigenuous sediments are currently supplied from a shallower slope while the amount of melage may be constant in each inner slab (Fig. 4). In this model, there may be a seaward shift of trench axis rather fixed axis associated with zonal subduction of the accretionary prism under the deeper part of the arc (KARIG and SHERMAN, 1976; SEELY, 1978; HONZA, 1981a).

Some of trenches are postruated to have no accretionary prism on the foot of the inner trench slope as is suggested on the Mariana Trench (HUSSONG *et al.*, 1977; YUASA *et al.*, 1982) and a part of the Chili Trench (KULM and SCHWELLER, 1977) which may be categorized consumption or tectonic erosion type arc (KULM and SCHWELLER, 1977; HONZA, 1981a). A possible tectonic erosion is also noticed in the Japan Trench during the earlier stage of its development where subduction complex developes in the lower part of the inner trench slope during the later stages. There may have been a wide shallow sea on the forearc of the Tohoku area before subduction begun, which is suggested in the unconformity developing most of the forearc area and in the coarser material lying on the unconformity examined by drilling of IPOD Leg 57 (SCIENTIFIC PARTY, 1980).

These facts may suggest that there is an alternative development of subduction complex, one the transgressive stage, development of the subduction complex with migration of trench axis seaward and the other the regressive stage, tectonic erosion

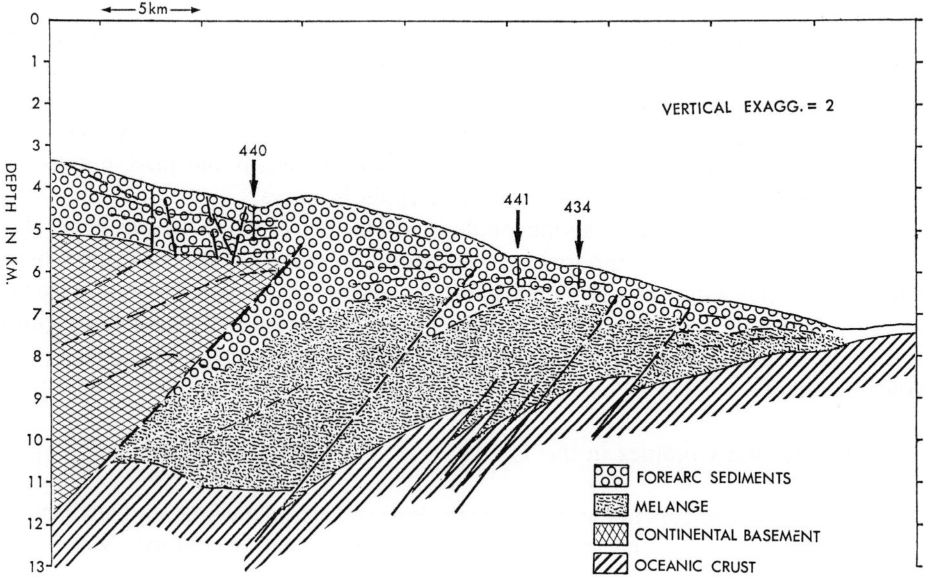

Fig. 4.　Diagrammatic structural interpretation of a multi-channel seismic profile in the Japan Trench (HONZA, 1981a).

with migration of trench axis landward.

The regressive stage may be taking place in the Mariana Arc while there may be modern spreading of the Mariana Trough in the backarc side (HUSSONG *et al.*, 1978). The transgressive stage may be taking place in the Tohoku Arc while no modern spreading is suggested in the Japan Sea.

It is inferred from these facts that there may be a spreading of marginal sea while tectonic erosion is taking place in the trench, and subduction complex may begin to develop when spreading of marginal sea ceased.

7.   A Model for Arc Volcanism and Marginal Sea Spreading

Several models or hypotheses have been proposed for arc volcanism and for marginal sea spreading. Most of them were discussed independently on arc volcanism and on origin of marginal sea. Those models for spreading of marginal sea are proposed on the relation with subduction of oceanic plate under an arc in which some of them are based on the convection current in the upper mantle formed by the subduction of the oceanic plate (MCKENZIE, 1969; UYEDA, 1977; KARIG, 1971; TOKSÖZ and BIRD, 1977; HONZA, 1978). In cases without any heat supply from subducted slab or that from upper mantle under arc, convection current might occur downwards by mechanical drag along the upper surface of subducted slab followed by arc ward horizontal current near the surface on the backarc side (MCKENZIE, 1969; TOKSÖZ and BIRD, 1977). In case where frictional heating occurred along the upper boundary of the subducted slab, convection current might be welling upwards turning to the both sides on the surface with relatively strong current towards the inner side of arc (horizontally away from the descending plate) and a weak counter current towards the outer side (ICHIE, 1971; HONZA, 1978). Geology of the arc seems to support a current towards a continent in the backarc basin, especially on double arcs such as seen in the Ogasawara and the Ryukyu Arcs intervened by the Shikoku Basin.

Convection current toward continent seems to imply arc to drift apparently towards a continent. However, when a continent forms a barrier to horizontal components of the current near surface, the arc may drift seaward adding new material along upwelled sites of the current near the surface. There may be a drift towards a continent, when a new subduction occurs along the margin of the continent as is seen on the both sides of the Shikoku Basin, and when supply rate of new material along upwelled sites on the surface to form a backarc basin is not sufficient to recover the total loss occurred by consumption.

Arc volcanism needs a heat supply under an arc. A calculated heat pattern under an arc suggests that a lower temperature takes place along the subducted oceanic slab as compared with that in surrounding asthenosphere which may be resulted in the subduction of the cooled plate (HASEBE *et al.*, 1970) in which it seems difficult to supply any heat energy for arc volcanism along the subducted slab.

One possible supply of heat energy for arc volcanism is a frictional heating along the upper surface of the subducting oceanic plate which is the same model examined for the origin of marginal sea spreading (Fig. 5). Upwelling component of the convection current may roll a heat supply for arc volcanism, although theoretical distance of

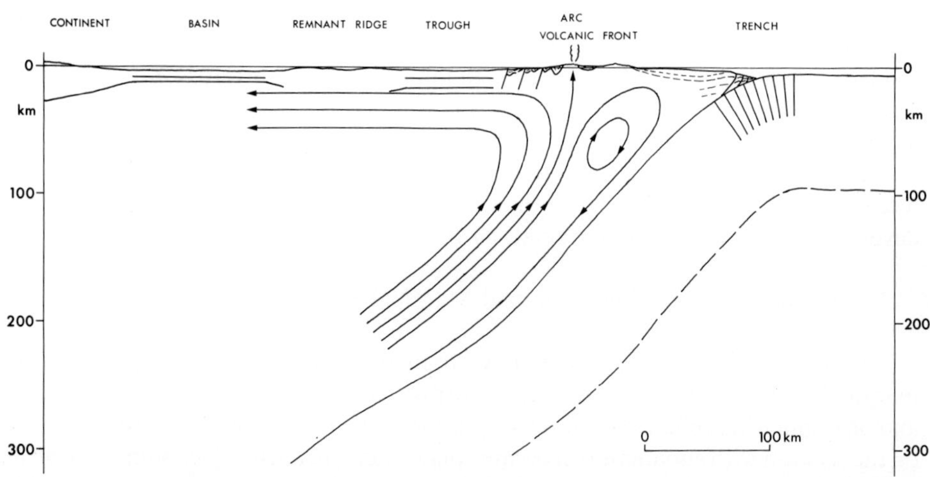

Fig. 5. A convection current model under an arc illustrated in the Tohoku Arc (Honza, 1978). The model is based on the calculated result by Ichie (1971) in which frictional heating may occur on the boundary between the subducted oceanic slub and the asthenosphere under the arc.

upwelling is rather backarc side than that of volcanic chain according to the calculated result by Ichie (1971). In this case, there may be a heat supply whenever subduction occurs. A difficulty of this model may be on the mechanism of deep earthquakes which occur at the heated boundary between a subducted slab and an upper mantle under the arc.

8. Discussion and Concluding Remarks

Geomorphologic variations of arcs makes it difficult to understand the fundamental frameworks which constitute arcs. Some of them occur within an arc. Further difficulty comes from the temporal variations of geomorphology discussed above. In modern arcs it is difficult to distinguish whether transgressive or regressive plases are taking place in the forearc development.

Here, forearc and backarc features are joined as a related phenomenon to constitute an arc. However, there may be another possible interpretation, as is shown in the Aleutian area, where the volcanism was active during the same time as the formation of accretionary wedge within a long-range geological age since Triassic (Raymond, 1980). There may be another possibility that the mechanism to form arc volcanism is independent from the formation of marginal sea, although both of them may need heat supply under the arc.

A quite different mechanism is suggested for mid-oceanic volcanism and for continental volcanism which may be attributed to heat process in a deeper mantle area than that of the arc area.

Here, a possible mechanism for arc volcanism is suggested as an alternative activity with marginal sea spreading and also a related activity to the subduction complex in trench on the basis of convection current under the arc formed by the frictional heating along the upper surface of the subducted oceanic plate.

Intense volcanism may occur when marginal sea spreading is slightly active. It is also inferred that the period of formation of subduction complex on the trench is correlated to the period of volcanism in the arc. Low level volcanism may occur when marginal sea is spreading. The period of consumption on the trench is also correlated to the period of low level volcanism (Fig. 6).

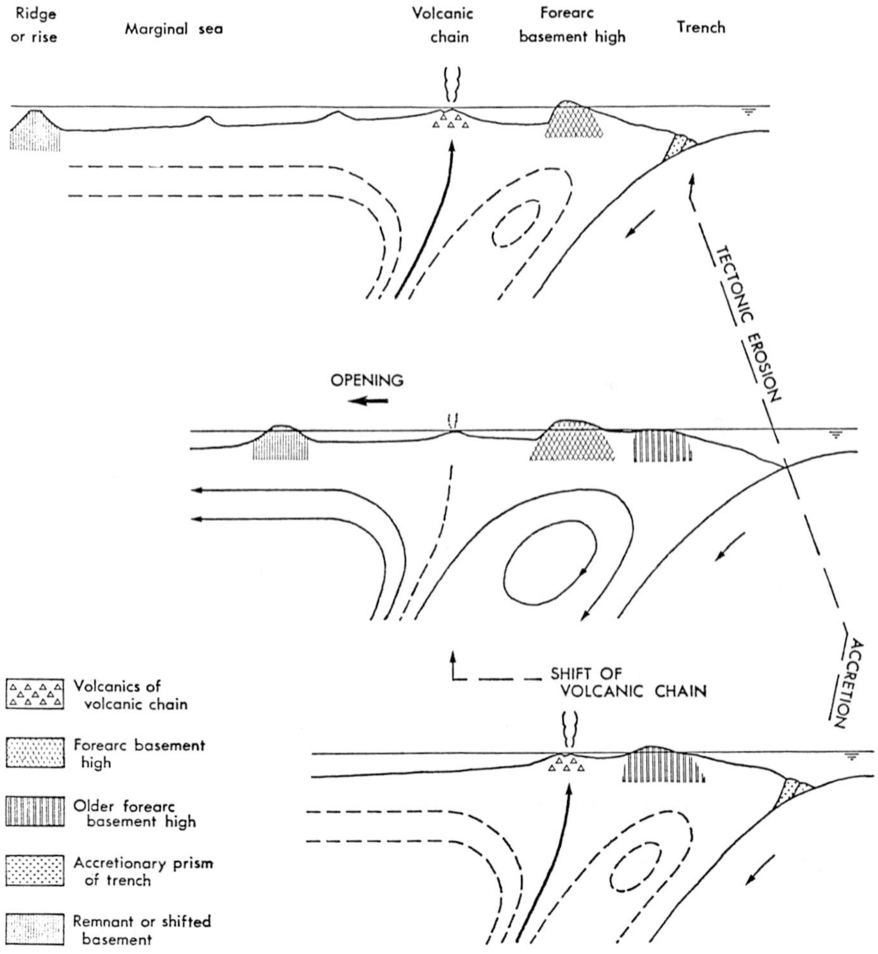

Fig. 6.   A model for evolution of fundamental framework of an arc based on a convection current model (HONZA, 1981). Slight spreading may occur in the marginal sea when the volcanism is active on the volcanic chain associated with the development of subduction complex at the trench. Low level volcanism may occur in the volcanic chain while the marginal sea is opening sssociated with development of tectonic erosion and horizontal shortening of distance between forearc basement high and trench.

In convection current model, it is inferred that the upwelling heat energy might provide the energy for arc volcanism in the volcanic chain and might be consumed by the horizontal component of current which results in the opening of the marginal sea and a possible consumption on the trench.

I wish to thank Prof. K. Nakamura and Dr. K. Ono for critical reading of the manuscript and for helpful suggestions, Dr. K. Shibata for his helpful suggestions on radiometric age determinations of arc volcanism, and Prof. D. Shimozuru for his stimulating to publish.

REFERENCES

DEVRIES KLEIN, G. and K. KOBAYASHI et al., Initial Reptorts of Deep Sea Dribling Project, V. 48, U.S. Govt. Printi. Office) Washington.

DICKINSON, W. R., Widths of modern arc trench gaps proportional to past duration of igneous activity in associated magma arcs, J. Geophys. Res., 78, 3376–3389, 1973.

HANZAWA, S., Eocene foraminifera from Haha-jima (Hillsborough Island), J. Paleont., 21, 254–259, 1947.

HASEBE, K., N. FUJII and S. UYEDA, Thermal processes under island arcs, Tectonophysics, 10, 335–355, 1970.

HERMAN, B. M., R. N. ANDERSON, and N. TRUCHAN, Extensional tectonics in the Okinawa Trough, Mem. Am. Assoc. Petrol. Geol. Bull., 199–208, 1978.

HESS, H. H., Major structural features of the western north Pacific, an interpretation of H. O. 5485 bathymetric chart, Korea to New Guinea, Bull. Geol. Soc. Am., 59, 417–446, 1948.

HONZA, E. (ed.), Ryukyu Island (Nansei-shoto) Arc, Cruise Rept. 6, Geol Surv. Japan, 1976.

HONZA, E., Geological map around Ryukyu Arc, in Marine Geol. Map Ser. 7, Geol. Surv. Japan, 1977.

HONZA, E., Geological history of the Kuril Basin and the Tartary Trough, in Geological Investigation of the Okhotsk and Japan Seas off Hokkaido, Cruise Rept. 11, edited by E. Honza, Geol. Surv. Japan, 1978.

HONZA, E., Subduction and accretion in the Japan Trench, Oceano, Acta Sp., vol., 251–258, 1981a.

HONZA, E., The geological settings of the Ogasawara and the northern Mariana Arcs, in Geological Investigation of the Ogasawara (Bonin) and Northern Mariana Arcs, Cruise Rept. 14, edited by E. Honza, I. Inoue, and T Ishihara, Geol. Surv. Japan, 1981b.

HONZA, E., E. INOUE, and T. ISHIHARA (eds.), Geological investigation of the Ogasawara (Bonin) and northern Mariana Arcs, in Cruise Rept. 14, Geol. Surv. Japan, 1981.

HONZA, E., K. KAGAMI and N. NASU, Neogene geological history of the Tohoku Island Arc System, J. Oceanog. Soc. Japan, 33, 197–310, 1977.

HUSSONG, D., S. UYEDA et al., Leg 60 ends in Guam, Geotimes Oct., 19–22, 1978.

ICHIE, T., A. note on the convection currents in the mantle, Tectonophysics, 11, 407–418, 1971.

ISSHIKI, N., Igneous activity, in Geology and Mineral Resources of Japan, 3rd edition, Vol. 1, Geology, Geol. Surv. Japan, 1977.

IWASAKI, Y. and M. AOSHIMA, Report on geology of the Bonin Islands, in Nature of the Bonin Islands, pp. 205–219, Educational Agency of Japan, 1970.

KANEOKA, I., N. ISSHIKI, and S. ZASHU, K-Ar ages of the Izu-Bonin Islands, Geophys. J., 4, 53–60, 1970.

KARIG, D. E., Origin and development of marginal basins in the Western Pacific, J. Geophys. Res., 76, 2542–2561, 1971.

KARIG, D. E., J. C. INGLE, Jr. et al., Initial Rept. DSDP 31, U.S. Govt. Print. Office, Washington, D.C., 1975.

KARIG, D. E. and G. F. SHARMAN, III, Subduction and accretion in trenches, Geol. Soc. Am. Bull., 86, 377–389, 1975.

KOBAYASHI, K. and M. NAKADA, Magnetic anomalies and tectonic evolution of the Shikoku inter-arc basin, in Geodynamics of the Western Pacific, edited by S. Uyeda, R. W. Murphy, and K. Kobayashi, pp. 391–402, 1979.

KONDA, T. and Y. UYEDA, K-Ar age of the Tertiary volcanic rocks in the Tohoku area, *J.*, *Japan. Assoc. Miner. Pet. Econ. Geol. Sp.*, vol., **2**, 343–346, 1980.

KULM, L. D. and W. J. SCHWELLER, A preliminary analysis of the subduction processes along the Andian continental margin, 6° to 45°S, in *Island Arcs, Deep Sea Trenches and Back-Arc Basins*, edited by M. Talwani and W. C. Pitman III, Maurice Ewing Ser., 1, pp. 285–301, Am. Geophys. Union, 1977.

KUNO, H., Differentiation of basalt magmas, in *Basalts, V. 2*, edited by H. H. Hess and A. Poldervaart, 1968.

LADD, H. S., Chitons and Gastropods (Haliotidae trough Adeorbidae) from the western Pacific islands, *U.S. Geol. Surv. Prof.* Paper, 531, 1966.

MATSUDA, T., K. NAKAMURA, and A. SUGIMURA, Late Cenozoic Orogeny in Japan, *Tectonophysics*, **4**, 349–366, 1967.

MATSUDA, T. and S. UYEDA, On the Pacific-type orogeny and its model-Extension of the paired belts concept and possible origin of marginal seas, *Tectonophysics*, **11**, 5–27, 1970.

MATSUMARU, K., The transition of the larger forminiferal assemblages in the western Pacific Ocean-Especially from the Tertiary Period, *J. Geogr. Soc. Japan*, **83**, 281–301, 1974.

McKENZIE, D. P., Speculations on the consequences and causes of plate motions, *Geophys. J. R. Astron. Soc.*, **18**, 1–32, 1969.

MIYASHIRO, A., Volcanic rock series in island arcs and active continental margins, *Am. J. Sci*, **274**, 321–355, 1974.

NAKAMURA, K., Island arc tectonics, a hypothesis, in *papers for sympo.*, *Problems Concerning Green Tuff*, *1969 Meeting*, Geol. Soc. Japan, pp. 31–38, 1969.

ONO, K., T. SOYA, and K. MIMURA (eds.), *Volcanoes of Japan*, 2nd ed. 1 :2,000,000 map ser. 11, Geo. Surv. Japan, 1981.

RAYMOND, L. A., Episodic accretion and plutonism in southwestern Alaska, *Earth Planet. Sci. Lett.*, **49**, 29–33, 1980.

SCIENTIFIC PARTY, *Initial Reports of Deep Sea Dribling Project V. 56–57*, U.S. Govt. Print. Office), Washington, D.C., 1980.

SEELY, D. R., The evolution of structural highs bordering major forearc basins, in *Geological and Geophysical Investigations of Continental Margins*, edited by J. S. Watkins, L. Montadert, and P. W. Dickerson, *Am. Assoc. Pet. Geol. Mem.*, **29**, pp. 245–260, 1979.

SIMKIN, T. and L. SIEBERT (eds.), Holocene volcanoes, in *Plate-Tectonic Map of the Circum-Pacific Region*, *NW Quad.*, edited by C. Nishiwaki, Circum-Pacific Council Energ. Min. Reso., 1981.

TANAKA, K. and T. NOZAWA (eds.), *Geology and Mineral Resources of Japan*, 3rd edition, Vol. 1, Geology, Geol. Surv. Japan, 1977.

TOKSÖZ, M. N. and P. BIRD, Formation and evolution of marginal basins and continental plateaus, in *Island Arcs Deep Sea Trenches and Back-Arc Basins*, edited by M. Talwani and W. C. Pitman III, Maurice Ewing Ser. 1, pp. 379–393, Am. Geophys. Union, 1977.

TSUCHI, R. (ed.), Fundamental data on Japanese Neogene bio- and chronostratigraphy, IGCP-114, National Working Group of Japan, 1979.

UJIIÉ, H. and K. MATSUMARU, Stratigraphic outline of Haha-jima (Hillsborough Island), Bonin Islands, *Memo. Nat. Sci. Museum, Japan*, **10**, 5–18, 1977.

UYEDA, S., Some basic problems in the trench-arc-backarc system, in *Island Arc, Deep Sea Trenches and Back-Arc Basins*, edited by M. Talwani and W. C. Pitman III, Maurice Ewing Ser, 1, pp. 1–14, Am. Geophys. Union, 1977.

WATTS, A. B. and J. K. WEISSEL, Tectonic history of the Shikoku marginal basin, *Earth Planet. Sci. Lett.*, **25**, 239–250, 1975.

YUASA, M., E. HONZA, K. TAMAKI, M. TANAHASHI, and A. NISHIMURA, Geological map of the southern Ogasawara and northern Mariana Arcs, in *Marine Geol. Map Ser. 18*, Geol. Surv. Japan, 1982.

Arc Volcanism: Physics and Tectonics, edited by D. Shimozuru and I. Yokoyama, 191–224.
Copyright © 1983 by Terra Scientific Publishing Company (TERRAPUB), Tokyo.

Magmatism and Subduction in
the Eastern Aleutian Arc

J. Kienle, S. E. Swanson, and H. Pulpan

*Geophysical Institute, University of Alaska,
Fairbanks, Alaska 99701, U.S.A.*

Volcanism and tectonism in the eastern Aleutian arc are controlled by the subduction of the Pacific plate beneath the North American plate. Worldwide earthquake data and data from local seismic networks in Cook Inlet, on the Alaska Peninsula and on Kodiak Island have defined the arcuate plate boundary and the Wadati-Benioff zone. A calc-alkaline volcanic arc of approximately 20 volcanic centers is well developed above the subduction zone.

Calc-alkaline volcanism and lithospheric plate subduction are intimately linked, even though the details of magma generation are still controversial. The volcanoes in the Cook Inlet region (N of 59° N) have a regular spacing of 60 ± 19 km and follow a northerly trend which is parallel to the strike of the Wadati-Benioff zone. In contrast, the Katmai volcanic centers (S of 59° N) have a much closer spacing of 13 ± 7 km and follow a cross-cutting trend with respect to the strike of the Wadati-Benioff zone. A misorientation of 35° (from N 20° E to N 55° E) marks the change in trend of the volcanoes of Cook Inlet and Katmai. The pivot point lies south of Augustine Volcano. Other more subtle misorientations in the volcanic arc are found at Kaguyak and Katmai craters. This segmented nature of the volcanic arc is in contrast to the smoothly changing strike of the Wadati-Benioff zone.

Geochemical reconnaissance of the volcanic centers reveals two distinct types of patterns of magmatism based on major element chemistry. Volcanoes near subsegment boundaries (Kaguyak and Katmai) are characterized by relatively high abundances of dacite and rhyolite while volcanoes within arc segments are dominantly andesite. Augustine, dominated by andesitic lavas, is a typical intrasegment volcano. The transition from the Katmai segment to the Cook Inlet segment is not marked by any unusual geochemical variation in the lavas. Overall, however, the Katmai lavas are richer in SiO_2 and slightly richer in K_2O than the Cook Inlet lavas.

Volcanism in the eastern Aleutian arc apparently ends at Hayes Volcano, but seismically defined subduction continues as far northeast as Mt. McKinley. Andesitic volcanism has been linked to sediment subduction. The large accretionary wedge in the forearc of the eastern Aleutian subduction system suggests that sediments are not subducted and this may explain the cessation of volcanism in the easternmost Aleutian arc.

1.   Introduction

The Aleutian trench and volcanic arc define the northern margin of the Pacific lithospheric plate. The arc structure spans a distance of 3,800 km from Kamchatka to mainland Alaska and contains about 80 Quaternary volcanic centers of which about 40 have erupted since 1700 (Fig. 1). The Aleutian arc consists of two structurally continuous parts, a western arc of islands built on oceanic crust of the deep Bering Sea and an eastern arc constructed on the continental basement of the Alaska Peninsula and Cook Inlet region. While the seismicity associated with the subduction process is becoming fairly well defined (ENGDAHL, 1973, 1977; DAVIES, 1975; LAHR, 1975; JACOB *et al.*, 1977; ENGDAHL and SCHOLZ, 1977; PULPAN and KIENLE, 1979; DAVIES and HOUSE, 1979; LAFORGE and ENGDAHL, 1979; REYNERS and COLES, 1982), many volcanologic studies of the arc are still of a reconnaissance nature. This is reflected in the fact that new Late Quaternary and Holocene volcanic centers are still being discovered (MILLER and SMITH, 1976).

The purpose of this paper is to investigate the relationship of magmatism and subduction in the eastern Aleutian arc. The geometry of the subduction system is derived from our studies of seismicity, and magmatism is characterized by recent petrologic studies (KOSCO, 1981) and an analysis of samples collected from 18 volcanic centers that lie between Kejulik and Redoubt volcanoes, representing a 350 km long segment of the arc (see Fig. 1).

We will also investigate the relationship of magma chemistry to arc segmentation and compare the results to other parts of the Aleutian arc where tholeitic and calc-alkaline volcanism appears to be controlled by tectonic setting (KAY *et al.*, 1982).

2.   Pacific-North American Plate Interaction

Subduction under the Aleutian arc changes from normal to oblique, from east to west along the arc (Fig. 2). The transition from convergent to transform plate motion is completed near Buldir Volcano. The transform nature of the plate boundary is marked by the cessation of Quaternary magmatism and intermediate-depth seismicity in the Kommandorsky section of the arc, clearly demonstrating the intimate relationship between subduction and magmatism (McKENZIE and PARKER, 1967). Arc magmatism resumes when plate motion becomes again convergent in the Kamchatka-Kuril arc. Sheveluch Volcano marks the transition point (Fig. 2).

The volcanic front of the Aleutian arc generally lies about 100 to 150 km above the dipping seismic zone. Bogoslof and Amak Islands lie behind the volcanic arc and erupt more potash rich lavas (MARSH, 1979). This secondary arc has started to develop only during the past 7,000 years and its tectonic significance is not yet clear.

Figure 3 shows the position of Quaternary volcanic centers in the eastern Aleutian arc and illustrates current thoughts on the transition from plate convergence to transform motion between the Pacific and North American plates in the Gulf of Alaska. This transition is not sharp but is spread out in a zone about 600 km wide and the relationship of active volcanism to plate subduction is not as clear as at the western end of the arc. The kinematic model for Pacific-North American plate interaction shown in Fig. 3 is taken from LAHR and PLAFKER (1980) who based their model on the historical seismicity and known rates of relative motion along a series of important regional faults in the area. A similar model for the Gulf of Alaska transition from convergent to transform plate motion involving several microplates was recently

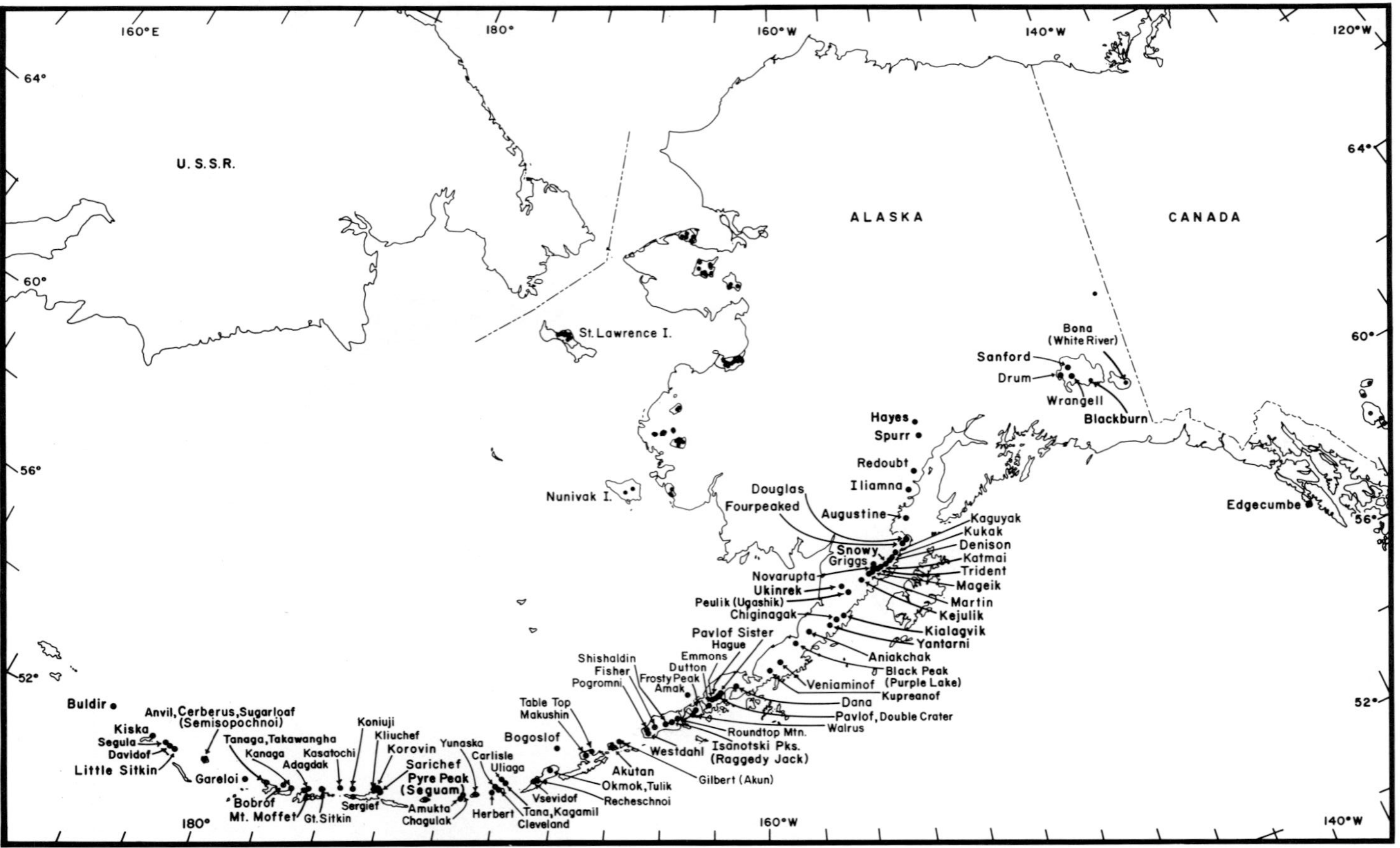

Fig. 1.   Quaternary volcanic centers in the Aleutian arc.

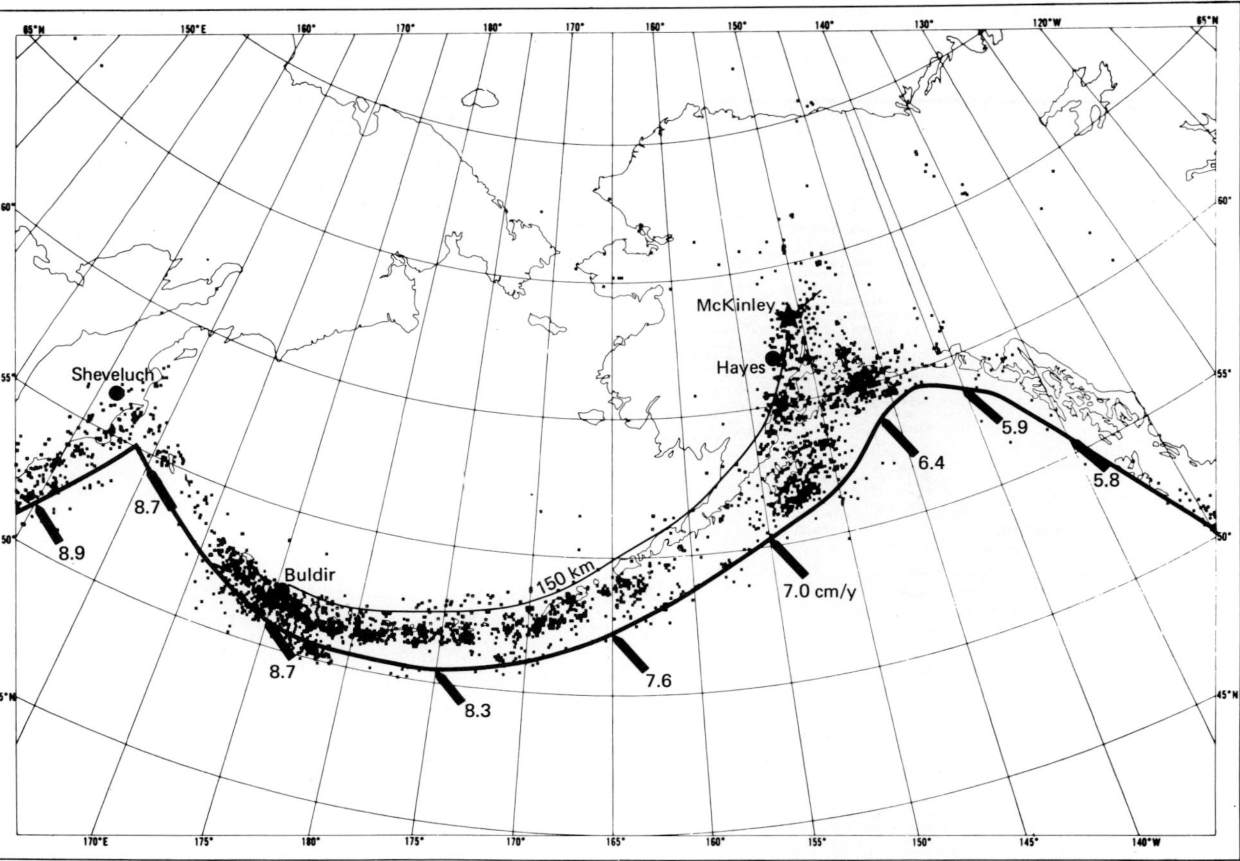

Fig. 2. Relative motion vectors between Pacific and North American plates after Minster and Jordan (1978), and seismicity of the Aleutian arc. Trench and 150 km depth contour to Wadati-Benioff zone shown by solid lines. Figure is modified from Jacob *et al.* (1977).

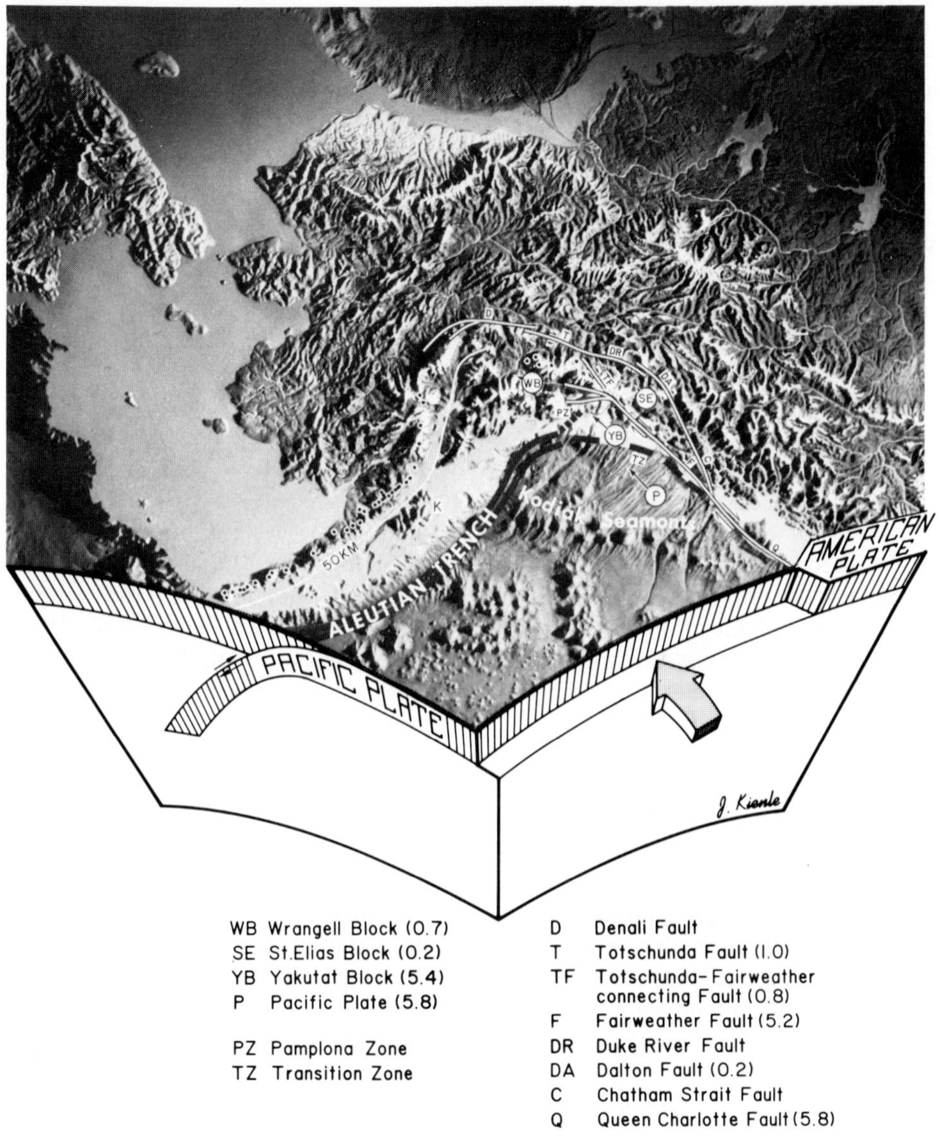

WB Wrangell Block (0.7)   D   Denali Fault
SE St.Elias Block (0.2)   T   Totschunda Fault (1.0)
YB Yakutat Block (5.4)   TF   Totschunda–Fairweather
P   Pacific Plate (5.8)   connecting Fault (0.8)
                                 F   Fairweather Fault (5.2)
PZ Pamplona Zone   DR   Duke River Fault
TZ Transition Zone   DA   Dalton Fault (0.2)
                                 C   Chatham Strait Fault
                                 Q   Queen Charlotte Fault (5.8)

Fig. 3.   Plate interaction in Alaska. Physiographic view shows transition from convergent to transform plate motion in the Gulf of Alaska. Motion vectors of St. Elias block (SE), Yakutat block (YB), and Wrangell block (WB) after LAHR and PLAFKER (1980). 50 km depth contour to Wadati-Benioff zone shown by solid white line. K, Kodiak Island. Black dots on white circles, volcanoes. Numbers in parenthesis after fault names indicate right lateral slip rate along given fault (cm/yr); numbers in parenthesis after block names give rates of motion relative to fixed North American plate (cm/yr).

developed by PEREZ and JACOB (1980) based on a new set of fault plane solutions for the area.

There are two apparent anomalies in the distribution of volcanoes at the eastern end of the Aleutian arc: Even though northwesterly subduction is defined by intermediate depth earthquakes well into the interior of of Alaska, extending to the vicinity of Mt. McKinley, no Quaternary volcanoes have yet been found north of Hayes Volcano (Figs. 1 and 3). On the other hand, the Wrangell volcanoes lying some 350 km east of the Aleutian trend (Figs. 1 and 3) are *not* underlain by a seismically defined Wadati-Benioff zone. LAHR and PLAFKER (1980) have suggested that the Wrangell volcanoes may be a dying magmatic system that has not been very eruptive for the past 3 m.y. and formed in an earlier, more northerly directed episode of subduction. PEREZ and JACOB (1980) take a different view and propose currently active, but seismically quiet, northerly subduction of the Pacific plate beneath the still active Wrangell volcanoes.

3.   Seismicity of the Easternmost Aleutian Arc

3.1   General discussion

The geometry of the subducting Pacific plate in the eastern Aleutian arc has recently become better defined through a series of unpublished theses on subduction seismicity (DAVIES, 1975; LAHR, 1975; ESTES, 1978; AGNEW, 1980) and various reports not available in the literature (PULPAN and KIENLE, 1979; LAHR *et al.*, 1974; and WOODWARD-CLYDE CONSULTANTS, 1980). In Fig. 3, the 50 km depth contour to the Wadati-Benioff zone (compiled from the literature cited above and published data taken from DAVIES and HOUSE (1979) and DAVIES *et al.* (1981)) fairly closely delineates the aseismic front defined by YOSHII (1975) as the seaward edge of the aseismic wedge beneath the active volcanic arc.

The volcano-trench gap measured along the slip direction widens dramatically from 170 km in the central Aleutians to 300 km on the Alaska Peninsula and 570 km in the Gulf of Alaska (JACOB *et al.*, 1977) suggesting that the trench in the Gulf of Alaska may have been forced to migrate as much as 200 km seaward by continental accretion. During this migration the underthrusting plate maintained a shallow subduction angle. Shallow subduction over hundreds of kilometers under southern Alaska results in strong plate coupling and very large rupture areas of great earthquakes such as the Alaskan earthquake of 1964 (a shallow thrust event), which had a magnitude of M_w 9.2 (KANAMORI, 1977).

A pronounced 35 degree bend occurs in the volcano line northwest of Kodiak Island (K in Fig. 3). This bend is also reflected in the strike of the Wadati-Benioff zone. In Cook Inlet, the strike of the dipping seismic zone is anomalous and distinctly more northerly than the adjacent zones to the northeast and southwest (see 50 km depth contour in Fig. 3). Plate plunge in Cook Inlet is $30°$ more westerly than the direction of plate convergence while in the dipping seismic zones north and south of Cook Inlet plate plunge and plate convergence are nearly parallel.

3.2 Seismicity of the Cook Inlet-Katmai-Kodiak region

Members of the seismology group of the Geophysical Institute of the University of Alaska have been studying the regional seismicity, focal mechanisms and seismotectonics in the Cook Inlet-Katmai-Kodiak region. Since 1976, the University of Alaska has operated a high resolution seismographic network with over 30 stations in the area. All stations have single component, vertical, short-period (natural period 1 sec) seismometers. Figure 4 shows the current configuration of the seismic network and the location of an additional four stations, operated by other agencies, which are routinely used for our earthquake locations. Data from all stations are radio-telemetered to three data gathering points, located in King Salmon, Kodiak, and Homer. Homer is the recording site for all the data which arrive there either via direct radio links or leased satellite transmission links from King Salmon and Kodiak. The data is recorded on 16 mm film. Arrival times are read from the film with a precision of 0.05 sec.

In our routine procedure for hypocentral location procedure we use a computer program developed by the U.S. Geological Survey. The program uses a horizontally layered velocity structure, which for the shallow crustal layers is similar to that

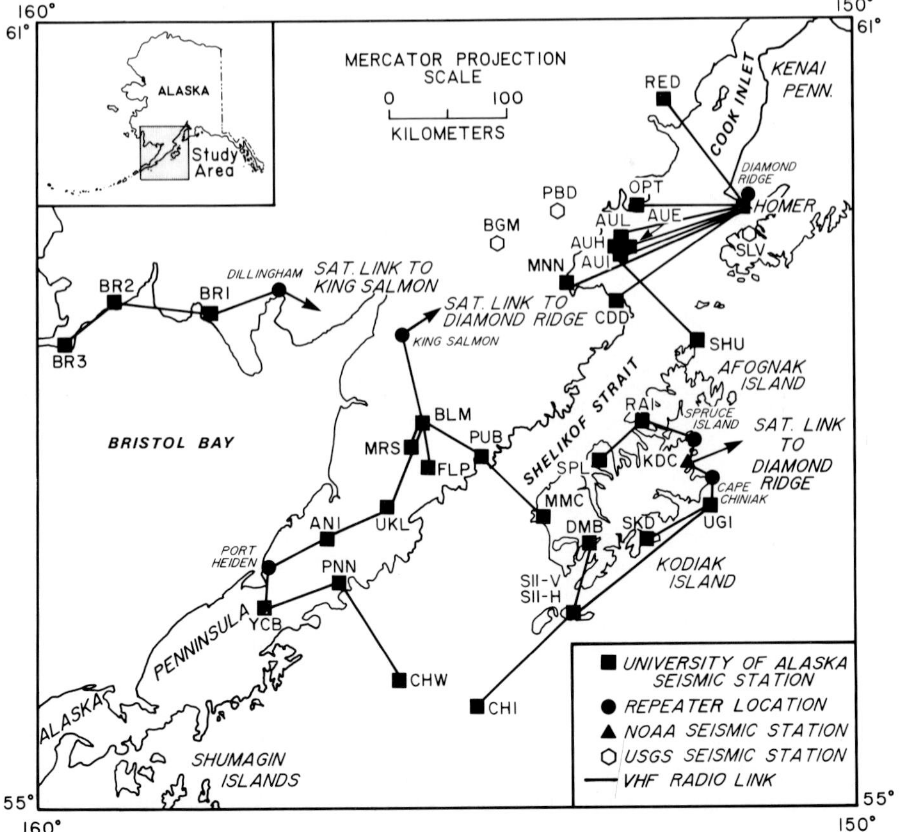

Fig. 4. Cook Inlet-Alaska Peninsula-Bristol Bay-Kodiak seismic network, 1981 configuration.

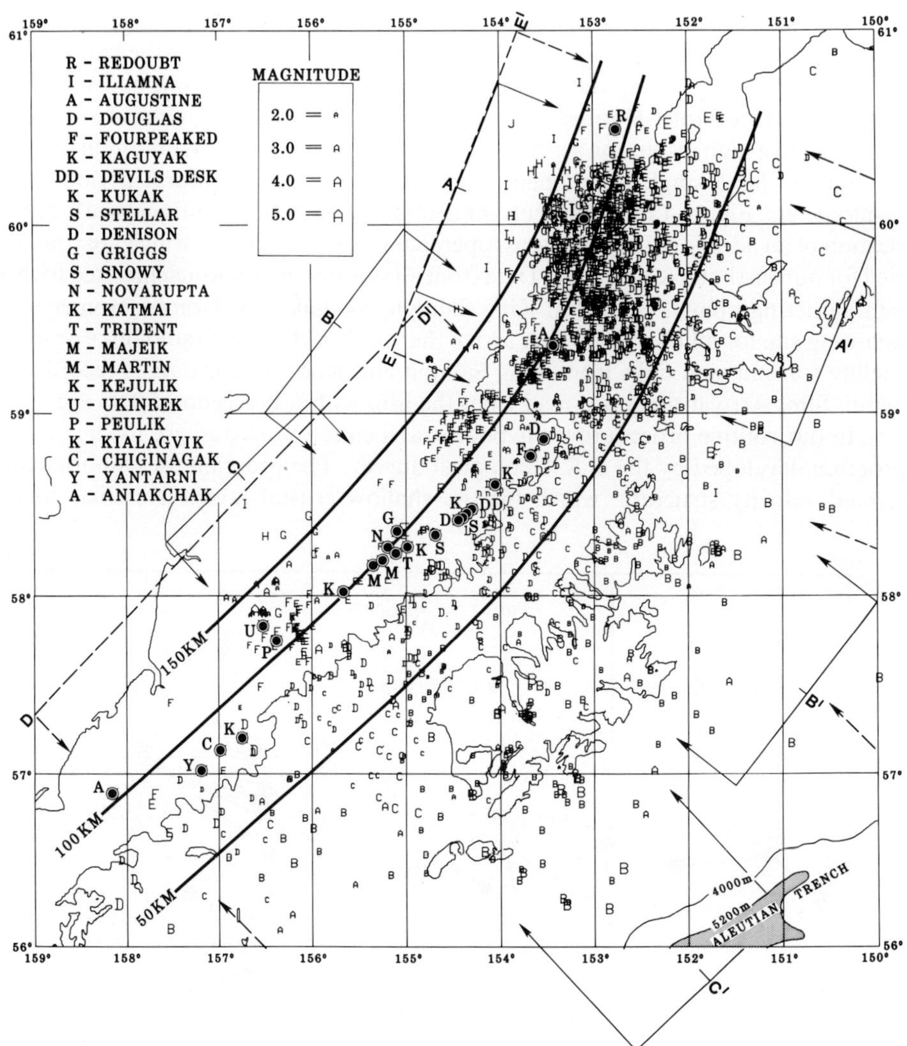

Fig. 5. Epicenters, volcanoes, depth contours to top of Wadati-Benioff zone (50, 100, 150 km) and location of projection areas for cross sections and frontal views shown in Figs. 6 and 7. Epicenters from local network data July 1977 to June 1981; selection criteria: Recorded on at least 6 stations, RMS travel time residual ⩽ 0.4 sec, relative vertical and horizontal location error ⩽ 10 km. Epicenters are coded according to magnitude and depth range, A :0–25 km, B :26–50 km, C :51–100 km, etc.

obtained by Hasegawa (1973) for the central Aleutians. For the mantle layers we use the standard P-wave velocity structure of Herrin (1968).

The earthquake detection threshold within the network is about magnitude 2, but can be as low as magnitude 1 in some portions of it.

Figure 5 is a map of epicenters of the area. For this figure we selected a file of 1,881 events from the period July 1977 to June 1981, which were recorded on at least 6 stations, had vertical and horizontal hypocentral location errors of less than 10 km, and had a root mean square travel time errors of less than 0.4 sec. The letters denote depth intervals A :0–25 km, B :26–50 km, etc. and the size of the letter is proportional to the magnitude.

The dominant source of earthquakes in our study area is the Wadati-Benioff zone, which consists of an upper shallow thrust portion extending from the Aleutian trench to about 50 km depth and a more steeply dipping portion reaching a maximum depth of 250 km behind the volcanic arc.

Shallow (less than 30 km deep) earthquakes occur in the overriding plate. These earthquakes are generally diffuse and in most cases cannot be correlated with known faults. We also routinely locate shallow swarms of earthquakes that are associated with the volcanoes.

3.3   Hypocentral cross sections

In Fig. 6 earthquake hypocenters within the three blocks shown on the epicenter map have been projected on vertical planes indicated by lines A–A', B–B' and C–C' in Fig. 5. The orientations of the planes were chosen by trial and error to minimize the thickness of the Wadati-Benioff zone and thus correspond to the best-fit plunge direction of the subducting plate. Two types of data are plotted on Fig. 6: hypocenters for events with a depth greater than 50 km are fairly closely constrained, and we required that the event was recorded on at least six stations, had a vertical (ERZ) and horizontal (ERH) location error of less than 10 km, had a RMS travel time error of less than 0.3 sec and the distance from the epicenter to the nearest station was within twice the depth to the hypocenter. Since these constraints filtered out too many of the shallow events (depths 0 to 50 km), we relaxed the selection criteria for shallow events requiring that the event be recorded on at least five stations and had an RMS travel time residual of less than 0.5 sec and an ERH and ERZ of less than 10 km.

Very noticeable in the three plots shown in Fig. 6 is the systematic widening of the arc-trench gap as one moves from southwest to northeast (bottom to top in Fig. 6).

Cross section A–A' samples the seismicity of a 133 km wide section of the arc in Cook Inlet between Redoubt and Augustine Volcanoes and shows a very well developed Wadati-Benioff zone. The zone is 20–30 km thick and dips about 35° to 40° beneath the Cook Inlet volcanoes. It reaches a depth of 200 km with a few events scattering no deeper than 250 km. As mentioned above, Cook Inlet is anomalous in that the best fit plunge direction of the subducting plate is 30° more westerly than the direction of plate convergence.

The shallow seismicity in the overriding plate is diffuse and cannot be correlated with tectonic or volcanic features. Had we used data recorded before 1977, extending back to 1975, a dense cluster of very shallow (0–5 km) seismicity would have been plotted directly beneath Augustine Volcano, which erupted explosively in 1976 (REEDER et al., 1977; JOHNSTON, 1978; KIENLE and SHAW, 1979; KIENLE and SWANSON, 1980).

The aseismic wedge landward of the dipping seismic zone is very clear. The aseismic front is located beneath the eastern shore of Cook Inlet, roughly following the

J. Kienle *et al.*

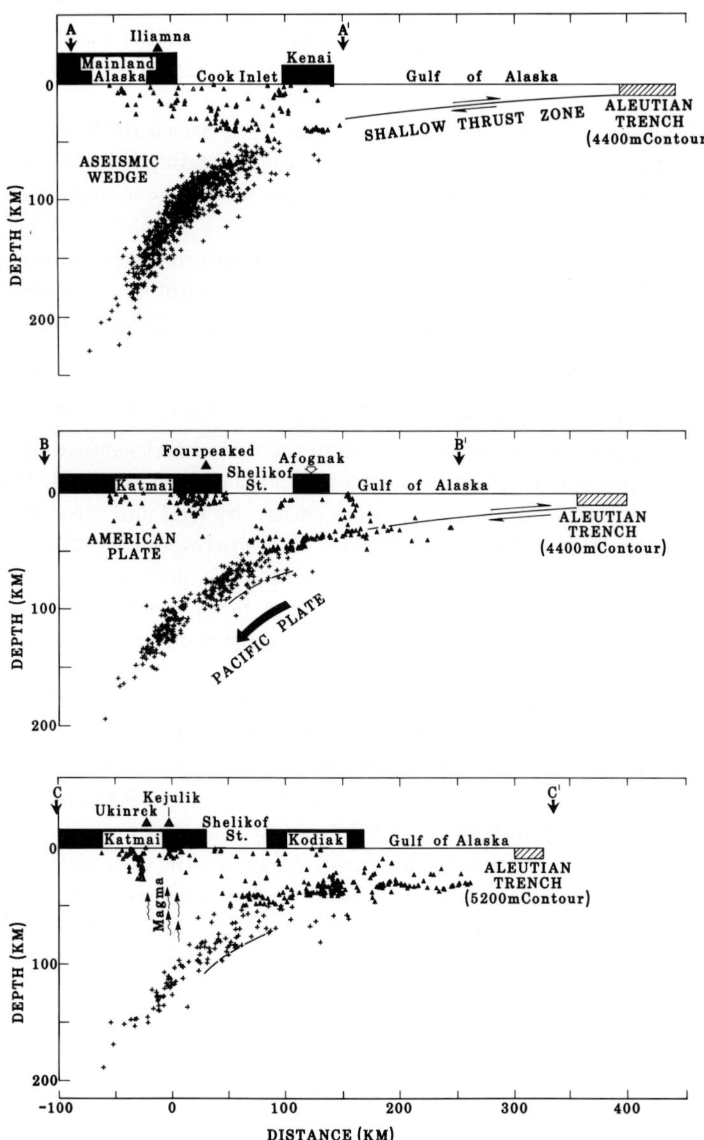

Fig. 6. Cross sectional views of Wadati-Benioff zone along lines A–A′, B–B′, C–C′ shown in Fig. 5. No vertical exaggeration. Landmasses (solid black), the position of the trench, and location of volcanoes are also shown. Shallow thrust zone is added schematically. Dashed lines show possible position of second seismic zone in cross sections B–B′ and C–C′. Selection criteria for events plotted in the 0–50 km depth range (triangles): Recorded on at least five stations (STA ≥ 5), relative horizontal and vertical location error (ERZ and ERH) ≤10 km, RMS travel time residual ≤0.5 sec. Selection criteria for events > 50 km depth (crosses): STA ≥ 6, ERH and ERZ ≤10 km, RMS ≤0.3 sec, and distance from epicenter to nearest station ≤2 × depth.

50 km depth contour to the dipping seismic zone shown in Fig. 3.

Augustine, Iliamna, and Redoubt Volcanoes are located on a linear trend that slighly cross-cuts the strike of the dipping seismic zone. The depth to the top of the dipping seismic zone beneath the volcanoes is 100–110 km (compare Fig. 5).

Cross section B–B′ samples the seismicity of a 138 km wide section of the arc between Augustine Volcano in lower Cook Inlet and Snowy Mt. in Katmai and shows very similar features to cross section A–A′.

The seismic zone plunges with a dip of 35° to 40° beneath the Katmai volcanoes but in contrast to middle and upper Cook Inlet plate plunge is nearly parallel to the direction of plate convergence. The dipping seismic zone below 50 km is 20 km thick and no events deeper than 250 km have been recorded. There is a hint of a second seismic zone below and subparallel to the main Wadati-Benioff zone (dashed line Fig. 6, section B–B′). Such double-planed dipping seismic zones have been observed elsewhere in the Aleutians (ENGDAHL and SCHOLZ, 1977—but questioned by TOPPER, 1978; REYNERS and COLES, 1982), and in northern Honshu, Japan (HASEGAWA et al., 1978).

The shallow seismicity in this section is more clustered than in section A–A′. A pronounced cluster of shallow seismicity occurs in the vicinity of Katmai volcanoes Douglas and Fourpeaked Mtn. Even though we do not have the epicentral resolution needed to correlate the seismicity specifically with one or the other of the two volcanoes, the swarm character and very shallow depth of the earthquakes suggest that they are probably associated with hydrothermal activity beneath one or both of the volcanoes. The summit crater of Douglas contains a hot, acid lake.

The upper Katmai volcanoes in this section of the arc lie on a linear trend slightly oblique to the strike of the Benioff zone. The top of the dipping seismic zone lies at a depth of only about 80 km beneath the volcanoes, markedly shallower than for the Cook Inlet volcanoes.

Cross section C–C′ includes earthquakes within a 122 km wide section of the arc from Snowy Mtn. to Peulik/Ukinrek Volcanoes and shows a particularly well defined shallow-dipping thrust zone. Many of the events that map out the thrust are aftershocks of a M_L 6.5 earthquake which occurred offshore southern Kodiak on April 12, 1978 (PULPAN and KIENLE, 1979; HAMPTON et al., 1979; LAWTON et al., 1982). The other features of this section are very similar to the other two sections. There is again a hint of a double-planed dipping seismic zone. Plate plunge and the direction of plate convergence are parallel and the volcanoes line up with the strike of the Wadati-Benioff zone.

The depth of the seismic zone is about 100 km beneath the volcano line and somewhat deeper, about 125 km, beneath the Ukinrek Maars. These maars formed in April 1977 (KIENLE et al., 1980; SELF et al., 1980) and erupted a weakly undersaturated alkali olivine basalt ($Ne = 1.2\%$). The shallow seismicity beneath the Ukinrek Maars seen in this section is associated with their formation in 1977.

3.4 Frontal views

In Fig. 7 all the data shown in the hypocentral cross sections are projected onto planes oriented parallel to the strike of the Wadati-Benioff zone. We chose two planes, one approximating the strike of the dipping seismic zone in the Katmai region, the other approximating the strike in Cook Inlet (lines D–D' and E–E' in Fig. 5).

The shallow seismicity (0–50 km) is fairly intense in the overriding plate. Earthquake clusters at depths shallower than 5 km are located beneath several volcanoes (Ukinrek, Martin-Mageik-Trident, Snowy, Kaguyak, Douglas, Four-peaked, Augustine) and are probably associated with volcanic or hydrothermal activity.

An interesting and unexplained clustering of deep earthquakes occurs in the 100 to 200 km depth range beneath and behind certain volcanoes, as also observed in the central Aleutians by ENGDAHL (1977). Fairly well defined clusters of intermediate depth earthquakes occur beneath Peulik Volcano (100–150 km deep), behind Kaguyak-Douglas-Fourpeaked (100–150 km deep), and slightly off Augustine (100–150 km deep). The intense cluster of seismicity beneath Iliamna, some 100 km in diameter and extending from 70 to 180 km in depth, is a major unexplained anomaly, with frequent earthquakes in the magnitude 4–5 range.

4.   Segmentation in the Eastern Aleutian Arc

Volcanoes in subduction-related arcs, such as the Cascades or Central America, are often arranged in linear segments that are separated by transverse offsets (STOIBER and CARR, 1973; HUGHES *et al.*, 1980). A similar model of segmentation has been applied to the western Aleutian arc (KAY *et al.*, 1982). In addition to the intrasegment volcanoes that form the volcanic front, a second group of back arc volcanoes that erupt either alkaline magmas or bimodal basalt/rhyolite have been recognized in some arcs. These alkalic volcanoes are found to lie on the back-arc extension of the transverse offsets (HUGHES *et al.*, 1980; KAY *et al.*, 1982). Intrasegment volcanoes are characterized by calc-alkaline fractionation patterns and commonly erupt homogeneous two-pyroxene andesite (HUGHES *et al.*, 1980). Volcanoes at the ends of segments, adjacent to the offsets, erupt a variety of lavas that range from basalt to rhyolite (HUGHES *et al.*, 1980) and may show tholeiitic fractionation patterns (KAY *et al.*, 1982).

Segmentation of the volcanic front is also found in the eastern Aleutian arc (Fig. 8). Three major arc segments defined in the eastern Aleutian arc on the basis of the volcano alignment are the Semidi, Katmai and Cook Inlet segments. In Fig. 8 offsets between these segments have been drawn parallel to direction of convergence between the Pacific and North American plates (MINSTER and JORDAN, 1978). The boundary between the Semidi and Katmai segments also marks the limit of the rupture zone of the 1964 Alaskan earthquake.

Our studies have concentrated on the geophysics and volcanic petrology of the Kodiak and Kenai segments (termed collectively the eastern Aleutian arc). After a brief overview of the limited data base for the Semidi segment, we will present our data on the lavas of the eastern Aleutian arc and discuss the implications in terms of a segmentation model.

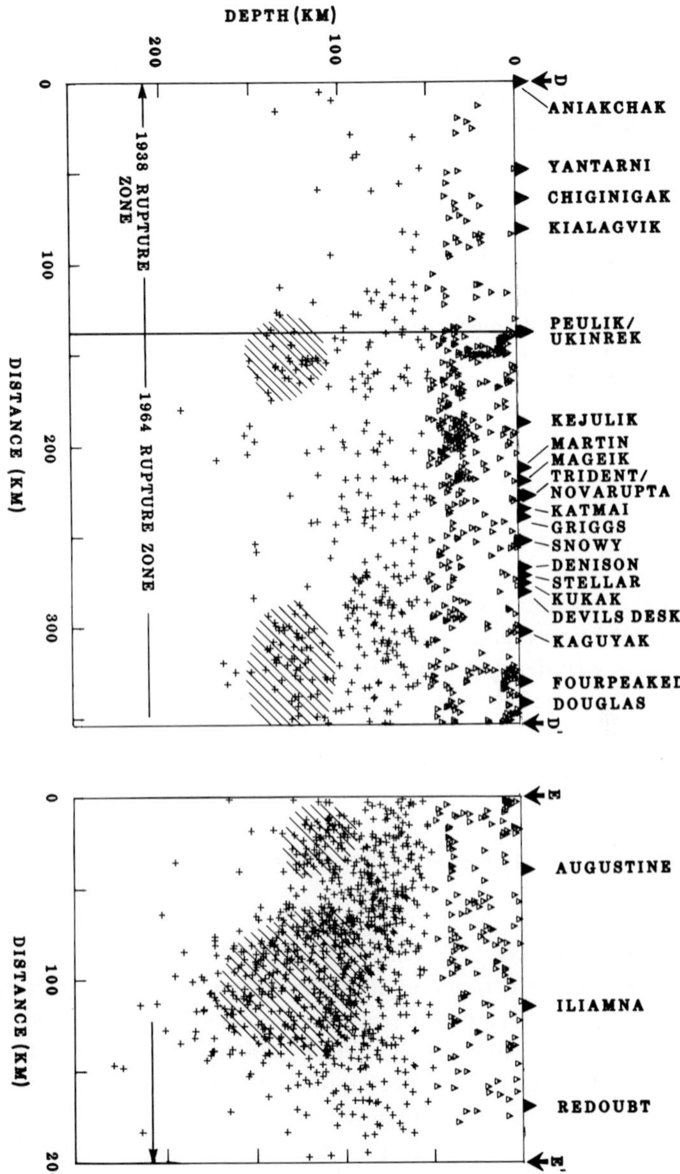

Fig. 7. Same data as in Fig. 6 projected on a vertical plane along the strike of the Wadati-Benioff zone (along lines D–D′ and E–E′ shown in Fig. 5). Position of volcanoes also shown. Deep clusters cross-hatched. Vertical line in section D–D′ marks rupture zone boundary of 1964 (M_w 9.2) and 1983 (M_w 8.2) shallow thrust earthquakes.

4.1 Volcanoes of the Semidi segment

The Semidi segment (Fig. 8) contains three intrasegment volcanoes (Yantarni, Chiginagak, and Kialagvik) and is bounded by transverse offsets, the northern one corresponding to the boundary between the rupture zone of the 1964 Alaskan earthquake and the 1938 Semidi earthquake rupture zone (KIENLE *et al.*, 1980). Volcanic centers found along the segment offsets include Aniakchak, Peulik and Ukinrek. Aniakchak is a large caldera complex characterized by dacite, but composition of the lavas ranges from basalt to rhyolite (SMITH, 1925; MILLER and SMITH, 1977). Petrographic data for the Peulik/Ugashik/Ukinrek volcanic system is, to the authors knowledge, unavailable and the basalt to rhyolite designation on Fig. 8 is based on a brief reconnaissance visit to the area by Kienle and information supplied by Thomas P. Miller (U.S. Geological Survey). Ukinrek is a recent (1977) maar-forming eruption of alkali basalt (KIENLE *et al.*, 1980; SELF *et al.*, 1980). Intrasegment volcanoes are poorly known. MILLER and SMITH (1976) describe Kialagvik as a dacite dome with associated ash flows. To the writers knowledge, nothing is known about the petrology of Yantarni or Chiginagak.

5. Volcanoes of the Katmai and Cook Inlet Segments

Most of the volcanoes in the eastern Aleutian arc (Fig. 8) are composite cones, typical of subduction-related volcanoes. Two of the volcanoes (Katmai and Kaguyak) have well-developed summit calderas filled with lakes. Augustine is composed of pyroclastic debris and domes or fragments of domes (typical eruptions at Augustine start by explosive removal of plug domes emplaced at the end of the previous eruption (KIENLE and SWANSON, 1980). Kaguyak (SWANSON *et al.*, 1981) also has a high proportion of pyroclastic material and domes. However, most of volcanoes in the eastern Aleutian arc are composed of lava flows and fragmental deposits (minor pyroclastics, major debris flows), that together form the characteristic composite cones.

5.1 Age and physiography

Glaciers are currently found on all of the volcanoes in the easternmost Aleutian arc with the exception of Kaguyak and Augustine. This is to be expected at this latitude given the elevation of the volcanoes (average 1,800 m in the Katmai area, 2,400 m in Cook Inlet) and the high rate of precipitation. Kaguyak apparently lacks sufficient accumulation areas at elevations required for glacier development while the active eruptive history of Augustine would preclude any ice accumulation. Glacial erosion has produced extensive dissection of Kejulik, Denison, Devils Desk, and Fourpeaked Volcanoes and this together with a lack of any known surface hydrothermal activity suggests these volcanoes have not been active since the last period of glacial activity (which ended about 10,000 y.b.p. in this region (PÉWÉ, 1975)). Glacial activity has also extensively modified Martin, Mageik, Trident, Katmai, Snowy, Kukak, Douglas, Iliamna and Redoubt, but surface hydrothermal activity or recorded historic eruptions (Mageik, Trident, Katmai, and Redoubt) indicate that all of these volcanoes are still active. Volcanoes that have a symmetric, undissected profile apparently related to a

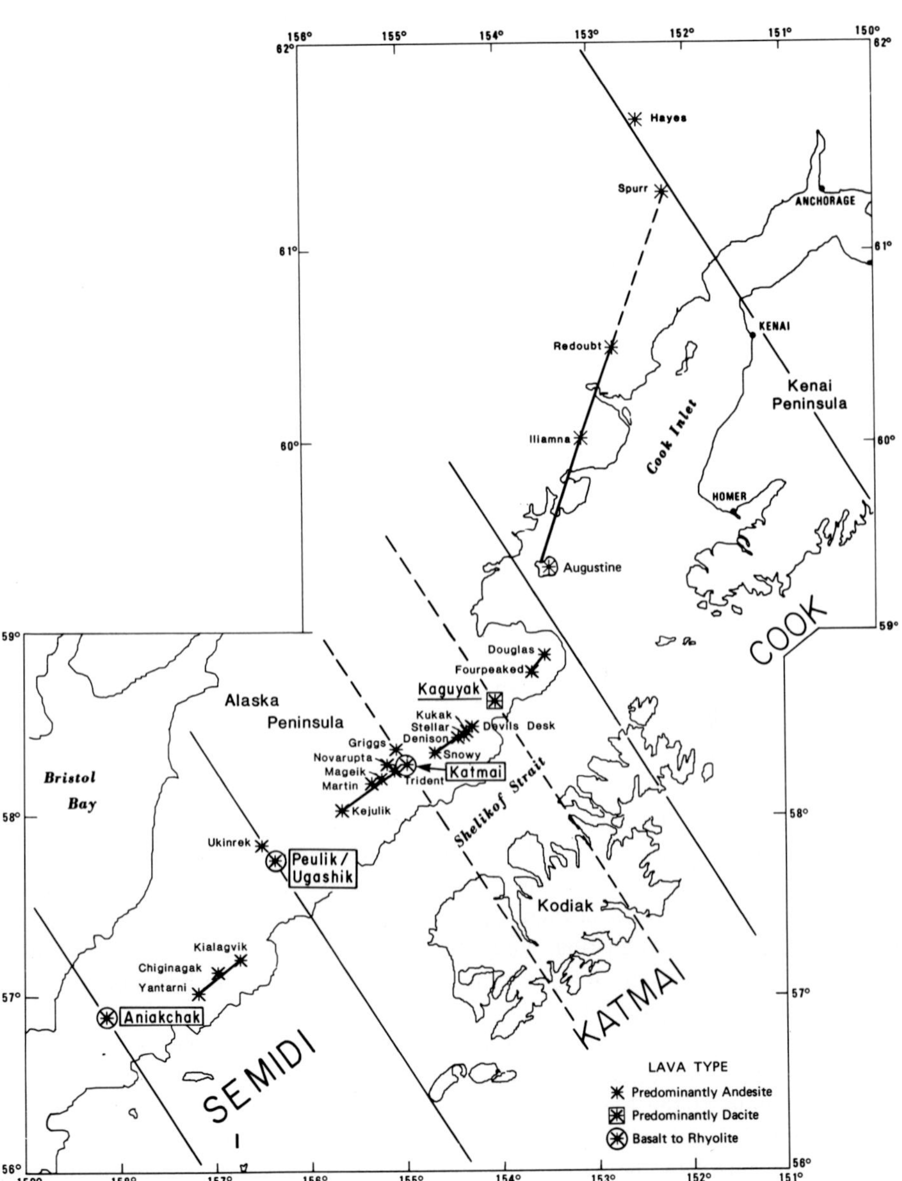

Fig. 8. Volcanoes of the eastern Aleutian arc grouped into three volcanic segments, Cook, Katmai, and Semidi. The volcanoes are classified according to their predominant lava types.

recent high rate of extrusion include Griggs, Kaguyak and Augustine.

Volcanoes in the eastern Aleutian Arc are developed on previously existing topographic highs consisting of Mesozoic sedimentary and plutonic rocks. KELLER and REISER (1959) recognized that volcanoes in the Katmai area were developed on the crest of a gentle anticline composed of sandstone and shale of the Jurassic Naknek and Cretaceous Kaguyak Formations. Augustine Volcano, located on an island in lower Cook Inlet, has been developed on an uplifted mass of Mesozoic sedimentary rocks that have been correlated with the Naknek and Kaguyak Formation (DETTERMAN and JONES, 1974). Despite the insular location of Augustine Volcano, only the early volcanics show evidence of deposition in a sub-aqueous environment (JOHNSTON, 1979). Iliamna Volcano is built upon a basement composed of Naknek sediments and granitic rocks of the Peninsula Batholith (REED and LANPHERE, 1973). Redoubt Volcano is built upon the granitic rocks of the Peninsula Batholith (MAGOON *et al.*, 1976). None of the volcanoes in the eastern Aleutian arc are especially large, much of the elevation of the more impressive cones is due to the prevolcanic topographic high upon which the volcano developed. For example, Iliamna Volcano at 3,000 m is an impressive structure that dominates much of the southern Cook Inlet. Yet, the outcropping of lavas that form Iliamna Volcano are generally confined to elevations above 1,500 m (JUHLE, 1955) thus giving a misleading picture of the size of the volcano.

6.   Petrology and Geochemistry

Porphyritic andesite ($SiO_2 = 53$ to 63 wt.%) is the most common rock type in the volcanoes of the eastern Aleutian arc (Figs. 9 and 10). Plagioclase is the most common phenocryst followed by hypersthene, augite and magnetite in decreasing order of abundance. The groundmass is typically holocrystalline and composed of the same mineralogy as the phenocrysts. Olivine is found as phenocrysts in a few of the samples from Martin, Snowy, Devils Desk, Douglas, Augustine, Iliamna, and Redoubt. Hornblende, both the common green variety and brown basaltic hornblende, is a minor constituent of the phenocryst assemblage at Snowy, Devils Desk, Fourpeaked, Augustine, and Redoubt. Hornblende is often partially to completely oxidized to an opaque mineral assemblage and can be recognized based only on the shape of pseudomorphs.

Dacite ($SiO_2 = 63$ to 70 wt.%) forms a minor portion of the lavas at Martin, Mageik, Griggs, Katmai, Snowy, Kukak, and Augustine (Figs. 9 and 10). Dacite is the rock type found at the Novarupta dome (emplaced during the 1912 eruption that formed the Valley of Ten Thousand Smokes (CURTIS, 1968) and other domes (Cerberus and Falling Mtn.) in the same area (KOSCO, 1981). Dacite is also an important constituent of the 1912 deposits in the Valley of Ten Thousand Smokes (FENNER, 1950). Dacite is the dominate lava at Kaguyak Crater (SWANSON *et al.*, 1981). Mineralogically the dacites are similar to the andesites, but hypersthene is generally more abundant than augite. Phenocrysts of quartz, often embayed with reaction rims of augite, are found in the dacites of Kaguyak (SWANSON *et al.*, 1981). Hornblende is commonly present as phenocrysts in the dacites. Reaction rims of hypersthene and plagioclase surround some of the hornblende in the dacites of Kaguyak Crater. Dacitic

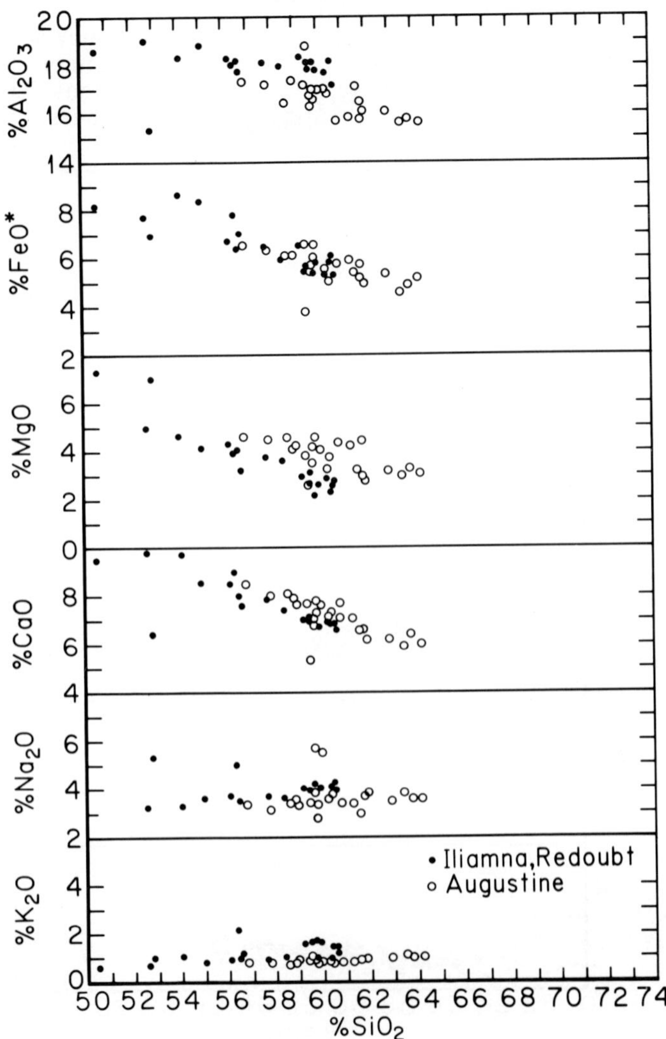

Fig. 9. Harker variation diagrams for Cook Inlet volcanoes. Data from BECKER (1898), JUHLE (1955). FORBES *et al.* (1965), DETTERMAN (1973), KIENLE and FORBES (1976), and this paper.

pumices from Augustine and Kaguyak plus the dacite of Novarupta contain a glassy groundmass.

Basalt ($SiO_2 < 53$ wt. %) has been identified only from Katmai (FENNER, 1950) and Iliamna (Fig. 10). Phenocrysts of olivine, hypersthene, plagioclase, and augite set in a holocrystalline groundmass of plagioclase, hypersthene and opaques characterize the mineralogy of the basalts.

Chemical analyses of samples from the eastern Aleutian arc are given in the

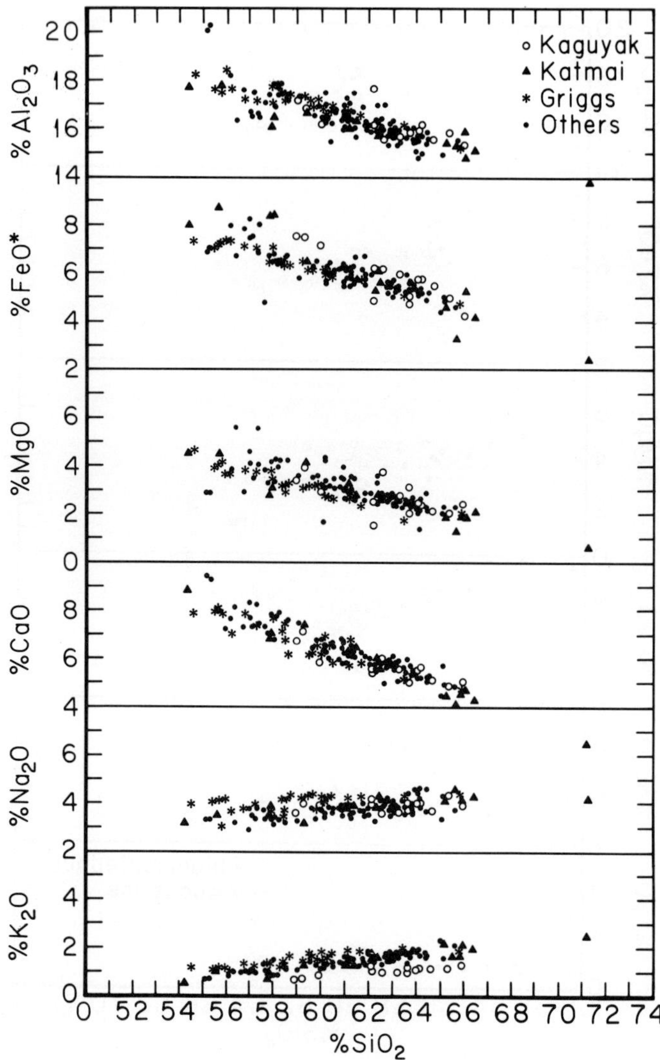

Fig. 10. Harker variation diagrams for Katmai volcanoes. Data from FENNER (1926, 1950), RAY (1967), KOSCO (1981), and this paper.

Appendix. This data together with the previously reported analyses from the area (BECKER, 1898; FENNER, 1926, 1950; JUHLE, 1955; DETTERMAN, 1973; FORBES *et al.*, 1965; RAY, 1967; KIENLE and FORBES, 1976; and KOSCO, 1981) provide a data base of over 200 analyses from the eastern Aleutian Arc (Table 1). Coverage for individual volcanoes ranges from good (Mageik, Trident, Griggs, and Augustine) to very poor (Denison, Snowy, and Martin). The compositions of the lavas show typical calc-alkaline patterns (Figs. 11 and 12). Slight increases in K_2O and Na_2O with increasing SiO_2 content are characteristic of all the volcanoes (Figs. 9 and 10). Griggs has a higher

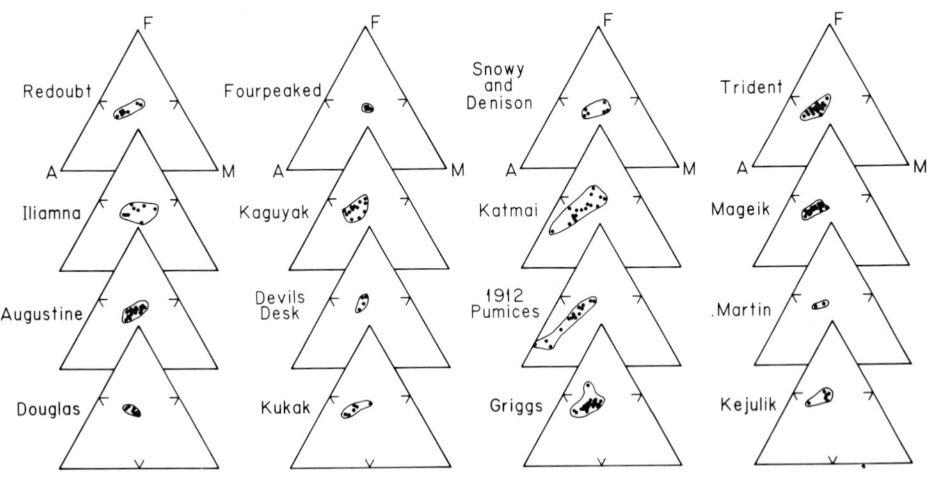

Fig. 11.   AFM diagrams for eastern Aleutian volcanoes. Data sources listed in Figs. 9 and 10.

Na$_2$O and K$_2$O content than other nearby volcanoes (Fig. 10), such as Katmai or Trident, probably due to the slightly back-arc position of Griggs. Kaguyak has a lower K$_2$O content than other nearby volcanoes. Most of the volcanoes show a decrease in Al$_2$O$_3$ with increasing SiO$_2$. Augustine is somewhat different than the other Cook Inlet volcanoes (Iliamna and Redoubt) as evidenced by the lower Al$_2$O$_3$, and high MgO and CaO (Fig. 9). Also the Augustine lavas do not show Fe-enrichment as do the Redoubt at Iliamna lavas (Fig. 12a). Katmai shows the greatest variation in SiO$_2$ (Fig. 10), from basalt (SiO$_2$ = 48 wt. %) to rhyolite (SiO$_2$ = 72 wt. %). Most of the lavas show slight FeO enrichment (Fig. 12). However, the extent of the FeO enrichment is not sufficient to classify the lavas as tholeiitic based on MIYASHIRO's (1974) classification (Fig. 12). Possible exceptions are Devils Desk (Fig. 12b) and Kejulik (Fig. 12c), both of which include lavas classified as tholeiitic and calc-alkaline.

Older volcanoes (identified on the basis of the large amount of glacial erosion) seem to show the same general patterns of chemical variation as the more recent volcanic cones. Heavily glaciated volcanic centers that show no sign of recent volcanic or hydrothermal activity, such as Fourpeaked or Denison, show the same type of compositional variation as other volcanoes (Fig. 11). However, some of these older volcanoes (Devils Desk and Kejulik) also show significantly different patterns (Figs. 12b and c) suggesting some change in the pattern of chemical variation with time.

Compositional differences between the Cook Inlet volcanoes and the Katmai volcanoes are not obvious. The Cook Inlet volcanoes tend to be lower in K$_2$O, FeO* and SiO$_2$ than the Katmai volcanoes (Figs. 9 and 10), while the CaO content is higher in the Cook Inlet.

6.1   Subduction geometry and K$_2$O variation

Variation of K$_2$O content in intermediate lavas deserves separate discussion in light of the petrotectonics implications that have been based on these patterns

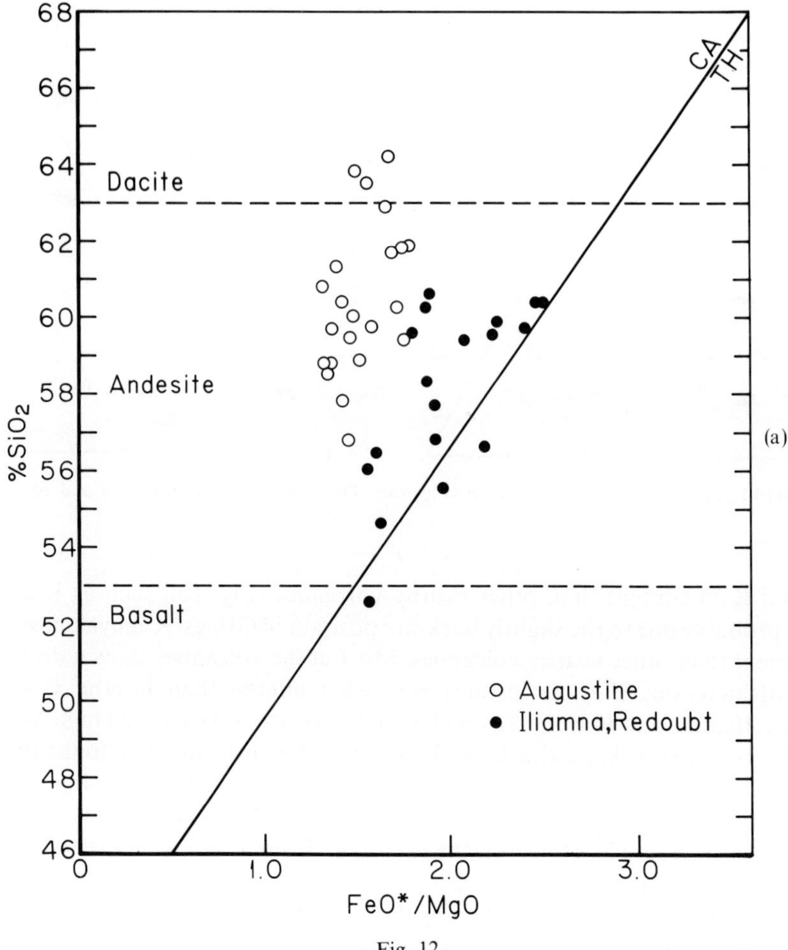

Fig. 12

(Dickinson, 1975). Variation of K_2O with SiO_2 for volcanoes in the eastern Aleutian arc (Figs. 9 and 10) forms a regular pattern related to fractionation. Since the lavas do not crystallize any phases that contain K_2O as a major component, the K_2O is concentrated in the residual liquid. Dominant phenocrysts within the lavas (hypersthene, augite, and plagioclase) contain approximately the same amount of SiO_2 and are uniformly low in K_2O, therefore fractionation of any combination of these phenocrysts will produce the observed variation. Fractionation of phenocrysts that contain more or less SiO_2 than the pyroxene-plagioclase assemblage might be expected to change the K_2O–SiO_2 relations, but the presence of quartz in the Kaguyak lavas and olivine in the Iliamna samples does not appear to significantly alter the K_2O–SiO_2 relations for these volcanoes relative to the other volcanic centers in the eastern Aleutian arc (Figs. 9 and 10).

Regional comparison of K_2O–SiO_2 variations requires standarization of the

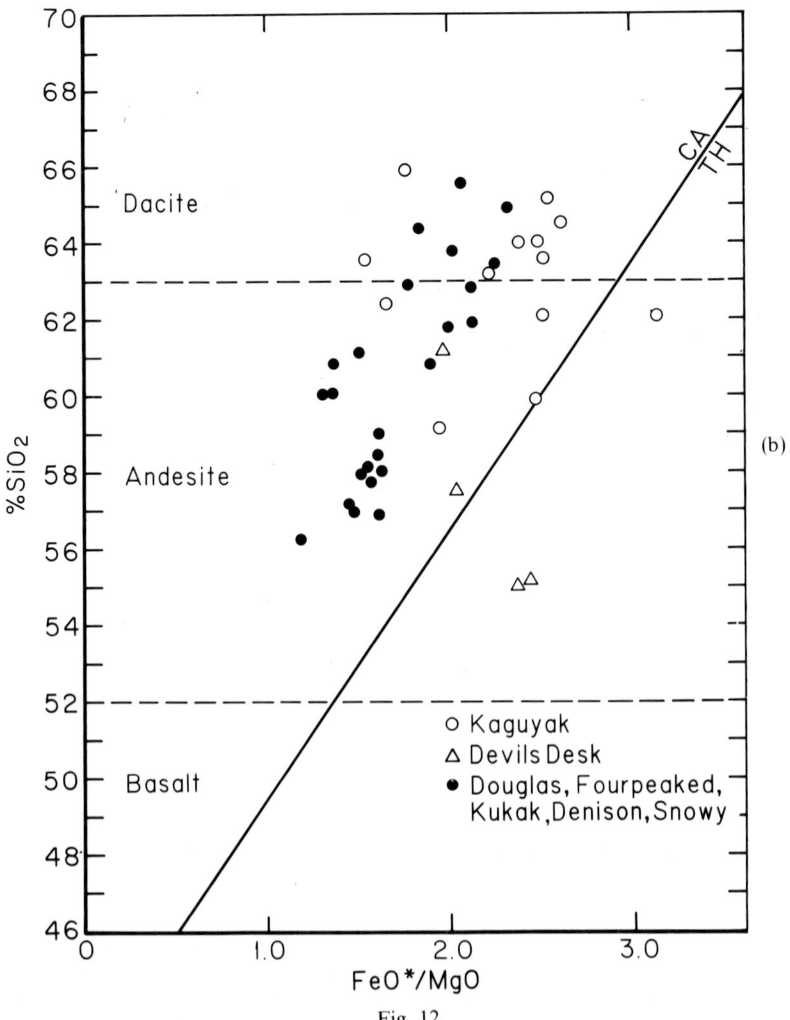

Fig. 12

fractionation pattern. One method of comparing K_2O and SiO_2 relations is to fit a straight line to the Harker variation diagram and then consider the K_2O content at some fixed SiO_2 content. This method takes advantage of the commonly observed linear variation pattern between K_2O and SiO_2 (Figs. 9 and 10). Dickinson and Hatherton (1967) utilized this approach and compared K_2O contents at 55 and 60 wt. % SiO_2 (K_{55} and K_{60}, respectively). CARR et al. (1979) followed a similar approach, but used 50 wt. SiO_2 %, in an attempt to minimize the effects of fractionation. In the present study, $K_2O–SiO_2$ variation diagrams were prepared for each of the volcanoes in the eastern Aleutian arc and then a straight line was fit to the data. A summary of these results is presented in Table 1. Most of the volcanoes in the eastern Aleutian arc have SiO_2 contents in the range of 60 wt. % (Figs. 9 and 10) and K_{60} is chosen as a basis

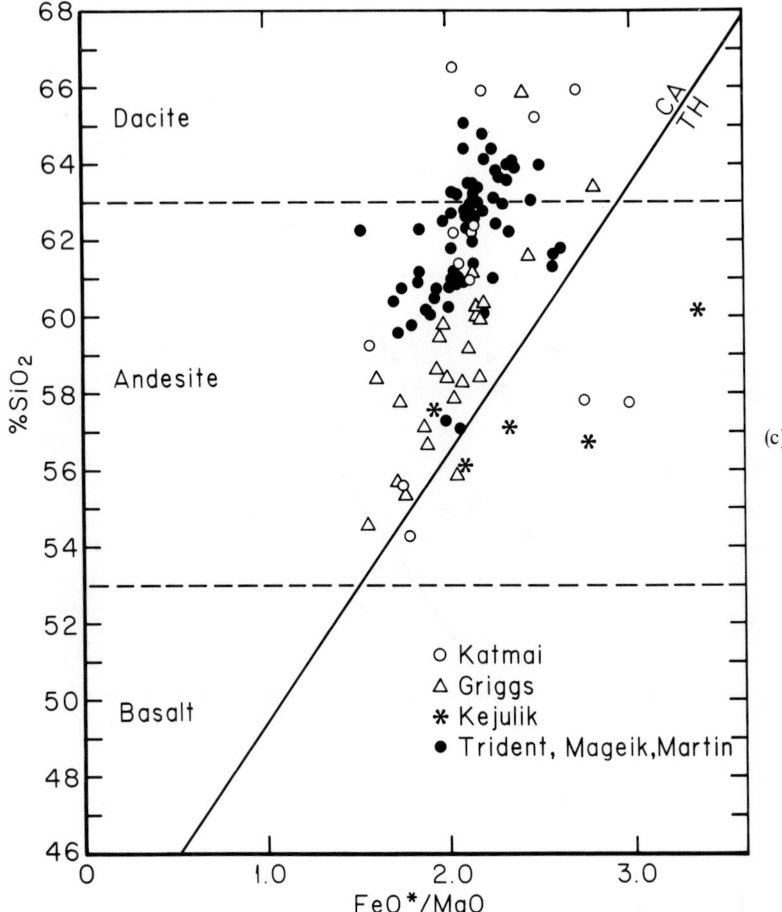

Fig. 12.   FeO*/MgO vs. SiO$_2$ variation diagram where FeO* = FeO + 0.899 Fe$_2$O$_3$. CA/TH is the calc-alkaline/tholeiitic trend for island arcs from MIYASHIRO (1974). Data sources given in Figs. 9 and 10. (a) Cook Inlet volcanoes. (b) Upper Katmai volcanoes, Douglas to Snowy. (c) Lower Katmai volcanoes, Katmai to Kejulik.

for comparison. Similar patterns of K$_2$O–SiO$_2$ variation for most of the eastern Aleutian volcanoes suggest that similar fractionation patterns must have operated in most of these volcanic centers. Thus the K$_{60}$ value should provide a measure of the *relative* difference between volcanoes in the eastern Aleutian arc.

The orientation of the Wadati-Benioff zone beneath the volcanoes of the eastern Aleutian arc is shown on Figs. 5 and 6. Depths to the top of the Wadati-Benioff zone under each of the eastern Aleutian arc volcanoes have been determined and the results are given in Table 1. For adjacent volcanoes in the Katmai area, depth differences to the Wadati-Benioff zone of only a few km are recorded (Table 1). Resolution of the seismicity does not warrent such a slight distinction, but the volcanoes in the Katmai area show a consistent linear pattern that cuts across the trend of the Benioff zone (Fig.

Table 1. Summary of Depth of Seismic Zone (H) and K_2O–SiO_2 Variation for Eastern Aleutian Volcanoes.

Volcano	Analyses	H (km)	K_{50}	K_{60}	Intercept	Slope	References
Cook Inlet							
Redoubt	9	112	0.18	1.52	−6.42	0.132	FORBES *et al.* (1965), this paper
Iliamna	11	113	0.69	1.08	−1.16	0.037	JUHLE (1955), this paper
Augustine	23	98	0.49	0.88	−1.31	0.036	BECKER (1898), DETTERMAN (1973), KIENLE and FORBES (1976), this paper
Katmai							
Douglas	7	78	0.47	1.25	−3.43	0.078	this paper
Foupeaked	5	79	0.82	0.89	+0.519	0.006	this paper
Kaguyak	14	83	0.19	0.83	−3.01	0.064	this paper
Devils Desk	4	84	0.01	1.39	−6.89	0.138	this paper
Kukak	6	85	−0.22	1.18	−7.22	0.140	this paper
Denison	2	86					this paper
Snowy	4	89	−0.51	1.29	−3.29	0.076	this paper
Katmai	18	95	0.25	1.32	−4.95	0.104	FENNER (1926, 1950), KOSCO (1981)
Griggs	28	104	0.66	1.64	−4.14	0.096	FENNER (1926), KOSCO (1981)
Trident	36	96	0.50	1.30	−3.45	0.079	RAY (1967), KOSCO (1981)
Mageik	33	99	0.04	1.31	−6.36	0.128	FENNER (1926), KOSCO (1981)
Martin	2	100					FENNER (1926), this paper
Kejulik	6	100	−1.17	2.09	−17.47	0.326	this paper

5). The slight differences in depth estimates for adjacent volcanoes in the Katmai area are consistent with this cross cutting pattern. Refinements in earthquake location procedures for the eastern Aleutian arc may change the Wadati-Benioff zone pattern shown in Fig. 5 somewhat and hence the depth estimates of Table 1. However, such changes will be systematic and will thus not change the relative relations between individual volcanoes.

Relations between the depth to Benioff zone and K_2O (at $SiO_2 = 60$ wt. %; K_{60}) are shown in Fig. 13. Volcanoes in the Katmai and Cook Inlet areas appear to form two distinct groups, each of which shows a crudely linear pattern. The pattern on Fig. 13 is such that, for any given depth to the Benioff zone, the K_2O content of the lava in the Katmai area should be greater than the lavas from the Cook Inlet volcanoes. Various authors have argued the K_2O-depth plots either show a poor linear relation or no relation at all (Carr *et al.*, 1979). However, in the eastern Aleutian arc there does seem to be a regional difference in K_2O-depth Wadati to Benioff zone relations for different segments of the volcanic arc albeit the relations are only crudely linear.

7. Discussion

The eastern Aleutian arc appears to be segmented into two blocks, based on the orientation and distribution of volcanoes. The Cook Inlet block (Fig. 8) contains the Cook Inlet volcanic segments with a trend of about N 20°E and an intervolcano spacing of 60 ± 19 km. The volcanoes on the Katmai block trend about N 55°E with

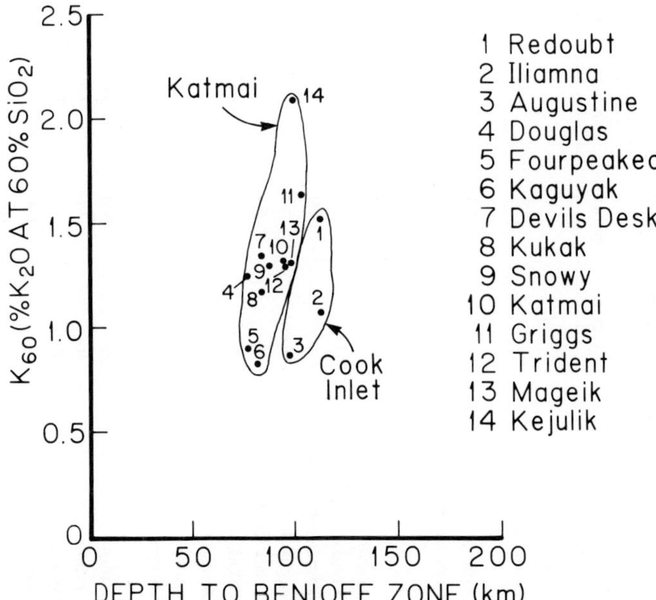

Fig. 13. K_2O contents at 60 wt. % SiO_2 for eastern Aleutian volcanoes. See text for explanation.

an intervolcano spacing of 13 ± 7 km. Within the Katmai segment slight misalignment of volcanoes suggests the presence of some subsegments, but the general trend of the Katmai volcanoes is still markedly different from the Cook Inlet segment. Volcanoes in the eastern Aleutian arc occupy three positions relative to the arc segments: 1) intrasegment volcanoes, 2) volcanoes at segment boundaries, and 3) back-arc volcanoes that lie behind the volcanic front.

Intrasegment volcanoes are large composite cones that erupt porphyritic two-pyroxene andesite of rather uniform composition and correspond to the "coherent" volcanoes of the Cascades that occupy a similar tectonic setting (McBirney, 1968; Hughes *et al.*, 1980). With few exceptions, most of the volcanoes in the eastern Aleutian arc fall into the intrasegment category. Volcanoes at segment boundaries erupt a more diverse suite of lavas and correspond to the "divergent" volcanoes identified in the Cascades (McBirney, 1968; Hughes *et al.*, 1980). Segment-boundary volcanoes are characterized by andesitic and dacitic lavas that often erupt as ash flows and form pumices with a glassy goundmass. More rarely basalt and rhyolite may be associated with these volcanic centers. Calderas and dacite domes are often found at segment boundary volcanoes. Two volcanoes, Katmai and Kaguyak within the Katmai segment appear to mark intrasegment boundaries within the Katmai segment. These volcanoes have a rather diverse suite of lavas, form calderas and appear to mark slight offsets in the volcano front within the Katmai segment. Back-arc volcanoes lie along segment or subsegment boundaries and erupt lavas that are more alkaline than lavas erupted along the volcanic front. Examples of back-arc volcanoes in the eastern Aleutian arc include Ukinrek and Griggs.

Tholeiitic differentiation trends appear to characterize segment-boundary volcanoes in the western Aleutian arc while intrasegment volcanoes show calc-alkaline fractionation patterns (Kay *et al.*, 1982). However, this pattern has not been found in our study of eastern Aleutian magmatism. Kejulik, Katmai, and Devils Desk contain some tholeiitic lavas (Figs. 12b and c) and would be classified as "transitional" by Kay *et al.* (1981). Other volcanoes (Iliamna and Griggs) have some lavas that barely fall in the tholeiitic field defined by Miyashiro (1974) on the basis of FeO*/MgO vs. SiO_2 (Figs. 12b and c) and are probably best termed calc-alkaline. Out of this group of transitional volcanoes, only Katmai lies along a segment or subsegment boundary. Thus, the characterization of segment-boundary volcanoes in the eastern Aleutian arc on the basis of tholeiitic differentiation trends does not appear valid.

The segment boundary between the Cook Inlet and Katmai blocks in the eastern Aleutian arc is not marked by any boundary-type volcanoes. Misalignment of the Cook Inlet and Katmai volcanic segments is one of the most striking features on a map of eastern Aleutian volcanoes (Fig. 8). In addition to the misalignment, volcanoes in the two segments show different relations to the underlying Wadati-Benioff zone. The Katmai segment of volcanoes strikes at an oblique angle to the 100 km Wadati-Benioff zone contour and ranges from 78 to 104 km above the inclined zone of seismicity (Fig. 6). Cook Inlet volcanoes are also aligned obliquely to the Benioff zone, but the depth to the Wadati-Benioff zone under the volcanoes is deeper and ranges from 98 to 113 km. The depth to the top surface of the subducting slab would be about 20 km shallower beneath the volcanoes in both Katmai and Cook Inlet if we assume that Stefani *et al.*'s

(1982) direct measurement of the distance from the top of the dipping seismic zone to the top of the slab is applicable to our area. Their measurement is based on a P-wave reflection from the upper surface of the slab for a 181 km deep earthquake in the seismic zone of the Kuril arc, which dips at about 45° beneath the volcanoes.

Misalignment of the Cook Inlet-Katmai volcanoes reflects a difference in orientation of the subducting slab in these areas. The subduction system appears to undergo a smooth change in trend, based on the seismicity. To date we have not found any evidence of a break in the plate (hinge fault) near the Katmai-Cook Inlet transition. One might speculate that the intense and unusual seismicity beneath Iliamna Volcano reflects stresses associated with the lateral bending of the subducting plate in that area.

The different positioning of the Cook Inlet and Katmai volcanoes over the subducting plate (Cook Inlet volcanoes having 20–30 km greater average depth to the Wadati-Benioff zone) is reflected in a plot showing K_{60} vs. depth to the seismic zone (Fig. 13). In addition to the subtle differences in K_2O content between the two segments, the lavas of the Katmai segment appear to be more SiO_2-rich as indicated by the relative abundance of dacites in the Katmai area (Fig. 12).

8.   Conclusions

Patterns of geochemical variation for the volcanoes of the eastern Aleutian arc are fairly similar. More extensively dissected (presumably older) volcanoes generally show the same calc-alkaline fractionation pattern as adjacent more symmetrical (younger) cones. Andesite is the most common rock type in both the Katmai and Cook Inlet segments, but dacite is more abundant in the Katmai volcanoes.

Volcanoes of the eastern Aleutian arc do not show the same geochemical relations to segment boundaries as volcanoes in the western Aleutian arc. The distinct break in the volcano alignment between the Cook Inlet and Katmai segments is not marked by any unusual differentiation trends. Douglas and Augustine, the two volcanoes adjacent to the transarc offset, appear typical of intrasegment volcanoes with dominant andesitic lavas and without a marked FeO* enrichment. Apparently the Cook Inlet-Katmai segment boundary is not marked by unusual petrochemical variation.

The most significant petrochemical variations have been identified at Mt. Katmai and Kaguyak crater, both within the Katmai segment. Kaguyak is characterized by an abundance of dacite, while Katmai lavas range from basalt to rhyolite in composition. Kaguyak and Katmai both occur at minor misorientations of the Katmai volcano line and these comparatively anomalous intrasegment volcanoes may mark subsegment boundaries within the Katmai segment.

The differences in volcano spacing and perhaps alignment between the Katmai and Cook Inlet segment may be due to localization of Katmai volcanoes along a crustal fracture. Most of the Katmai volcanoes are located along the crest of a gently-dipping anticline in the Jurassic sedimentary basement rocks (KELLER and REISER, 1959). Volcanoes in the Cook Inlet segment do not show any particular relation to crustal structure (MAGOON *et al*, 1976). Intravolcano spacing in the Katmai segment is much shorter than spacing in the Cook Inlet segment (13 ± 7 km vs. 60 ± 19 km). The 60 ± 19 km spacing in Cook Inlet is approximately the spacing found for other portions of the Aleutian arc and may be related to convection and diapiric rise of magma from the source region (MARSH, 1979). Closer spacing of the Katmai volcanoes may be related to magma leakage along crustal fractures. HILDRETH (1981) has suggested a caldera was formed near Novarupta simultaneously with the collapse of the Mt. Katmai summit during the 1912 eruption that formed the Valley of Ten Thousand Smokes. A seismically-identified magma body has been located at shallow levels in the Novarupta-Griggs area (MATUMOTO, 1971). The close spacing of volcanoes in this part of the Katmai segment may be related to this large magma body. Interconnection of some of the volcanoes with the same magma body is suggested by the events during the 1912 eruption (FENNER, 1920; HILDRETH, 1981).

The lack of volcanism in the Aleutian subduction system east of Hayes volcano may be related to the large accretionary wedge in the forearc region. Subduction-related seismicity in the eastern Aleutian arc can be traced as far as Mt. McKinley, but the volcanism apparently stops at Hayes. The accretionary prism in the Hayes area as measured by the volcanic arc-trench spacing along the slip direction is 570 km wide and increases further to the northeast. Various authors have linked the generation of andesitic magmas in subduction systems to sediment subduction (KAY *et al.*, 1978; MCCULLOCH and PERFIT, 1981), however the role of the subducted sediment in the petrogenesis of arc magmas is unclear. Sediments may be involved directly in the melting process (KAY, 1980) or they may contribute H_2O (via dehydration that accompanies subduction-related metamorphism) that aids partial melting in some other part of the subduction system (COATS, 1962). In any case, most sediment in the eastern Aleutian arc is apparently being accreted rather than subducted and this lack of sediment subduction may explain the cessation of volcanism east of Hayes.

We appreciate the critical review of the seismologic part of the manuscript by John Davies. We also thank Dianne Marshall who processed much of the seismic data and William Witte who calculated plate motion vectors for Fig. 2. Daniel Kosco made his analytical data on Griggs, Katmai, Trident, and Mageik available to the authors and helped to complete our data base for the Katmai volcanoes. This study was supported by the U.S. Bureau of Land Management (BLM) through an interagency agreement with the U.S. National Oceanic and Atmospheric Administration (Contract NOAA-03-4-022-55) under which a multi-year program responding to the need of petroleum development of the Alaska Continental Shelf is managed by the Outer Continental Shelf Environmental Assessment Program (OCSEAP) office. The operation of a part of the seismic network is supported by the office of Basic Energy Science of the U.S. Department of Energy (Contract EY-76-S-06-2229, Task No. 6).

APPENDIX

Bulk Compositions of Lavas from the Eastern Aleutian Arc.

	REDOUBT							ILIAMNA								
Sample Number	RD-2-1	RD-3-2	RD-1-3	RD-1-4	1	2	3	Sample Number I-1B	I-1C	I-1D	I-2A	I-2B	I-3B	I-3G	I-3H	I-3J
SiO_2	59.56	55.56	56.85	54.64	60.46	60.40	59.46	52.62 50.53	56.66	59.62	60.58	52.75	56.06	60.32	57.73
TiO_2	0.59	0.62	0.65	0.75	0.56	0.54	0.74	0.73 0.67	0.75	0.56	0.56	0.64	0.62	0.54	0.61
Al_2O_3	18.25	18.90	18.04	18.44	18.16	18.20	18.10	19.02 18.62	17.68	17.82	17.19	15.34	18.31	17.69	18.07
Fe_2O_3	2.78	3.09	3.19	4.80			2.75	3.28 3.35	3.34	4.33	3.70	3.87	2.79	2.29	3.19
FeO	3.32	4.64	4.08	3.40	6.14*	5.18*	2.91	4.76 5.20	4.02	1.76	1.99	3.36	4.23	3.38	3.56
MnO	0.14	0.16	0.15	0.16	0.15	0.15	0.14	0.14 0.14	0.13	0.12	0.12	0.08	0.14	0.12	0.13
MgO	2.62	3.78	3.62	4.72	2.51	2.34	2.60	4.95 7.25	3.22	3.14	2.81	6.99	4.31	2.91	3.75
CaO	6.90	8.35	8.79	9.61	6.80	6.82	7.14	9.77 9.48	7.60	7.01	6.60	6.43	8.47	6.87	7.90
Na_2O	3.96	3.49	4.86	3.24	4.19	4.08	3.94	3.18 2.87	3.42	4.15	3.90	5.31	3.66	4.07	3.73
K_2O	1.47	0.71	1.05	0.98	1.40	1.41	1.58	0.72 0.62	1.20	0.91	1.23	0.99	0.85	1.02	0.89
P_2O_5	0.22	0.15	0.15	0.14	0.23	0.22	0.20	0.12 0.10	0.18	0.15	0.15	0.13	0.14	0.13	0.13
H_2O^+								0.48 0.51	0.22	0.07	0.49	1.78	0.07	0.47	0.06
H_2O^-								0.19 0.41	0.31	0.03	0.05	0.79	0.08	0.06	0.12
Total	99.81	99.45	101.43	100.88	100.60	99.97	99.56	99.96 99.75	98.73	99.67	99.37	98.46	99.73	99.87	99.87

*Total Fe as FeO

	AUGUSTINE									ILIAMNA						
Sample Number	AU-79-8B	AU-79-8C	AU-79-8D	AU-79-8F	AU-79-10A	AU-79-13	74-AU-JK1	74-AU-JK4	74-AU JK6	Sample Number K-4B	K-4C	K-4D	K-4E	K-10A	K-10C	K-3I
SiO_2	60.31	61.78	61.88	59.70	59.74	59.97	59.43	56.78	58.91	60.85	61.18	60.10	60.10	60.89	61.78	58.98
TiO_2	0.59	0.54	0.53	0.58	0.57	0.59	0.62	0.67	0.64	0.60	0.60	0.61	0.61	0.61	0.59	0.59

Al$_2$O$_3$	17.08	16.60	16.24	16.45	16.87	17.09	17.30	17.40	17.47	Al$_2$O$_3$	16.39	16.39	16.38	16.30	16.95	16.45	17.31
Fe$_2$O$_3$	2.91	2.09	1.69	2.03	3.73	2.28	2.97	2.47	3.11	Fe$_2$O$_3$	1.69	1.74	1.92	1.85	2.44	2.66	1.90
FeO	2.98	3.35	3.50	3.80	2.44	3.95	3.26	4.40	3.36	FeO	3.99	3.89	3.99	4.05	3.43	3.07	4.13
MnO	0.13	0.12	0.12	0.12	0.13	0.13	0.13	0.13	0.13	MnO	0.10	0.10	0.10	0.10	0.10	0.10	0.11
MgO	3.27	2.99	2.83	3.54	4.22	4.06	3.75	4.57	4.09	MgO	4.04	3.64	4.31	4.38	2.98	2.76	3.62
CaO	7.18	6.57	6.20	6.82	7.14	7.62	7.70	8.48	7.92	CaO	6.74	6.55	6.98	7.01	6.60	6.09	7.52
Na$_2$O	3.62	3.66	3.76	3.86	5.67	5.53	3.47	3.42	3.53	Na$_2$O	3.46	3.42	3.36	3.40	3.48	3.49	3.32
K$_2$O	0.92	0.95	1.02	0.92	0.80	0.83	0.86	0.75	0.83	K$_2$O	1.29	1.33	1.31	1.25	1.20	1.41	1.41
P$_2$O$_5$	0.13	0.13	0.14	0.14	0.12	0.14	0.16	0.15	0.14	P$_2$O$_5$	0.12	0.11	0.11	0.10	0.10	0.09	0.09
										H$_2$O$^+$	0.32	0.15	0.16	0.08	0.28	0.24	0.70
										H$_2$O$^-$	0.12	0.06	0.17	0.19	0.41	0.25	0.14
Total	99.12	98.78	97.91	97.96	101.43	102.19	99.65	99.22	100.13	Total	99.71	99.16	99.50	99.42	99.47	8.98	99.55

	DEVILS DESK				KUKAK						FOURPEAKED				
Sample Number	K-7D	K-8A	K-8B	K-9A	K-1B	K-1D	K-1H	K-1J	K-1L	K-1Q	K-3B	K-3C	K-3F	K-3H	K-3I
SiO$_2$	57.66	55.05	55.19	61.19	58.51	63.53	61.96	64.95	63.83	65.61	58.11	57.85	58.24	58.09	56.98
TiO$_2$	0.67	0.71	0.71	0.60	0.64	0.68	0.66	0.54	0.60	0.55	0.60	0.61	0.61	0.61	0.60
Al$_2$O$_3$	17.36	20.18	20.27	16.17	17.18	15.99	16.08	14.93	15.51	15.47	17.43	17.79	17.87	17.74	17.85
Fe$_2$O$_3$	3.25	2.57	3.16	2.28	3.15	2.62	2.73	2.47	2.65	2.09	2.52	2.79	2.61	2.54	2.66
FeO	3.82	4.59	4.13	3.88	3.86	3.53	3.43	2.21	3.10	2.75	4.19	3.96	4.15	4.28	4.18
MnO	0.13	0.12	0.12	0.11	0.12	0.10	0.11	0.10	0.10	0.09	0.13	0.12	0.13	0.13	0.12
MgO	3.33	2.94	2.93	3.04	4.18	2.64	2.79	1.93	2.74	2.26	4.00	4.15	4.23	4.29	4.50
CaO	7.01	9.40	9.31	6.18	7.60	5.33	5.71	4.36	5.30	4.83	7.75	7.75	7.88	7.85	7.79
Na$_2$O	3.42	3.25	3.30	3.44	3.18	3.49	3.50	3.29	3.53	3.74	3.36	3.37	3.41	3.36	3.29
K$_2$O	1.05	0.68	0.71	1.53	1.09	1.82	1.75	2.28	1.89	1.97	0.91	0.87	0.86	0.87	0.87
P$_2$O$_5$	0.12	0.10	0.09	0.10	0.22	0.20	0.22	0.18	0.14	0.13	0.14	0.13	0.13	0.13	0.11
H$_2$O$^+$	1.24	0.27	0.11	0.19	0.49	0.52	0.32	2.04	0.21	0.13	0.15	0.17	0.11	0.20	0.27
H^{2}O$^-$	0.57	0.15	0.13	0.20	0.83	0.49	0.66	0.44	0.30	0.31	0.08	0.21	0.14	0.14	0.46
Total	99.63	100.01	100.16	98.91	101.05	100.94	99.92	99.72	99.90	99.93	99.37	99.77	100.37	100.14	99.68

Appendix (contn'd)

J. KIENLE *et al.*

KAGUYAK

Sample Number	KC-1B	KC-2A	KC-3B	KC-6	KC-7	KC-8A	KC-9B	KC-10B	KC-14	KC-16	KC-18B	KC-19A	KC-20A	KC-20C
SiO_2	59.22	63.19	65.90	64.61	64.11	59.94	62.07	63.97	58.86	62.45	65.29	62.05	63.55	63.59
TiO_2	0.69	0.60	0.42	0.54	0.60	0.79	0.69	0.60	0.76	0.60	0.54	0.66	0.43	0.54
Al_2O_3	16.77	15.62	15.30	15.50	16.05	16.09	16.14	15.85	17.11	15.47	15.79	17.64	15.80	15.80
Fe_2O_3	3.00	2.72	1.89	2.34	2.98	2.88	2.75	2.69	3.05	2.50	3.52	4.80	1.68	2.16
FeO	4.73	3.50	2.52	3.34	2.98	4.48	3.76	3.27	4.80	3.87	1.78	0.46	3.19	3.04
MnO	0.14	0.13	0.09	0.12	0.11	0.14	0.14	0.12	0.15	0.13	0.12	0.10	0.11	0.11
MgO	3.87	2.70	2.40	2.10	2.31	2.89	2.50	2.40	3.43	3.73	1.97	1.54	3.07	2.06
CaO	7.11	5.48	5.03	5.07	5.64	5.83	5.51	5.50	6.73	6.03	4.83	5.42	5.16	5.02
Na_2O	3.98	3.61	3.90	3.70	3.96	3.91	4.08	3.99	3.61	3.59	4.27	3.97	3.69	3.95
K_2O	0.68	0.99	1.18	1.07	1.08	0.87	0.95	1.10	0.69	0.97	1.14	1.24	0.98	1.09
P_2O_5	0.11	0.11	0.09	0.11	0.11	0.13	0.14	0.12	0.13	0.11	0.13	0.17	0.09	0.12
H_2O^+	0.06	0.11	1.25	0.74	0.17	0.93	1.07	0.26	0.25	0.47	0.15	1.31	1.26	1.73
H_2O^-	0.10	0.18	0.31	0.17	0.10	0.13	0.18	0.13	0.26	0.13	0.09	0.80	0.99	0.44
Total	100.46	98.94	100.28	99.41	100.20	99.01	99.98	100.00	99.83	100.05	99.62	100.17	100.00	99.59

	DENISON		SNOWY				MARTIN	KEJULIK					
Sample Number	K-2B	K-2C	K-5A	K-5B	K-5C	K-6B	K-13A	K-11B	K-11H	K-12B	K-12C	K-12F	K-12H
SiO_2	62.86	56.90	62.89	56.28	64.42	57.22	61.29	56.14	57.55	60.12	57.98	56.67	57.11
TiO_2	0.62	0.65	0.53	0.66	0.54	0.66	0.67	0.72	0.65	0.72	0.72	0.86	0.57
Al_2O_3	15.62	16.70	16.33	16.34	15.59	16.56	16.03	18.15	17.05	17.50	16.85	17.63	17.45
Fe_2O_3	1.91	3.30	2.18	2.25	2.25	2.13	2.22	4.19	2.15	1.86	3.25	4.35	3.00
FeO	3.95	4.40	3.07	4.70	3.20	4.83	4.36	4.23	5.40	3.92	3.87	3.99	4.79
MnO	0.11	0.14	0.11	0.12	0.11	0.13	0.12	0.15	0.14	0.15	0.12	0.16	0.15

MgO	2.70	4.60	2.86	5.63	2.89	5.54	3.48	3.82	3.80	1.68	3.37	2.88	3.52
CaO	5.79	8.25	5.84	8.13	5.23	8.17	6.33	7.65	7.32	6.16	6.79	7.25	7.29
Na_2O	3.45	2.87	3.73	3.17	3.72	3.28	3.56	3.64	3.66	3.63	3.13	3.64	3.46
K_2O	1.83	1.01	1.52	1.00	1.57	1.02	1.46	0.88	1.25	2.15	1.49	1.03	1.11
P_2O_5	0.12	0.13	0.11	0.13	0.09	0.11	0.11	0.16	0.15	0.20	0.14	0.17	0.14
H_2O^+	0.36	0.43	0.15	0.71	0.30	0.33	0.15	0.37	0.32	1.08	1.25	1.16	0.53
H_2O^-	0.12	0.16	0.11	0.20	0.31	0.18	0.11	0.32	0.12	0.30	0.49	0.28	0.33
Total	99.44	99.54	99.43	99.32	100.22	100.16	99.89	100.42	99.56	99.47	99.45	100.07	99.63

REFERENCES

AGNEW, J. D., Seismicity of the central Alaska Range, 1904–1978, M. S. Thesis, University of Alaska, Fairbanks, 88 pp., 1980.

BECKER, G. F., Reconnaissance of the goldfields of southern Alaska, with some notes on the general geology, *U.S. Geol. Surv., 18th Ann. Rept.*, pt. 3, 28–31, 50–58, 1898.

CARR, M. J., W. I. ROSE, and D. G. MAYFIELD, Potassium content of lavas and depth to the seismic zone in Central America, *J. Volc. Geotherm. Res.*, **5**, 387–401, 1979.

COATS, R., Magma type and crustal structure in the Aleutian arc, in *The Crust of the Pacific Basin*, edited by G. Macdonald and H. Kuno, Am. Geophys. Union, Mon. 6, pp. 92–109, 1962.

CURTIS, G. H., The stratigraphy of the ejecta of the 1912 eruption of Mount Katmai and Novarupta, Alaska, *Geol. Soc. Am. Mem.*, **116**, 153–210, 1968.

DAVIES, J. N., Seismological investigations of plate tectonics in southcentral Alaska, Ph. D. Thesis, University of Alaska, Fairbanks, 193 pp., 1975.

DAVIES, J. N. and L. HOUSE, Aleutian subduction zone seismicity, volcano-trench separation, and their relation to great thrust-type earthquakes, *J. Geophys. Res.*, **84**, 4583–4591, 1979.

DAVIES, J. N., L. SYKES, L. HOUSE, and K. JACOB, Shumagin seismic gap, Alaska Peninsula: History of great earthquakes, tectonic setting, and evidence for high seismic potential, *J. Geophys. Res.*, **86**, 3821–2855, 1981.

DETTERMAN, R. L., Geologic map of the Iliamna B-2 quadrangle, Augustine Island, Alaska, *U.S. Geol. Surv. Map*, GQ-1068, 1973.

DETTERMAN, R. L. and D. L. JONES, Mesozoic fossils from Augustine Island, Cook Inlet, Alaska, *Am. Assoc. Pet. Geol. Bull.*, **58**, 868–870, 1974.

DICKINSON, W. R., Potash-depth (K-h) relations in continental margin and intra-oceanic magmatic arcs, *Geology*, **3**, 53–56, 1975.

DICKINSON, W. R. and T. HATHERTON, Andesitic volcanism and seismicity around the Pacific, *Science*, **157**, 801–803, 1967.

ENGDAHL, E. R., Relocation of intermediate depth earthquakes in the central Aleutians by seismic ray tracing, *Nature, Phys. Sci.*, **245**, 23–25, 1973.

ENGDAHL, E. R., Seismicity and plate subduction in the central Aleutians, in *Island Arcs, Deep Sea Trenches, and Back-Arc Basins*, Maurice Ewing Series, vol. 1, edited by M. Talwani and W. C. Pitman III, pp. 259–272, Am. Geophys. Union, Washington, D. C., 1977.

ENGDAHL, E. R. and C. H. SCHOLZ, A double Benioff zone beneath the central Aleutians: An unbending of the lithosphere, *Geophys. Res. Lett.*, **4** (10), 473–476, 1977.

ESTES, S. A., Seismotectonic studies of lower Cook Inlet, Kodiak Island and the Alaska Peninsula areas of Alaska, M.S. Thesis, University of Alaska, Fairbanks, 142 pp., 1978.

FENNER, C. N., The Katmai region, Alaska, and the great eruption of 1912, *J. Geol.*, **28**, 569–606, 1920.

FENNER, C. N., The Katmai magmatic province, *J. Geol.*, **35**, 673–772, 1926.

FENNER, C. N., The chemical kinetics of the Katmai eruption, *Am. J. Sci.*, **248**, 593–627, 1950.

FORBES, R. B., D. K. RAY, T. KATSURA, H. MATSUMOTO, and M. S. FURST, The comparative chemical composition of continental vs. island arc andesites in Alaska, *Proc. And. Conf., Ore. Dept. Geol. and Minl. Ind. Bull.*, **65**, 111–120, 1965.

HAMPTON, M. A., A. H. BOUMA, H. PULPAN, and R. VONHUENE, Geoenvironmental assessment of the Kodiak shelf, in *Proc. 11th Offshore Tech. Conf., April 30-May 3, 1979, Houston, Texas*, pp. 365–376, 1979.

HASEGAWA, H. S., Surface and body-wave spectra of Cannikin and shallow Aleutian earthquakes, *Bull. Seismol. Soc. Am.*, **63**, 1201–1225, 1973.

HASEGAWA, A., N. UMINO, and A. TAKAGI, Double-planed deep seismic zone and upper-mantle structure in the northeastern Japan arc, *Geophys. J. Roy. Astron. Soc.*, **54**, 281–296, 1978.

HERRIN, E., Seismological tables for P, *Bull. Seismol. Soc. Am.*, **58**, 1196–1219, 1968.

HILDRETH, W., The 1912 eruption in the Valley of Ten Thousand Smokes, Katmai National Monument, Alaska, *Int. Assoc. Vol. Chem. Ear. Int. Sym. Arc Vol.*, 126–127, 1981.

HUGHES, J. M., R. E. STOIBER, and M. J. CARR, Segmentation of the Cascade volcanic chain, *Geology*, **8**, 15–17, 1980.

JACOB, K. H., K. NAKAMURA, and J. N. DAVIES, Trench-volcano gap along the Alaskan-Aleutian arc: Facts, and speculations on the role of terrigeneous sediments, in *Island Arcs, Deep Sea Trenches, and Back-Arc Basins*, Maurice Ewing Series, vol. 1, edited by M. Talwani and W. C. Pitman III, pp. 243–258, Am. Geophys. Union, Washington, D.C., 1977.

JOHNSTON, D. A., Volatiles, Magma mixing, and the mechanism of eruption of Augustine Volcano, Ph.D. Thesis, University of Washington, Seattle, 177 pp., 1978.

JOHNSTON, D. A., Onset of volcanism at Augustine Volcano, Lower Cook Inlet, *U.S. Geol. Surv. Circ.*, **804B**, B78–B80, 1979.

JUHLE, W., Iliamna Volcano and its basement, in *U.S. Geol. Surv. Open File Report*, 74 p., 1955.

KANAMORI, H., The energy release in great earthquakes, *J. Geophys. Res.*, **82**, 2981–2988, 1977.

KAY, R. W., Volcanic arc magmas: implications of a melting-mixing model for element recycling in the crust upper-mantle system, *J. Geol.*, **88**, 497–522, 1980.

KAY, R. W., S. S. SUN, and C. N. LEE-HU, Pb and Sr isotopes in volcanic rocks from the Aleutian Islands and Pribilof Islands, Alaska, *Geochim. Cosmochim. Acta*, **42**, 263–273, 1978.

KAY, S. M., R. W. KAY, and G. P. CITRON, Tectonic controls on tholeiitic and calc-alkaline magmatism in the Aleutian arc, *J. Geophys. Res*, **87**, 4051–4072, 1982.

KELLER, A. S. and H. N. REISER, Geology of the Mount Katmai area, Alaska, *U.S. Geol. Surv. Bull.* **1058-G**, 261–298, 1959.

KIENLE, J. and R. B. FORBES, Augustine—evolution of a volcano, *Geophys. Inst. Ann. Rept.* 1975–1976, 26–48, University of Alaska, Fairbanks, 1976.

KIENLE, J. and G. E. SHAW, Plume dynamics, thermal energy and long-distance transport of vulcanian eruption clouds from Augustine Volcano, Alaska, *J. Volc. Geotherm. Res.*, **6**, 139–164, 1979.

KIENLE, J. and S. E. SWANSON, Volcanic hazards from future eruptions of Augustine Volcano, Alaska, in *Geophys. Inst. Rept. UAG R-275*, 122 pp., Appendix and Map, University of Alaska, Fairbanks, 1980.

KIENLE, J., P. R. KYLE, S. SELF, R. J. MOTYKA, and V. LORENZ, Ukinrek Maars, Alaska, I. April 1977 eruption sequence, petrology and tectonic setting, *J. Volc. Geotherm. Res.*, **7**, 11–37, 1980.

KOSCO, D. G., Part I: the Edgecumbe volcanic field, Alaska: an example of tholeiitic and calc-alkaline volcanism; Part II: characteristics of andesitic to dacitic volcanism at Katmai National Park, Alaska, Ph.D. Thesis, University of California, Berkeley, 249 pp., 1981.

LAFORGE, R. and E. R. ENGDAHL, Tectonic implications of seismicity in the Adak Canyon region, central Aleutians, *Bull. Seismol. Soc. Am.*, **69**, 1515–1532, 1979.

LAHR, J. C., R. A. PAGE, and J. A. THOMAS, Catalog of earthquakes in south central Alaska, April-June 1972, in *Open File Report*, U.S. Geol. Surv., Menlo Park, California, 1974.

LAHR, J. C., Detailed seismic investigation of Pacific-North American plate interaction in southern Alaska, Ph.D. Thesis, Columbia University, New York, 140 pp., 1975.

LAHR, J. C. and G. PLAFKER, Holocene Pacific-North American plate interaction in southern Alaska: Implications for the Yagataga seismic gap, *Geology*, **8**, 483–486, 1980.

LAWTON, J., C. FROHLICH, H. PULPAN, and G. LATHAM, Earthquake activity at the Kodiak continental shelf, Alaska, determined by land and ocean bottom seismometer measurements, *Bull. Seismol. Soc. Am.*, **72**, 207–220, 1982.

MAGOON, L. B., W. L. ADKISON, and R. M. EGBERT, Map showing geology, wildcat wells, Tertiary plant fossil localities, K-Ar age dates, and petroleum operations, Cook Inlet area, Alaska, *U.S. Geol. Surv. Map*, I-1019, 1976.

MARSH, B. D., Island-arc volcanism, *Am. Scientist*, **67**, 161–172, 1979.

MATUMOTO, T., Seismic body waves observed in the vicinity of Mt. Katmai, Alaska, and evidence for the existence of molten chambers, *Geol. Soc. Am. Bull.*, **82**, 2905–2920, 1971.

MCBIRNEY, A. R., Petrochemistry of Cascade andesite volcanoes, in *Andesite Conf. Gdbk., Ore. Dept. Geol. and Minl. Ind. Bull.*, **62**, 101–107, 1968.

MCCULLOCH, M. T. and M. R. PERFIT, $^{143}Nd/^{144}Nd$, $^{87}Sr/^{86}Sr$ and trace element constraints on the petrogenesis of Aleutian island arc magmas, *Earth Planet. Sci. Lett.* **56**, 167–179, 1981.

224   J. KIENLE *et al.*

MCKENZIE, D. and R. L. PARKER, The North Pacific: An example of tectonics on a sphere, *Nature*, **216**, 1276–1280, 1967.

MILLER, T. P. and R. L. SMITH, "New" volcanoes in the Aleutian volcanic arc, *U.S. Geol. Surv. Circ.*, **733**, 11, 1976.

MILLER, T. P. and R. L. SMITH, Spectacular mobility of ash flows around Aniakchak and Fisher calderas, Alaska, *Geology*, **5**, 173–176, 1977.

MINSTER, J. B. and T. H. JORDAN, Present-day plate motions, *J. Geophys. Res.*, **83**, 5331–5354, 1978.

MIYASHIRO, A., Volcanic rock series in island arcs and active continental margins, *Am. J. Sci.*, **274**, 321–355, 1974.

PEREZ, O. J. and K. H. JACOB, Tectonic model and seismic potential of the eastern Gulf of Alaska and Yagataga seismic gap, *J. Geophys. Res.*, **85**, 7132–7150, 1980.

PÉWÉ, T. L., Quaternary geology of Alaska, *U.S. Geol. Surv. Prof. Paper*, **835**, 145 pp., 1975.

PULPAN, H. and J. KIENLE, Western Gulf of Alaska seismic risk, in *Proc. 11th Offshore Tech. Conf., April 30-May 3, 1979, Houston, Texas*, pp. 2209–2218, 1979.

RAY, D. K., Geochemistry and petrology of the Mt. Trident andesites, Katmai National Monument, Alaska, Ph.D. Thesis, University of Alaska, Fairbanks, 198 pp., 1967.

REED, B. L. and M. A., LANPHERE, Alaska Aleutian Range batholith: geochronology chemistry, and relation to circum-Pacific plutonism, *Geol. Soc. Am. Bull.*, **84**, 2583–2610, 1973.

REEDER, J. W., J. C. LAHR, J. THOMAS, S. CONERS, and M. BLACKFORD, Seismological aspects of the recent eruption of Augustine Volcano, *Trans. Am. Geophys. Union, EOS*, **58**, 1188, 1977.

REYNERS, M. and K. S. COLES, Fine structure of the dipping seismic zone and subduction mechanics in the Shumagin Islands, Alaska, *J. Geophys. Res.*, **87**, 356–366, 1982.

SELF, S., J. KIENLE, and J.-P. HUOT, Ukinrek Maars, Alaska II. Deposits and formation of the 1977 craters, *J. Volc. Geotherm. Res.*, **7**, 39–65, 1980.

SMITH, W. R., Aniakchak Crater, Alaska Peninsula, *U.S. Geol. Surv. Bull.* **132-J**, 139–149, 1925.

STEFANI, J. P., R. J. GELLER, and G. C. KROEGER, A direct measurement of the distance between a hypocenter in a Benioff-Wadati zone and the slab-asthenosphere contact, *J. Gephys. Res.*, **87**, 323–328, 1982.

STOIBER, . E. and M. J. CARR, Quaternary volcanic and tectonic segmentation of Central America, *Bull. Volc.*, **37**, 304–325, 1973.

SWANSON, S. E., J. KIENLE, and P. M. FENN, Geology and petrology of Kaguyak Crater, Alaska, *Trans. Am. Geophys. Union, EOS*, **62**, 1062, 1981.

TOPPER, R. E., Fine structure of the Benioff zone beneath the central Aleutian arc, M.S. Thesis, University of Colorado, Boulder, 148 pp., 1978.

WOODWARD-CLYDE CONSULTANTS, *Interim report on seismic studies for Susitna hydroelectric project, for Alaska Power Authority*, Anchorage, 1980.

YOSHII, T., Proposal of the 'aseismic front', *Zishin*, **28**, 365–367, 1975.

Arc Volcanism: Physics and Tectonics, edited by D. Shimozuru and I. Yokoyama, 225–241.
Copyright © 1983 by Terra Scientific Publishing Company (TERRAPUB), Tokyo.

Volcanism and the Evolution of the Ensimatic
Solomon Islands Arc

P. N. DUNKLEY

Geological Survey Division, Honiara, Solomon Islands

Since the Eocene the Solomon Islands have been the site of complex interaction between the Pacific and Australian plates, resulting in two separate episodes of subduction and volcanism. At the onset of subduction in U.Eocene-L.Oligocene times the oceanic basement of the arc was tectonised and metamorphosed contemporaneously with the tectonic emplacement of ultrabasic complexes. During this first episode the Pacific plate was subducted beneath the Australian plate resulting in the eruption of island arc tholeiites of Oligocene-L.Miocene age. By M. Miocene times the Ontong Java Plateau had collided with the arc, stopping subduction and volcanism. Further convergence between the plates was accommodated by the formation of a new trench on the south-west side of the arc, along which the Australian plate is being subducted beneath the Pacific plate. U.Miocene-Recent volcanism associated with this second subduction zone is generally calc-alkaline in character although tholeiitic volcanics have also been erupted. Since the Pliocene, the actively spreading Woodlark Basin has been subducted beneath the New Georgia Group, resulting in rapid and large scale uplift of this part of the arc, and the voluminous production of magmas of unusual composition.

1.   Introduction

The Solomon Islands are part of a complex of Melanesian island arcs and marginal basins that extend in a southeasterly direction across the South-West Pacific from New Ireland and New Britain through to Fiji and Tonga (Fig. 1). The Solomons occupy the central part of this complex and occur as a double linear chain of islands stretching for over 1,800 km from Buka-Bougainville in the northwest to Tikopia and Fatutaka in the southeast (Fig. 2).

Since Eocene times the region has been the site of complex interaction between the Pacific and Australian plates, resulting in two distinct and separate episodes of subduction and arc volcanism. Throughout this time the Solomons have evolved in an entirely oceanic environment. Several of the main events recorded in their evolution are broadly recognisable in other parts of the Melanesian arc complex, although here detailed discussion is confined to the Solomon Islands sector only.

2.   Oceanic Basement and the Initiation of Arc Tectonism

The oldest rocks of the region are ocean floor tholeiites of Cretaceous to

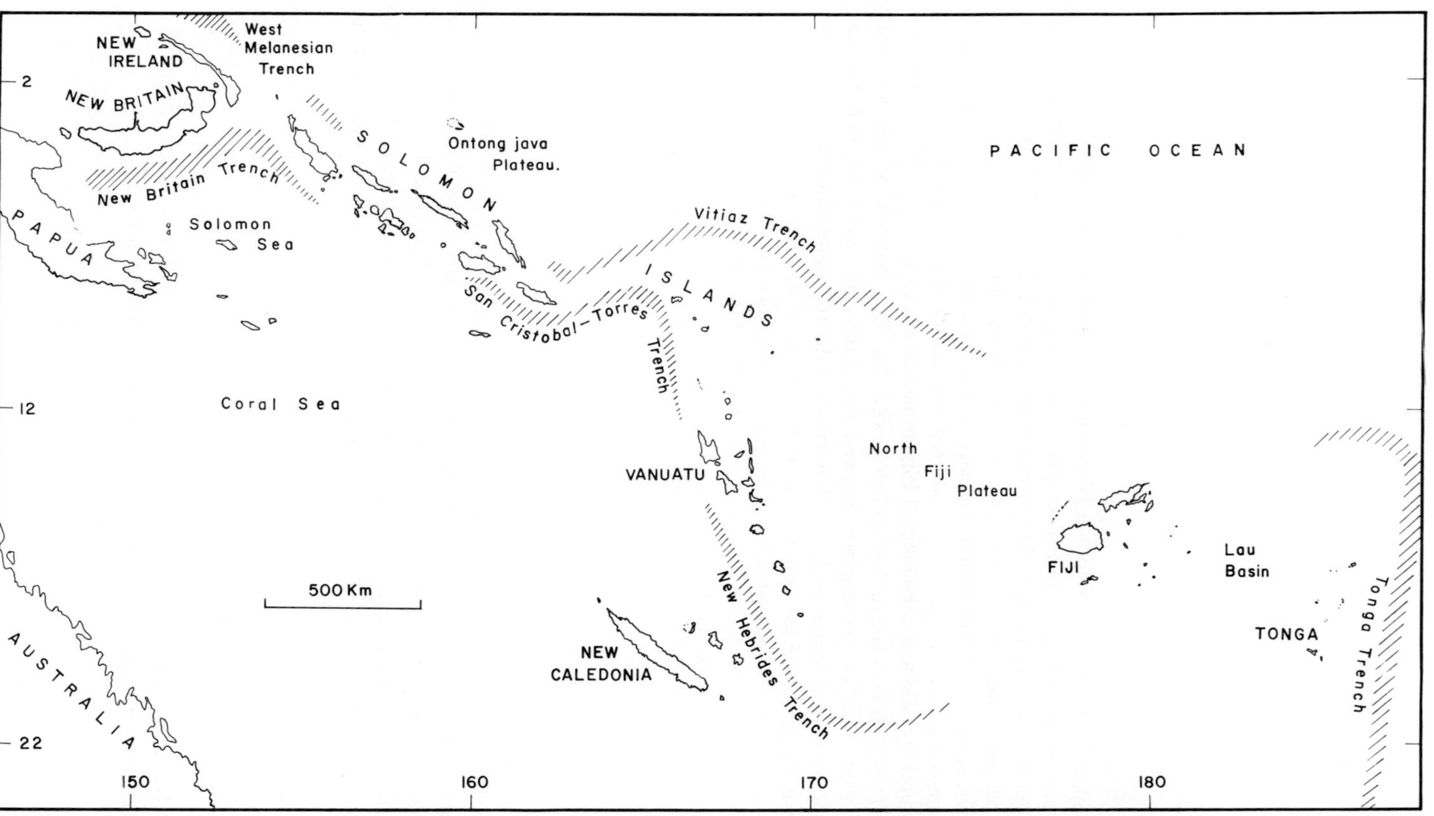

Fig. 1.　Regional map of the South West Pacific, showing the location of the Solomon Islands. Deep sea trenches are shown by hachuring.

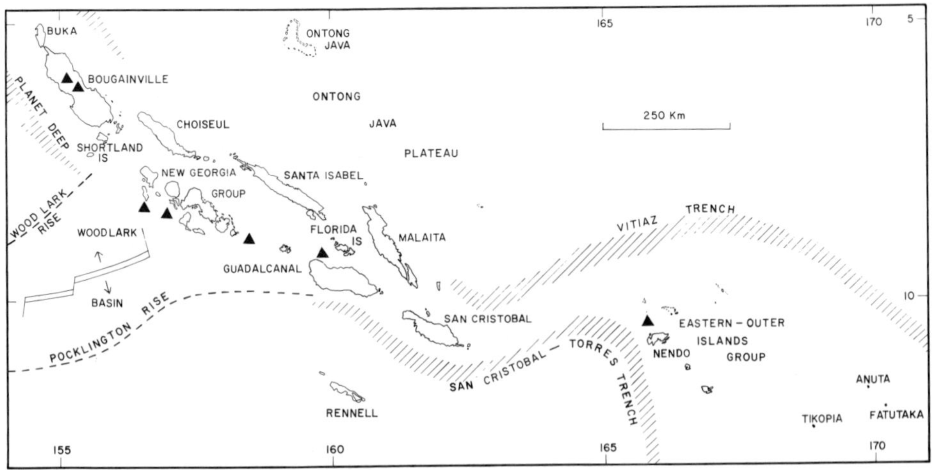

Fig. 2.   Map of the Solomon Islands. Deep trenches are shown by hachuring and active and dormant volcanoes as solid triangles. The Woodlark Rise and Pocklington Rise forming the margins of the Woodlark Basin are represented by dashed lines, and the Woodlark spreading centre is shown by two parallel lines with arrows illustrating the approximate sense of extension.

Palaeocene age and these form a basement on which later arc volcanism and sedimentation has taken place. On the islands of Choiseul, Santa Isabel, Florida, Guadalcanal, and San Cristobal, ultrabasic complexes consisting predominantly of harzburgites and serpentinised harzburgites have been tectonically emplaced, and the basaltic basement has been tectonised and metamorphosed to greenschist and amphibolite facies (e.g. STANTON and RAMSAY, 1975; ARTHURS, 1981). Radiometric dating of this metamorphism on Choiseul gives an age range of 32–51 Ma with a mean of 44 Ma (WEBB *et al.*, 1966), and on the Florida Islands an age of 35–45 Ma (NEEF and McDOUGALL, 1976), corresponding to the U.Eocene-basal Oligocene; similarly a basal Oligocene age of 35 Ma has been reported for the metamorphism of the ophiolitic basement complex of Pentecost Island, Vanuatu (MALLICK and NEEF, 1974). It is believed that this widespread tectono-metamorphic event took place in response to the initiation of subduction of the Pacific plate in a southwesterly or westerly direction beneath the Australian plate. The Vitiaz Trench to the east of the Solomon Islands and the West Melanesian Trench to the north of Bougainville (Fig. 1) are remnants of this first subduction system which is no longer active.

3.   U.Oligocene-L.Miocene Volcanism

Arc volcanism associated with this first phase of subduction had commenced by the U.Oligocene and continued until the L.Miocene. Volcanics of this age are preserved throughout the Solomons chain and include the Kieta Volcanics of Bougainville (BLAKE and MIEZITIS, 1967), the Suta and Marasa Volcanics of Guadalcanal (HACKMAN, 1980), the Soghonara Lavas (TAYLOR, 1977) and Ghumba Beds (PLIMER and NEEF, 1980) of the Florida Islands, the Mengalu and Nolua Volcanics of Nendo (HUGHES *et al.*, 1981) and probably the Alu Basalts and Masamasa Volcanics of the Shortland Islands (TURNER, in press).

The volcanics of this episode consist mainly of lavas with intercalations of volcaniclastics and carbonate sediments. They are mostly submarine in origin with pillow lavas being common, although subaerial sequences may exist as is thought to be the case for part of the Kieta Volcanics of Bougainville (BLAKE and MIEZITIS, 1967).

Within this suite basalts and basaltic andesites predominate (Fig. 3) with andesites and more salic rocks accounting for only a small proportion of the total volume (80% of the population contains $< 55\%$ SiO_2). The suite is tholeiitic (Fig. 3) with low TiO_2 ($< 1\%$), and is considered to be typical of the island arc tholeiite series (c.f. JAKES and GILL, 1970) in that it exhibits LREE depleted to chondritic REE patterns (Ce_N/Yb_N of 0.4–1.4, mean 0.7) (Fig. 4).

On Guadalcanal there are two intrusive complexes of U.Oligocene age (HACKMAN, 1980), and intrusions thought to be of similar age occur on Bougainville (BLAKE and MIEZITIS, 1967). These are generally more siliceous than their contemporaneous volcanics, ranging in composition from gabbro to tonalite with quartz diorite predominating, and in contrast to the volcanics they are calc-alkaline in character and exhibit weak porphyry copper style mineralisation.

This Oligocene-L.Miocene igneous event is not confined to the Solomon Islands, but is a common feature throughout the Melanesian complex. Volcanics and intrusions of this age occur to the southeast of the Solomons in Vanuatu, Fiji, and Tonga (e.g. GILL and GORTON, 1973) and to the northwest in New Ireland (HOHNEN, 1978) and New Britain (DAVIES, 1973). It is considered that at this time the whole Melanesian complex was represented by a single continuous easterly facing arc, which subsequent to the L.Miocene was disrupted and fragmented by collision with oceanic plateaus and

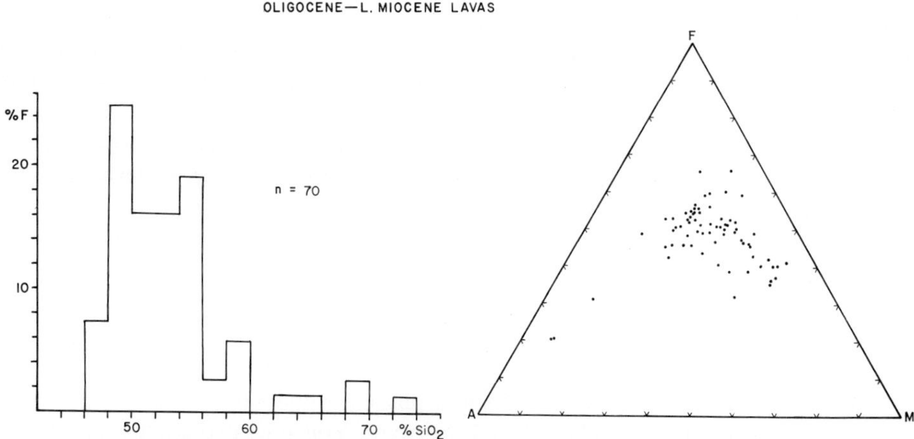

Fig. 3. Silica frequency histogram and F.M.A. diagram for Oligocene-L. Miocene lavas. Data from CHIVAS, 1977; HACKMAN, 1980; HUGHES *et al.*, 1981; PLIMER and NEEF, 1980; TAYLOR, 1977 and TURNER, 1982.

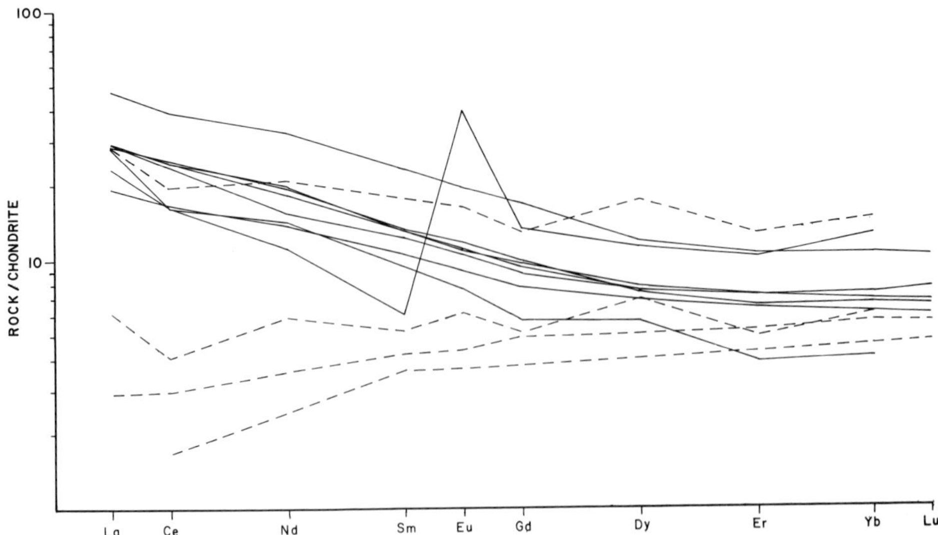

Fig. 4.   Rare earth element abundance patterns normalised against the Leedey chondrite (MASUDA *et al.*, 1973). Dashed lines represent Oligocene-L. Miocene volcanics, and continuous lines represent U. Miocene-Recent volcanics. Data taken from BULTITUDE *et al.*, 1978; JAKES and GILL, 1970; MASUDA pers. comm. and TAYLOR, 1977.

by the growth of marginal basins. The time of commencement of this first episode of volcanism in the Solomons is not precisely known, but it was certainly underway by the U.Oligocene. Furthermore, a L.Oligocene-M.Oligocene age is assigned to the Jaulu Volcanics of New Ireland (HOHNEN, 1978), and a radiometric date of 36.7 ± 1 Ma corresponding to the uppermost Eocene-basal Oligocene has been obtained from intrusive rocks of the Torres Islands (GREENBAUM *et al.*, 1975), which occur immediately to the south of the Solomons in northern Vanuatu. These ages suggest that this first episode of arc volcanism closely followed or possibly even overlapped the deformation and metamorphism of the oceanic basement and the tectonic emplacement of the ultrabasic complexes described in the previous section.

4.   Collision of the Ontong Java Plateau and the Cessation of Volcanism

By M.Miocene times the first episode of volcanism had ceased completely and the entire Solomons region was subject to extensive reef growth and deposition of carbonate sediments. The abrupt cessation of volcanism at this time was caused by the arrival at the Vitiaz-West Melanesian trench system of the Ontong Java Plateau, an anomalous area of the Pacific plate with abnormally thick oceanic crust (> 40 km) and lithosphere (COLEMAN and KROENKE, 1982). This thickened portion of the Pacific plate blocked and obliterated parts of the trench and consequently stopped subduction and its related volcanism. Deformation resulting from the collision folded and faulted the sediments upon the margin of the plateau (KROENKE, 1972) and culminated in obduction of its leading edge onto the arc (Fig. 5c). Subsequent uplift has exposed deformed pelagites and ocean floor basalts of the obducted slab on the island of Malaita (Hughes and Turner, 1977), and on northeast Santa Isabel (Coleman et al., 1978) where the Kaipito-Korigole Thrust (Stanton, 1961) may represent a major dislocation at the base of the slab.

5. Miocene-Recent Subduction and Volcanism

With the Vitiaz-West Melanesian trench system blocked by the Ontong Java Plateau further convergence between the two plates was accommodated by the formation of a second trench on the southwest side of the arc, along which the Australian plate and minor Solomon Sea plate are currently being subducted in a northeasterly direction beneath the Pacific plate (Fig. 5d). The rate of convergence between the two major plates is high, and is generally thought to exceed 10 cm/yr (e.g. Johnson and Molnar, 1972). However, not all of this convergence has necessarily been taken up by subduction alone; it has been suggested that part may have been accommodated by sinistral strike slip shearing along the length of the arc (e.g. Coleman, 1975). Although it is difficult to assess the importance of this transcurrent movement, Turner and Hughes (1979) consider it to be small.

The currently active trench system is well developed to the southwest of Bougainville and the Shortland Islands where it is represented by the Planet Deep (Fig. 2) which at its northern end curves sharply westwards into the New Britain Trench (Fig. 1). The Planet Deep reaches over 9 km in depth but dies out in a southeasterly direction, and along the central part of the arc between the New Georgia Group and western Guadalcanal the trench is absent or very poorly developed. Further to the southeast off eastern Guadalcanal and San Cristobal the trench reappears and passes via a transform structure into the Torres Trench which extends southwards towards Vanuatu (Fig. 2). Intense seismic activity characterises the northwestern and southeastern parts of the arc where the trench system is well developed (Fig. 6), and here clearly defined Benioff zones dip steeply northeastwards beneath the arc (Figs. 7a, c, and d). In contrast, along the central part of the arc where there is a gap in the trench system, seismic activity decreases dramatically and there are very few deep earthquakes (Fig. 6 and Fig. 7b). This reduced seismicity and lack of a trench along the central part of the arc are considered to be related to several other unusual features, including anomalous volcanism in the New Georgia Group, which are discussed in a later section.

Another feature of the seismicity illustrated in Fig. 7 is that beneath Bougainville and the Eastern Outer Islands there are regions of very deep earthquakes (500–700 km) which appear to be separate from the Benioff zones. It has been suggested that these may originate from a detached lithospheric slab derived from the earlier episode of southwesterly subduction along the Vitiaz Trench system (e.g. Halunen and von Herzen, 1973).

Volcanics related to the second episode of subduction have been erupted from U.Miocene times until the present day along the south-west side of the arc. Lavas and epiclastic breccias derived mainly from lavas appear to be more important than primary pyroclastic deposits, although intermediate ash-flow tuffs and pumice flows of

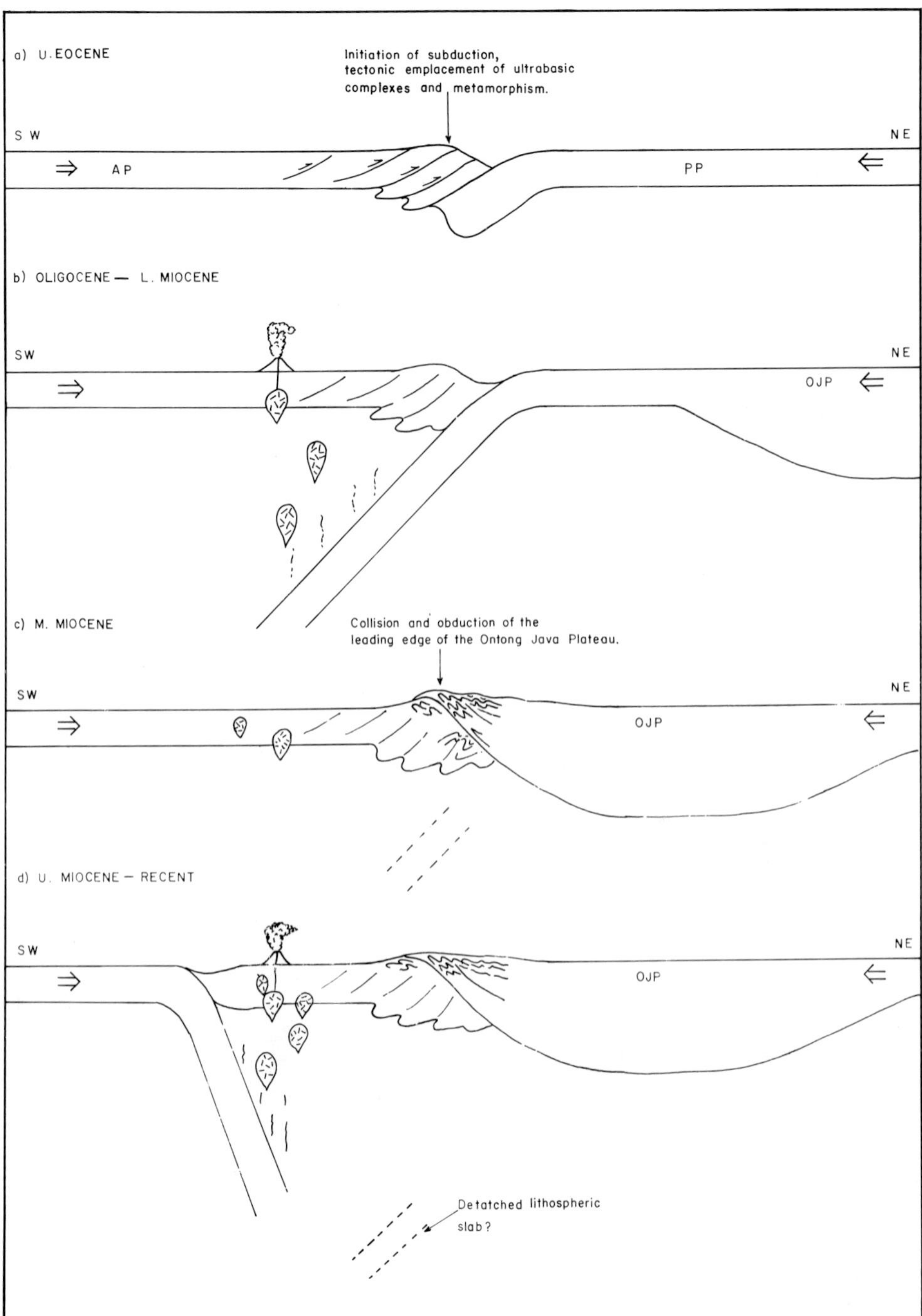

Fig. 5.   Schematic cross sections illustrating the main stages in the evolution of the arc. AP = Australian plate, PP = Pacific plate and OJP = Ontong Java Plateau.

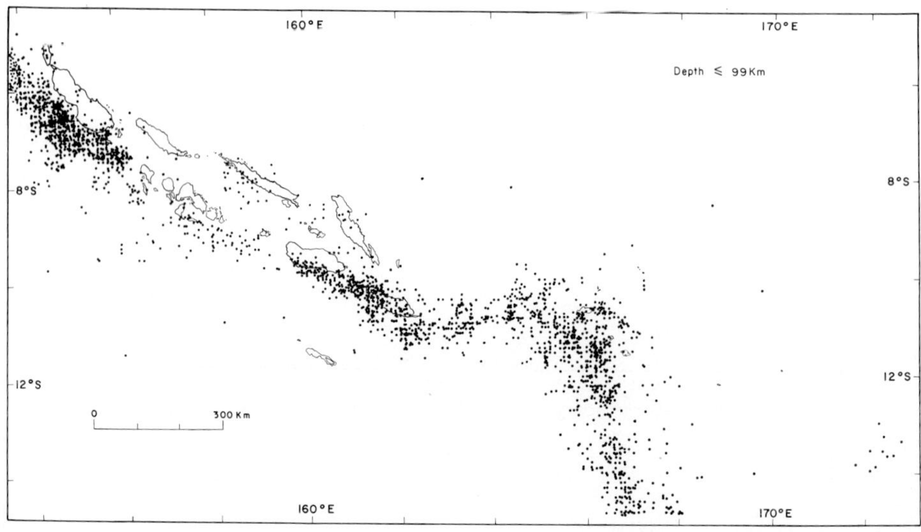

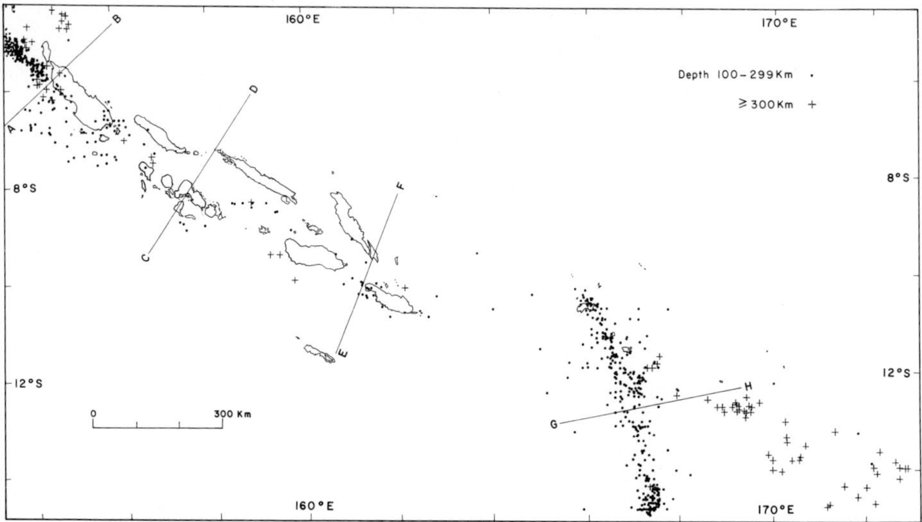

Fig. 6.   Earthquake epicentres for the Solomon Islands from 1962 to 1977. Top map shows the distribution of shallow earthquakes (< 99 km). Bottom map shows intermediate earthquakes (100–299 km) represented by dots and deep earthquakes represented by crosses. Lines of section appertain to Fig. 7.

limited aerial extent do occur, which on some recently active volcanoes are associated with andesite and dacite domes.

Compositionally the volcanics of this second episode are of variable character. In contrast to the earlier Oligocene-L.Miocene volcanics, those of the main part of the arc are typically calc-alkaline (Fig. 8a) and have LREE enriched patterns (Ce_N/Yb_N, 1–4)

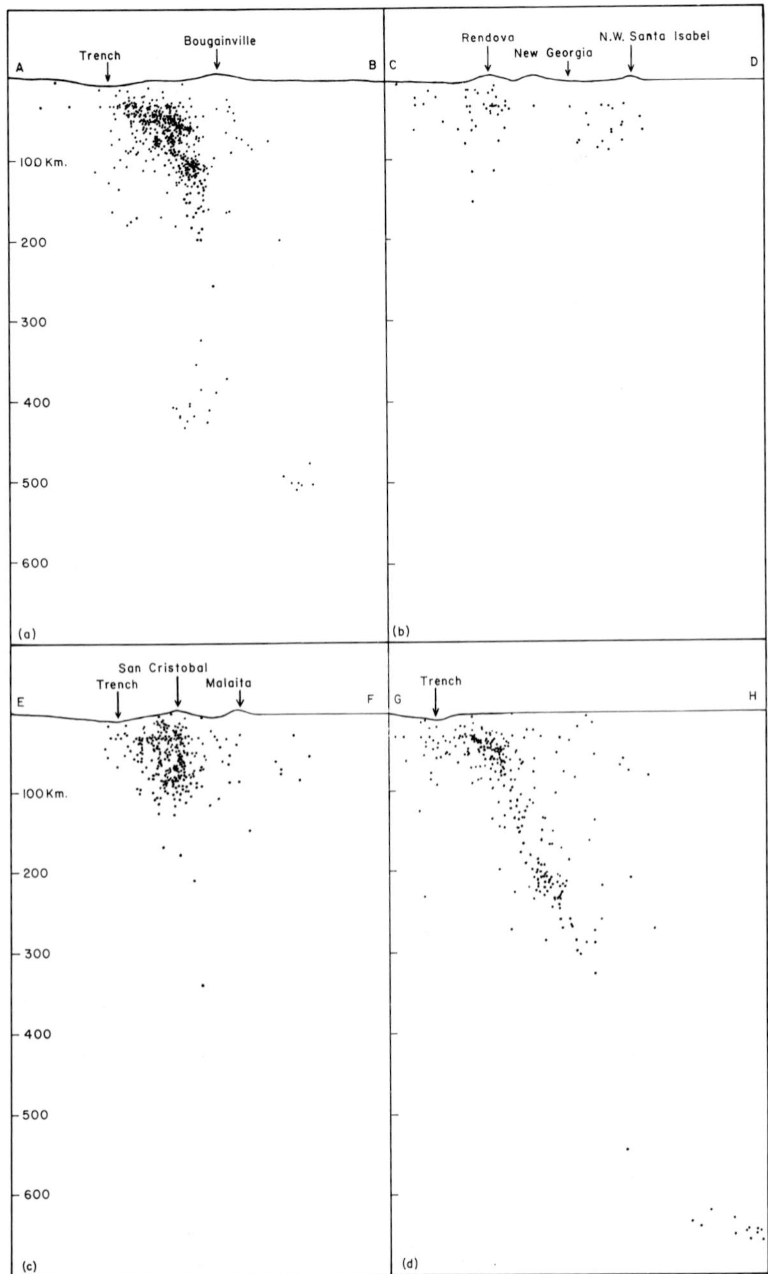

Fig. 7.   Distribution of earthquake depths. Positions of sections AB, CD, EF, and GH are given in Fig. 6. All earthquakes within 100 km either side of each line of section were projected normally onto the line of section. No vertical exaggeration.

(Fig. 4), whereas those of the Eastern Outer Island Group are tholeiitic (Fig. 8b) with low TiO$_2$ and belong to the island arc tholeiite series. A further variation occurs in the New Georgia Group, where large volumes of olivine rich basalts and picrites have been erupted; these are not considered to be related to normal subduction processes and are therefore discussed in a separate section. Excluding this anomalous volcanism of the New Georgia Group, the U.Miocene-Recent volcanics are generally more salic than those of the earlier Oligocene-L.Miocene episode and range in composition from basalt to rhyodacite with basaltic andesites and andesites predominating (Fig. 8c).

Along the main part of the arc, on Bougainville, the Shortland Islands, New Georgia, and Guadalcanal there are a number of intrusive complexes related to this second episode of volcanism, several of which have yielded Pliocene to Pleistocene radiometric ages (Webb in BLAKE and MIEZITIS, 1967; PAGE and MCDOUGALL, 1972; CHIVAS and MCDOUGALL, 1978). The complexes occur as irregular shaped stocks up to approximately 7 km in diameter, and range in composition from gabbro to grano-diorite and quartz monzonite, with quartz diorite and tonalite predominating. They are calc-alkaline in character and exhibit late stage sodium enrichment with the intrusion of trondhjemite dykes. Copper-gold porphyry style mineralisation is as-sociated with these complexes, notably at Panguna on Bougainville.

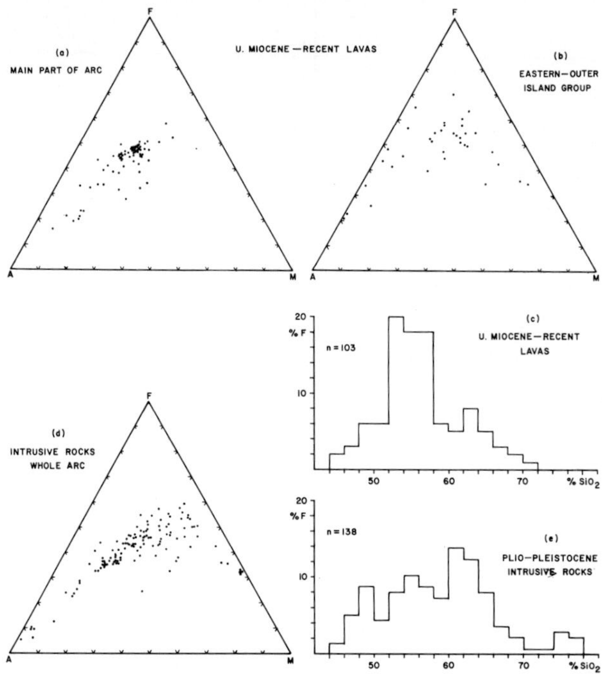

Fig. 8.   F.M.A. diagrams and silica frequency histograms for U. Miocene-Recent lavas and intrusions (excluding the New Georgia Group). Data from BLAKE and MIEZITIS, 1967; BULTITUDE *et al.*, 1978; CHIVAS, 1977; HACKMAN, 1980; HUGHES *et al.*, 1981; TURNER 1982, and unpublished records of the Solomon Islands Geological Survey.

6.   Pliocene-Recent Volcanism in the New Georgia Group and Subduction of the Woodlark Basin

From Pliocene to Recent times the New Georgia Group, which is situated very close to the projected line of the trench, has been the site of voluminous eruption of volcanics of unusual composition, some of which are among the most basic lavas known from island arcs. They were initially described by STANTON and BELL (1969) and are currently the subject of detailed mapping and petrological studies by the Solomon Islands Geological Mapping Project and others. These volcanics consist of large volumes of highly porphyritic olivine basalt and picrite basalt lavas and breccias, with minor hornblende basaltic-andesites and andesites. The basalts and picrites are hypersthene normative and on an FMA diagram show a trend of magnesium enrichment (Fig. 9), whereas the minor hornblende basaltic andesites and andesites, with which they are spatially and temporally associated, exhibit a broadly calc-alkaline trend more akin to that illustrated by the 'normal' volcanics occurring elsewhere in the main part of the arc. The picrites contain up to 55% modal olivine of highly forsteritic composition (Fo_{88}–Fo_{89}) and as much as 30% MgO, and are believed to have been produced by low pressure cumulus enrichment of olivine and clinopyroxene (COX and BELL, 1972). Recent mapping has discovered rare but widespread occurrences of fine grained aphyric high magnesium basalts (10–12% MgO) which could represent the parental or near parental liquids from which the picrites accumulated. It is difficult to define the petrogenetic character of the suite, except to say that it is subalkaline and in keeping with arc volcanics has low TiO_2 ($< 1\%$). However, the suite is high in potassium with the basalts containing up to 2.6% K_2O. This is unusually high when it is considered how close to the projected line of the trench these volcanics have been erupted.

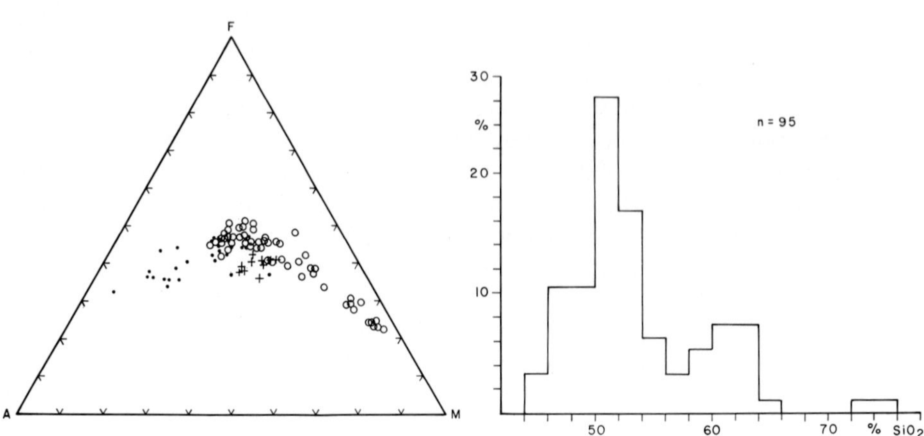

Fig. 9.   F.M.A. diagrams and silica frequency histogram for Pliocene-Recent lavas from the New Georgia Group. Open circles represent olivine basalts and picrites, dots represent hornblende bearing andesites and crosses represent two pyroxene magnesian andesites from Simbo volcano. Data by the author and from STANTON and BELL (1969).

Apart from the picrites, basalts and minor hornblende andesites, there are also minor occurrences of two pyroxene andesites. These are relatively magnesian (4.7–6.4% MgO at 58–62% SiO_2) and contain abundant orthopyroxene phenocrysts and noritic microxenoliths, and rare olivine. An unusual feature of these magnesian andesites is that almost all of them have been erupted from the dormant volcano of Simbo, which is the most southwesterly island of the group and is situated virtually upon the projected line of the trench.

To the west and southwest of the New Georgia Group lies the Woodlark Basin which is currently undergoing active extension and sea-floor spreading (Milsom, 1970; Luyendyke et al., 1973). This basin is bounded along its northern and southern margins by two arcuate ridges, the Woodlark Rise and Pocklington Rise respectively (Fig. 2). Based upon magnetic lineations and anomaly patterns Weissel et al. (1982) suggest that the basin is oldest in the east having commenced formation at least 3.5 Ma ago, and that the spreading centre has propagated westwards with time. Although there are uncertainties regarding the precise position of the spreading axis and the nature of movement along transform structures, they conclude that the spreading centre has always been subducted in the vicinity of the New Georgia Group; it is the subduction of this spreading centre which is considered to be the principal cause of the abnormal volcanism of this region.

In addition to the production of large volumes of mafic magmas of unusual composition in the New Georgia Group, the subduction of the Woodlark spreading centre and adjacent young crust has had substantial physical and structural influence upon the arc. The most obvious feature in this respect is the marked reduction in seismicity and virtual absence of intermediate and deep earthquakes between Guadal-canal and the Shortland Islands (Fig. 6 and Fig. 7b). This is believed to be the result of subducting very young, comparatively warm and ductile oceanic lithosphere which is likely to deform plastically. Evidence supporting the relatively warm nature of the subducting slab is provided by Halunen and Von Herzen (1973) who reported high heat flow values along the projected line of the trench adjacent to the Woodlark Basin. An equally striking feature is the absence of a trench in the region where the basin is being subducted (Fig. 2), and this is thought to be due to shoaling as a consequence of subducting young, buoyant ocean lithosphere. With respect to these features it is very significant that to the north of the Woodlark Rise and south of the Pocklington Rise (the margins of the basin) where much older and presumably colder and denser ocean crust is being subducted, there is an immediate and marked increase in seismicity with the development of steeply inclined Benioff zones and clearly defined trenches (c.f. Fig. 2, Fig. 6, and Fig. 7).

DeLong and Fox (1977) have discussed the consequences of subducting spreading centres and young buoyant oceanic lithosphere, and propose that this should produce uplift of the arc and forearc. The absence of a trench adjacent to the Woodlark Basin is considered to result from shoaling and other evidence indicates considerable uplift of the forearc and arc. In the New Georgia Group, Pleistocene to Recent coral reefs are raised to 800 m above sea level. This uplift increases southwestwards towards the site of subduction and micropalaeontological studies of foraminiferal depth assemblages on those islands nearest the projected line of the trench indicate that basal

Pleistocene sediments now elevated to 800 m above sea level were originally deposited in water depths of 1.5–2 km (HUGHES, 1981), implying uplift of 2–3 km during and since the Pleistocene.

ENGLAND and WORTEL (1980) propose that during subduction young oceanic lithosphere will, because of its buoyancy, exhibit increased coupling with the overriding plate giving rise to the development of compressive stresses in this plate. Evidence for such compression is seen along the south-western margin of the New Georgia Group, where several uplifted islands are composed of Plio-Pleistocene forearc sediments that have been folded along axes parallel to the projected line of the trench. If increased coupling does occur between the two plates then it should be possible for stresses within the subducted slab to be translated into the overriding plate. Therefore, on subducting a spreading centre one might expect to find tensional features within the overlying arc of the same orientation as the centre itself. It follows that the subduction of the Woodlark spreading centre should result in north-easterly trending tensional structures within the arc. Such structures of this orientation are in fact prominent in the New Georgia Group and take the form of normal faults, graben and dykes.

7.   Extension of the North Fiji Plateau

One other important post M.Miocene event that has affected the evolution of the arc is the formation of the North Fiji Plateau (Fig. 1), which is an extensional back-arc basin stretching from the southeastern Solomons to Fiji. Growth of this basin (or 'plateau') about a northwesterly spreading axis commenced during the U.Miocene and has produced clockwise rotation of the Vanuatu arc away from the Vitiaz Trench (FALVEY, 1978). The northern part of this basin which is now inactive extends into the Eastern Outer Island Group of the Solomons. On Tikopia, Fatutaka and Anuta, the most southeasterly of the Solomon Islands, high alumina tholeiitic basalts occur in addition to the arc volcanics produced by subduction. On Fatutaka JEZEK et al. (1977) report an U.Pliocene age for these basalts (2.2 Ma) concluding that their origin is related to subduction of the Pacific plate along the Vitiaz Trench. However, as the Vitiaz Trench was no longer active at this time such an origin is untenable. Furthermore, it is considered that these basalts are not likely to be related to subduction processes since they contain much higher TiO_2 (1.4–1.9%) and Al_2O_3 (up to 20.5%) and noticeably lower K_2O than the arc volcanics found on the same islands. A more feasible explanation for the origin of these high alumina tholeiites is that they were produced during extension of the North Fiji Plateau; this is plausible when it is considered that the islands on which they occur are situated upon or close to the axis of the basin.

8.   Discussion and Summary

Several authors have considered that the Solomon Islands do not represent a typical island arc (e.g. COLEMAN, 1976; HACKMAN, 1980), and have used terms such as "fractured arc" (COLEMAN, 1970; HACKMAN, 1973) and "non-arc" (COLEMAN, 1975) to describe the chain. Models put forward by these authors propose that the evolution of the islands may have been largely controlled by major transcurrent fractures along which magmas have risen, and upon which there has been substantial strike-slip movement to accommodate convergence between the Australian and Pacific plates. Features that have been cited in favour of such a non-arc origin include the absence of a trench and seismic zone together with the occurrence of high heat flow along the central part of the chain, and the eruption of abnormal volcanics in the New Georgia Group. However, in this paper an explanation has been put forward suggesting that all these anomalous features owe their origin to the subduction of the Woodlark Basin and its spreading centre. Furthermore, although strike-slip movements may have taken place along transcurrent faults these are not considered to be large.

COLEMAN (1976) also noted an absence of volcanism from eastern Guadalcanal through San Cristobal to the area where the San Cristobal and Torres trenches join. COLEMAN and KROENKE (1982) consider that this gap in volcanism occurs where subducted lithosphere of the Australian plate comes into direct contact beneath the arc with the thickened oceanic lithosphere and underlying mantle of the Ontong Java Plateau, which because of its cool and refractory nature is incapable of undergoing partial melting. However, there are a number of objections to this proposal. Firstly, much of the trench system east of San Cristobal to the junction with the Torres trench is in fact a transform structure characterised by shallow earthquakes only (Fig. 6); along this part of the trench there does not appear to be any subduction and therefore volcanism is not to be expected. Secondly, where subduction is taking place beneath eastern Guadalcanal and San Cristobal, the seismic zone dips so steeply beneath the arc that it is unlikely that the descending slab comes into direct contact with lithosphere of the Ontong Java Plateau, which presumably is confined to the region north and northeast of the Vitiaz Trench (Fig. 2). Furthermore, the lack of volcanism in eastern Guadalcanal and San Cristobal is not considered to be an unusual feature when the spacing of active or recently active volcanoes is taken into account for the entire arc. If the abnormal activity caused by ridge subduction in the New Georgia Group is excluded, then only two active and two possibly dormant volcanoes related to 'normal' subduction are found throughout the remainder of the arc; even within the New Georgia Group there is only one active and two possibly dormant volcanoes. By comparison with the volume of material erupted or intruded since the U.Miocene, it would appear from the paucity of Recent activity that volcanism within the arc is very much on the wane. This rapid reduction in activity during the past six million years can be accounted for by two related factors. Firstly, because the subducted lithosphere dips very steeply beneath the arc, the volume of the overlying mantle wedge available for partial melting will be comparatively small and therefore likely to undergo relatively rapid depletion in low melting point components. Secondly, this same portion of mantle beneath the arc would also have been involved in the earlier Oligocene-L.Miocene episode of subduction and volcanism, and therefore will already have been partially depleted prior to the second U.Miocene-Recent episode. As a result of the combination of these two factors the mantle beneath the arc is now probably quite refractory and thus unable to produce large volumes of magma.

In summary, the Solomon Islands have evolved since the Eocene in response to complex interaction between the Australian and Pacific plates. Initially, subduction of the Pacific plate caused deformation and metamorphism of the oceanic basement to the arc, and the eruption of Oligocene-L.Miocene island arc tholeiitic volcanics. The arrival of the Ontong Java Plateau during the M.Miocene stopped this subduction and produced a reversal in arc polarity. Subsequently, the Australian plate has been subducted resulting in a second episode of U.Miocene-Recent volcanism. During this episode, extensional opening of the North Fiji Plateau has affected the Eastern Outer Group giving rise to the eruption of high alumina tholeiites. The main part of the arc has also been complicated by Pliocene-Recent subduction of the Woodlark Basin and its spreading centre. This event has had profound effects upon the arc, resulting in the voluminous eruption of unusual magmas in the New Georgia Group, rapid and large scale uplift of the arc, the obliteration of the trench, high heat flow along the projected line of the trench and a marked reduction in seismic activity. Thus, although the Solomon Islands appear to have experienced a complicated geological history, it is concluded that their salient features are reconcilable with formation in an island arc environment, albeit a complex one.

The author is indebted to Julia Dunkley for drafting the diagrams and typing the manuscript, and to Wyn Hughes for stimulating discussions on the geology of the Solomon Islands. The paper is published with the permission of the Director of the Institute of Geological Sciences, NERC, United Kingdom.

REFERENCES

ARTHURS, J. W., Micropetrography of the Voza Lavas and Choiseul Schists, Choiseul, *Br. Tech. Co-op. W. Solomon Map Project, Rpt*, **16**, 1981.

BLAKE, D. H. and Y. MIEZITIS, Geology of Bougainville and Buka Islands, New Guinea, *Bull. Bur. Mineral. Resour. Geol. Geophys. Aust.*, **93**, 1967.

BULTITUDE, R. J., R. W. JOHNSON, and B. W. CHAPPELL, Andesites of Bagana volcano, Papua New Guinea: chemical stratigraphy, and a reference andesite composition, *BMR, J. Aust. Geol. Geophys.*, **3**, 281–295, 1978.

CHIVAS, A. R., Geochemistry, geochronology and fluid inclusion studies of porphyry copper mineralization at the Koloula igneous complex, Guadalcanal, Solomon Islands, Unpbl Ph.D. Thesis, University of Sydney, 1977.

CHIVAS, A. R. and I. McDOUGALL, Geochronology of the Koloula porphyry copper prospect, Guadalcanal, Solomon Islands, *Econ. Geol.*, **73**, 678–689, 1978.

COLEMAN, P. J., Geology of the Solomon and New Hebrides Islands, as part of the Melanesian re-entrant, *Pacif. Sci.*, **24**, 289–314, 1970.

COLEMAN, P. J., The Solomons as a non-arc, *Bull. Aust. Soc. Explor. Geophys.*, **6**, 60–61, 1975.

COLEMAN, P. J., A re-evaluation of the Solomon Islands as an arc system, in *Marine Geological Investigations in the Southwest Pacific and Adjacent Areas*, edited by G. P. Glasby and H. R. Katz *UNESCAP Tech. Bull.*, 2, pp. 134–139, 1976.

COLEMAN, P. J. and L. W. KROENKE, Subduction without volcanism, *Geomarine Lett.*, 1982.

COLEMAN, P. J., B. McGOWRAN, and R. W. RAMSAY, New early Tertiary ages for basal pelagites, northeast Santa Isabel, Solomon Islands, *Bull. Aust. Soc. Explor. Geophys.*, **9**, 110–114, 1978.

COX, K. G. and J. D. BELL, A crystal fractionation model for basaltic rocks of the New Georgia Group, British Solomon Islands, *Contrib. Mineral. Petrol.*, **37**, 1–13, 1972.

DAVIES, H. L., Gazelle Peninsula, New Britain, 1:250,000 Geological Series, *Bur. Mineral. Resour. Geol. Geophys. Aust.*, explan. Notes SB/56–2, 1973.

DELONG, S. E. and P. J. FOX, Geological consequences of ridge subduction, in *Island Arcs, Deep Sea Trenches and Back-Arc Basins*, edited by M. Talwani and W. G. Pitman, III, pp. 221–228, Am. Geophys. Union, Washington, 1977.

ENGLAND, P. and R. WORTEL, Some consequences of the subduction of young slabs, *Earth Planet. Sci. Lett.*, 47, 403–415, 1980.

FALVEY, D. A., Analysis of palaeomagnetic data from the New Hebrides, *Bull. Aust. Soc. Explor. Geophys.*, 9, 117–123, 1978.

GILL, J. and M. GORTON, A proposed geological and geochemical history of eastern Melanesia, in *The Western Pacific: Island Arcs, Marginal Seas, Geochemistry*, edited by P. J. Coleman, pp. 543–566, *University of Western Australia*, 1973.

GREENBAUM, D., D. I. J. MALLICK, and N. W. RADFORD, Geology of the Torres Islands, *Regional Rpt. New Hebrides Geol. Surv.*, 1975.

HACKMAN, B. D., The Solomon Islands fractured arc, in *The Western Pacific: Island Arcs, Marginal Seas, Geochemistry*, edited by P. J. Coleman, pp. 179–191, University of Western Australia, 1973.

HACKMAN, B. D., The geology of Guadalcanal, Solomon Islands, in *Overseas Mem. Inst. Geol. Sci.*, No. 6, H.M.S.O. London, 1980.

HALUNEN, A. J. Jr, and R. P. VON HERZEN, Heat flow in the western equatorial Pacific, *J. Geophys. Res.*, 78, 5195–5208, 1973.

HOHNEN, P. D., Geology of New Ireland, Papua New Guinea, *Bull. Bur. Mineral. Resour. Geol. Geophys. Aust.*, 194, 1978.

HUGHES, G. W., Micropalaeontological report for Tetepare and Rendova Islands, New Georgia Group, *Internal Rpt. Br. Tech. Co-op. W. Solomon Map. Project*, (Unpbl.), 1981.

HUGHES, G. W. and C. C. TURNER, Upraised Pacific Ocean floor, southern Malaita, Solomon Islands, *Geol. Soc. Am. Bull.*, 88, 412–424, 1977.

HUGHES, G. W., P. M. CRAIG, and R. A. DENNIS, Geology of the Eastern Outer Islands, *Bull. Geol. Surv. Solomon Isl.*, 4, 1981.

JAKES, P. and J. GILL, Rare earth elements and the island arc tholeiitic series, *Earth Planet. Sci. Lett.*, 9, 17–28, 1970.

JEZEK, P. A., W. B. BRYAN, S. E. HAGGERTY, and H. P. JOHNSON, Petrography, petrology and tectonic implications of Mitre Island, northern Fiji Plateau, *Mar. Geol.*, 24, 123–148, 1977.

JOHNSON, T. and P. MOLNAR, Focal mechanisms and plate tectonics of the southwest Pacific, *J. Geophys. Res.*, 77, 5000–5032, 1972.

KROENKE, L. W., Geology of the Ontong Java Plateau, *Hawaii Inst. Geophys. Rpt.*, HIG-72-5, 1972.

LUYENDYK, B. P., K. C. MACDONALD, and W. B. BRYAN, Rifting history of the Woodlark Basin in the southwest Pacific, *Geol. Soc. Am. Bull.*, 84, 1125–1134, 1973.

MALLICK, D. I. J. and G. NEEF, Geology of Pentecost, *Regional Rpt. New Hebrides Geol. Surv.*, 1974.

MASUDA, A., N. NAKAMURA, and T. TANAKA, Fine structures of mutually normalised rare-earth patterns of chondrites, *Geochim. Cosmochim. Acta*, 37, 239–248, 1973.

MILSOM, J. S., Woodlark Basin, a minor centre of sea-floor spreading in Melanesia, *J. Geophys. Res.*, 75, 7335–7339, 1970.

NEEF, G. and I. MCDOUGALL, Potassium-argon ages on rocks from Small Nggela Island, British Solomon Islands, *Pacific Geol.*, 11, 81–85, 1976.

PAGE, R. W. and I. MCDOUGALL, Geochronology of the Panguna porphyry copper deposit, Bougainville Island, New Guinea, *Econ. Geol.*, 67, 1065–1074, 1972.

PLIMER, I. R. and G. NEEF, Early Miocene extrusives and shallow intrusives from Small Nggela, Solomon Islands, *Geol. Mag.*, 117, 565–578, 1980.

STANTON, R. L., Explanatory notes to accompany a first geological map of Santa Ysabel, British Solomon Islands Protectorate, *Overseas Geol. Mineral. Resour.*, 8, 127–149, 1961.

STANTON, R. L. and J. D. BELL, Volcanic and associated rocks of the New Georgia Group, British Solomon Islands Protectorate, *Overseas Geol. Mineral. Resour.*, 10, 113–145, 1969.

STANTON, R. L. and W. R. H. RAMSAY, Ophiolite basement complex in a fractured island chain, Santa Isabel, British Solomon Islands, *Bull. Aust. Soc. Explor. Geophys.*, 6, 61–64, 1975.

TAYLOR, G. R., The ophiolite terrain and volcanogenic mineralisation of the Florida Islands, Solomon Islands, Unpbl. Ph. D. Thesis University of New England, 1977.

TURNER, C. C., Geology of the Shortland Islands, *Rpt. Brit. Tech. Co-op. W. Solomon Map. Project*, **13**, in press, 1982.

TURNER, C. C. and G. W. HUGHES, A review of the sedimentary facies of Solomon Islands, and implications concerning the evolution of this island arc, in *3rd Southwest Pacific Workshop Symposium* (abstract), University of Sydney, 1979.

WEBB, R. J. R., A. W. COOPER, and P. J. COLEMAN, Potassium-argon measurements of the age of basal schists in the British Solomon Islands, *Nature*, **211**, 1251–1252, 1966.

WEISSEL, J. K., B. TAYLOR, and G. D. KARNER, The opening of the Woodlark Basin, subduction of the Woodlark spreading system, and the evolution of northern Melanesia since mid-Pliocene time, *Tectonophysics*, 1982.

Arc Volcanism: Physics and Tectonics, edited by D. Shimozuru and I. Yokoyama, 243–259.
Copyright © 1983 by Terra Scientific Publishing Company (TERRAPUB), Tokyo.

Paleomagnetic Studies of the Bonin and Mariana Island Arcs

B. KEATING,* K. KODAMA,** and C. E. HELSLEY*

**Hawaii Institute of Geophysics, University of Hawaii,
Honolulu, Hawaii, U.S.A.*
***Earthquake Research Institute, The University of Tokyo,
Tokyo, Japan*

Results of paleomagnetic and radiometric studies of the Bonin and Mariana Islands indicate that these islands have undergone rotations about a vertical axis in excess of 90 degrees. We interpret the rotation as a clockwise rotation of the island arcs subsequent to their formation during the Eocene. This rotation is much greater than those predicted by previous tectonic models. In addition, rotation is coupled with northward translation of the Bonin Islands from equatorial latitudes.

1. Introduction

The origin and evolution of marginal basins has long been a subject of debate. Of the marginal basins in the Pacific, the Mariana arc and back arc basins have been the most studied. As a result of these studies several tectonic models have been suggested (UYEDA and BEN-AVRAHAM, 1972; KARIG, 1975; KOBAYASHI and ISEZAKI, 1976) that are consistent with some rotation or change in the curvature of the Mariana arc subsequent to its formation. In order to document any rotations and to place age constraints on the development of the island arc, both paleomagnetic and radiometric studies were conducted, utilizing rocks from the Bonin and Mariana islands. The preliminary results of the Bonin studies are presented here.

2. Geological Setting

The Bonin Islands (or Ogasawara Islands), situated about 1,000 km south of Tokyo, form a part of the Izu-Mariana arc-trench complex. The Bonin island arc consists of three groups of islands lying about 50 km apart; from south to north, they are the Hahajima, Chichijima, and Mukojima islands, respectively. Each group is made up of a few inactive volcanic islands and small islets. Several of the islands are up to 20 km² in size; of these, Chichijima is the largest, followed by Hahajima.

Since the end of the 19th century, the Bonin Islands have been of interest to a number of scientists mainly from the petrological and paleontological viewpoints. KIKUCHI (1890) first reported the high content of MgO in certain rocks on Chichijima, and PETERSON (1891) named these rocks "Boninite." The boninite (a vesicular and feldspar-free glassy andesite with high (6 to 12%) content of MgO) is found on several of the Bonin Islands (SHIRAKI and KURODA, 1977). No volcanic rock exactly equivalent to boninite has yet been found elsewhere, although similar rocks have been reported in the fore-arc area of the Marianas (HUSSONG, UYEDA *et al.*, 1982). Early in the 20th century, some paleontological studies on *Nummulites* on Hahajima were made by such workers as YOSHIWARA (1902) and HANZAWA (1925). The results indicate a middle Eocene age for this island. A similar age was reported later by SAITO (1962) and Ujiie and Matsumaru (1977), from study of planktonic and larger foraminifera on Hahajima. A slightly younger age of 40 m.y.b.p. (late Eocene) was obtained for the two-pyroxene andesite of Hahajima by KANEOKA *et al.* (1970). They also reported the absolute age of 26 m.y.b.p. (late Oligocene) for the slightly altered andesite of Chichijima. More recently TAKAYANAGI and KATAYAMA (1979) discovered planktonic foraminifera of early Eocene age from the tuffaceous sediments in Chichijima. H. Tsunakawa (personal communication, 1981) obtained K-Ar ages of 38 and 42 m.y.b.p. from dacites and 41 m.y.b.p. from a boninite on Chichijima. Therefore, the true age of Chichijima is unlikely to differ significantly from that of Hahajima, although some differences may exist between fossil and absolute ages. The inferred age of the Bonin Islands (Early Tertiary), thus, is comparable to the oldest ages for the Mariana Islands such as Guam and Saipan (INGLE, 1975).

3.   Regional Setting

Geophysical studies have shown that the Bonin Islands and the surrounding areas are characterized by large positive Bouguer and free-air gravity anomalies. KARIG and MOORE (1975) reported a free-air gravity anomaly with peak value of over 300 mgal across the Bonin Ridge. TANAKA *et al.* (1974) also observed free-air and Bouguer anomalies of over 300 mgal in the central and northern parts of Chichijima. These positive anomalies are the largest known in the oceans and island arcs.

In spite of these observations, except for the two major islands of Chichijima and Hahajima, the geology of the Bonin Islands has never been investigated thoroughly. Detailed geologic descriptions covering the whole of both islands have not been accomplished. We summarize briefly the geology of the Bonin Islands, based on the limited bits of geologic information so far reported and upon our own observations made during the paleomagnetic field work.

3.1   *Hahajima*

Hahajima, also known as Hillsborough Island, is comprised mainly of andesitic lava and volcanic breccia together with a number of dikes. Andesitic lavas and tuff breccias are exposed in the northern area, and tuffaceous sandstone and sandy limestone are intercalated in the volcanics in the south (IWASAKI and AOSHIMA, 1970). Volcanic layers are interrupted by several gently dipping faults (UJIIE and MATSUMARU, 1977). Marine sedimentary layers distributed in the southern half of the island contain fossils of both planktonic and larger foraminifera. The fossils yield a late middle Eocene age (SAITO, 1962; UJIIE and MATSUMARU, 1977).

3.2 Chichijima

Chichijima is composed mainly of boninitic and andesitic pillow lavas, hyaloclastites, dikes and subordinate sedimentary rocks such as sandstone and limestone. Volcanic breccias are composed of boninite, hypersthene andesite, and dacite. The lower successions are comprised dominantly of pillow lavas that are cut by numerous dikes of both boninitic and andesitic composition. The middle to upper horizons are composed of an alternation of pillow lavas and glassy hyaloclastites. In the uppermost horizon, andesitic and dacitic lavas are locally present. Boninitic and andesitic dikes are especially common in the eastern part of the island (S.Maruyama, personal communication, 1981). The dips of volcanic layers are gentle, generally less than 20° toward the west, except for a few local areas where the rock units dip strongly to the east or west. Sedimentary rocks dominated by reef limestone of Oligocene to early Miocene age are present in the southwestern part of the island near Minamijima, where they unconformably overlie the older volcanics of Chichijima (e.g., HANZAWA, 1925). The age of the basal volcanics is likely to be early to middle Eocene, as suggested by the K-Ar ages and the fossil age derived from intercalated calcareous layers within the volcanics.

3.3 Mukojima

Mukojima is comprised of boninitic and andesitic pillows, hyaloclastites and a number of dikes in the same manner as Chichijima. No fossil-bearing sediments have been found.

3.4 Otootojima and Anijima

A description of the geology of Otootojima and Anijima is in preparation by MARUYAMA et al. (1982). The geology of these islands is similar to that of Chichijima. The rock types are predominantly andesitic and boninitic pillow lavas, dikes, and hyaloclastite. The dike swarm, observed on the eastern margin of Chichijima, cuts across the island trend and continues along the southern and eastern portion of Anijima.

4. Sampling and Measurement

During the summer of 1980, multidisciplinary studies were conducted in the Bonin Islands. Rock samples were collected for paleomagnetic, geochronologic, petrologic, and paleontologic studies. In addition, field studies were conducted by Maruyama and Matsumoto, for the purpose of constructing a geological map of the islands. Exposures in the Bonin Islands are excellent. Vertical sea cliffs several hundred feet high expose large volumes of pillow basalts, dikes, and hyaloclastites. Invariably the freshest exposures occurred at the shoreline, so most of the sampling was concentrated there (see Fig. 1). Access to the northern islands was by boat. Samples for paleomagnetic study were collected from seven islands and were divided between the participating groups from the University of Tokyo and University of Hawaii. The combined results of these two coordinated studies are presented here. Samples were collected by means of a gasoline-powered hand-held drill (a few ancillary samples were collected by hand

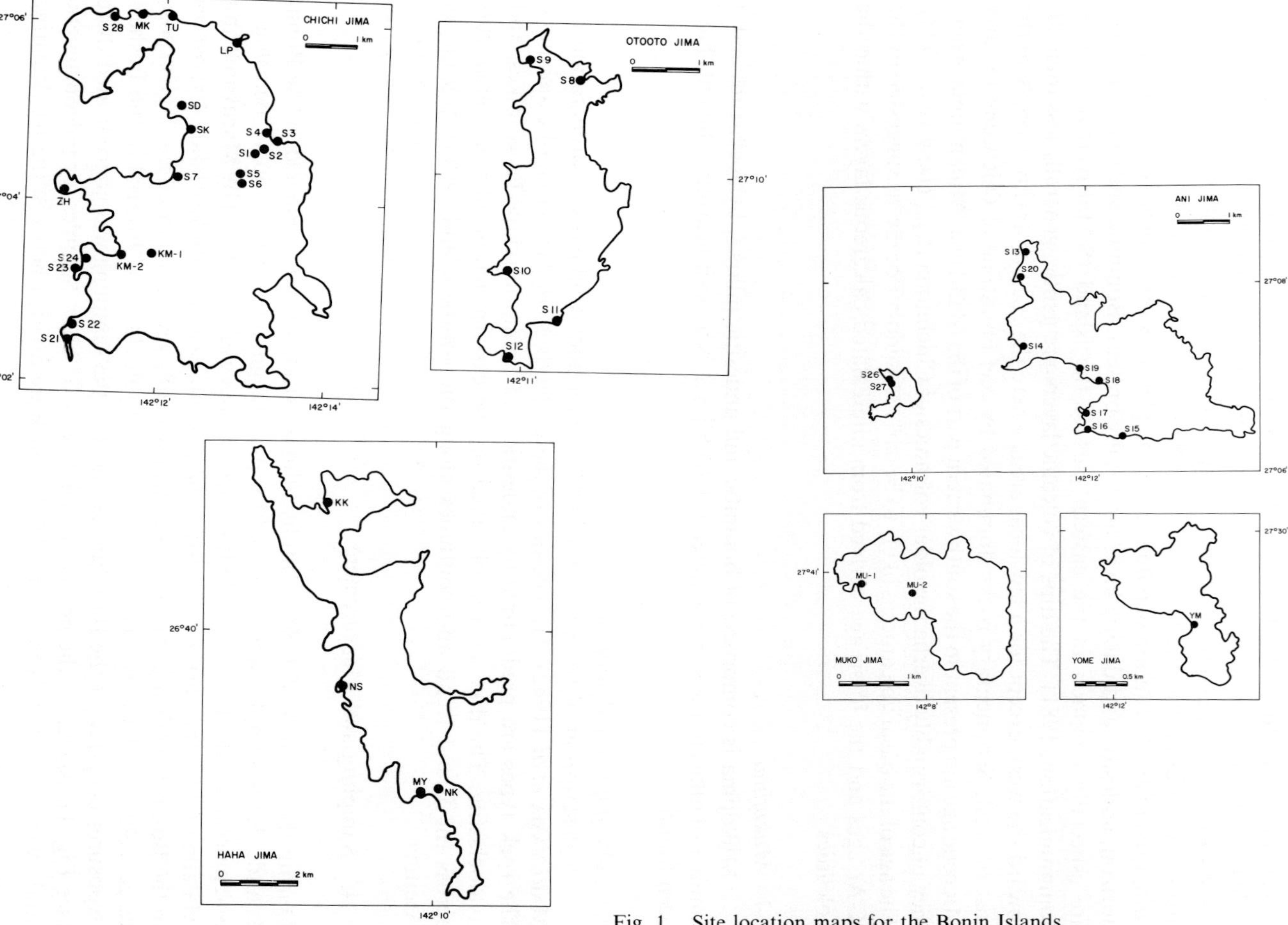

Fig. 1. Site location maps for the Bonin Islands.

from inland outcrops). The majority of the samples were oriented by means of a sun compass. A total of nine sites as well as portions of five other sites were oriented by magnetic compass (see Table 1). The igneous rock samples were measured on Schonstedt spinner magnetometers at both universities and were demagnetized by using alternating fields up to several hundred Oersted in multi-axis tumbler apparatus.

4.1 Paleomagnetic measurements

After the measurements of natural remanent magnetization (NRM) of all samples, test samples from each site were subjected to increasing alternating magnetic fields in order to determine the optimum cleaning field. The optimum demagnetization field was defined in the studies by the Hawaii group using the SYMONS and STUPAVSKY (1974) stability index. The Tokyo group demagnetized specimens at levels chosen on the basis of the cluster of magnetic directions and minimum change in direction on demagnetization. The results by either method appear to be fully comparable.

In addition to the alternating field demagnetization studies, thermomagnetic studies were conducted on specimens from Chichijima and Hahajima. The studies included heating and cooling in nitrogen gas in an ambient field of less than 100 nT, and heating in a vacuum ($\sim 10^{-5}$ Torr) and in a strong magnetic field (~ 4000 Oe). In addition, hysteritic behavior studies were conducted with a vibrating sample magnetometer at Goddard Space Center. These studies were conducted at ambient temperature and peak fields of 10 KG.

4.2 Magnetic behavior

In general, the demagnetization studies indicated that the lavas were very stably magnetized; they showed less than 5 degrees of direction change after demagnetization to 700 or 800 Oe (Fig. 2). Thermal demagnetization studies carried out on additional test specimens also revealed stable directional behavior up to 550°C. The results from representative samples are illustrated in Fig. 3. The changes of relative intensities after demagnetization indicate that these specimens are characterized by blocking temperatures in excess of 300°C (Fig. 4), indicative of stable primary magnetization acquired upon cooling. Only at site NS on Hahajima and 18 on Anijima were samples found that carried a large component of secondary magnetization.

Curie temperature analyses were conducted on specimens from the two primary rock types, the boninitic lavas and andesitic dikes. The results of these studies are shown in Fig. 5. The thermomagnetic curves were found to be irreversible. The saturation magnetization in each case increases after a heating-cooling cycle; however, the change is relatively small. The Curie temperature curves suggest that a low-titanium titanomagnetite produced by the subsolidus exsolution of ilmenite from an originally homogeneous titanomagnetite is present. High-temperature oxidation has occurred during initial cooling. Sample B is distinctly irreversible, as is typical when oxidation occurs during cooling.

In order to study the coercive force of the titaniferous magnetite, hysteresis experiments were conducted on boninitic lava and andesitic lavas. Examples of the hysteresis loops are shown in Fig. 6. The values of the coercive force, Hc, determined from these samples were 225 Oe and 150 Oe. Hysteresis parameters such as

Table 1. Fisherian Analysis and Paleomagnetic Direction.

Island	Site	Rock* unit	Demag. level	Declination	Inclination	Orientation**	Alpha$_{95}$	Number***
Chichijima	1		0100	139.8	13.0	S	4.4	14
Chichijima	2	P	0100	127.8	−43.9	S	2.7	8
Chichijima	2	P	0150	108.2	12.8	S	4.1	7
Chichijima	2	D	0200	137.6	7.4	S, M	4.1	9
Chichijima	3	D	0150	167.2	−53.8	M	3.2	7
Chichijima	3	P	0100	114.6	2.4	S	3.7	7
Chichijima	3	D	0150	252.7	16.0	S	5.4	4
Chichijima	4		0150	97.5	13.7	M	3.1	9
Otootojima	5		0200	125.3	30.3	S, M	7.5	14
Otootojima	6	T	0100	140.4	25.5	S	3.1	12
Otootojima	6	D	0100	132.5	15.3	S	1.8	4
Otootojima	7		0150	284.3	39.9	M	3.2	6
Otootojima	8		0125	113.1	−7.4	S	6.4	13
Otootojima	9	D	0125	149.5	−83.0	S	15.8	6
Otootojima	9	P	0125	95.7	11.0	S	8.5	11
Otootojima	11	T	0175	156.3	−36.2	S, M	22.8	18
Otootojima	11	B	0175	77.3	−73.1	S	20.65	5
Otootojima	12		0075	46.9	−20.3	M	17.0	20
Anijima	13		0075	275.3	29.5	S	3.5	15
Anijima	14	D1	0300	78.9	16.6	S	4.5	6
Anijima	14	D2	0300	80.6	0.4	S, M	3.6	7
Anijima	15		0250	116.4	6.9	S	21.2	11
Anijima	16		0150	344.0	15.5	S	15.5	29
Anijima	17	D1	0100	129.2	33.7	M	6.6	7
Anijima	17	D4	0100	97.0	18.8	S	4.2	3
Anijima	17	P1	0100	88.9	−6.8	M	4.3	9
Anijima	17	D5	0100	100.3	16.3	M	36.3	4
Anijima	17	P2	0100	83.6	28.6	M	5.2	8
Anijima	17	D6	0100	323.2	21.3	S	30.0	4

Anijima	17	D7	0100	314.7	31.5	S	11.7	6
Anijima	18	D	0100	156.0	54.1	S	65.7	11
Anijima	18	P	0100	33.0	−37.4	S	10.8	7
Anijima	19	P	0175	70.1	−15.1	S	11.8	36
Anijima	19	D	0175	82.5	9.4	S	15.5	18
Anijima	19	D1	0175	94.0	34.1	S	12.0	6
Anijima	19	D2	0175	101.3	−40.8	S	3.4	5
Anijima	19	D3	0175	100.6	−41.7	S	8.2	5
Anijima	19	D4	0175	83.8	20.3	M	14.3	4
Anijima	19	D5	0175	70.1	43.9	M	6.8	7
Anijima	19	D6	0175	67.0	12.1	M	5.3	9
Anijima	20		0200	244.5	57.0	S	3.0	19
Chichijima	22		0100	106.5	6.3	S	7.9	15
Chichijima	23	P	0100	149.5	60.3	M	2.1	13
Chichijima	23	D1	0200	110.0	−2.7	S	10.0	6
Chichijima	23	D2	0200	84.8	−31.9	S	3.5	4
Chichijima	24		0150	299.1	−7.5	S	6.5	14
Nishijima	26		0100	110.9	2.9	M	7.9	15
Nishijima	27		0150	154.3	4.4	M	5.5	20
Chichijima	28	0D	0200	203.2	−3.3	M	8.3	13
Chichijima	28	ID	0200	261.1	54.6	M	3.6	5
Chichijima	28	D3	0200	78.3	25.2	M	41.8	2
Chichijima	28	P1	0150	79.5	24.5	M	7.8	8
Chichijima	28	P2	0100	88.6	1.3	M, S	4.1	11
Chichijima	LP		0150	87.9	53.7	M, S	5.1	13
Chichijima	ZH		0100	125.8	47.9	M, S	6.3	5
Chichijima	KM-1		0100	88.2	−7.2	M, S	3.6	7
Chichijima	KM-2		0100	90.6	−15.2	M, S	9.2	7
Chichijima	MK		0250	75.4	31.0	M, S	9.2	6
Chichijima	TU		0250	90.4	27.4	M, S	5.1	7
Chichijima	SD		0150	261.8	−41.5	M, S	5.1	5
Chichijima	SK		0250	266.5	−32.5	M, S	4.8	9

Table 1 (continued).

Island	Site	Rock* unit	Demag. level	Declination	Inclination	Orientation**	Alpha$_{95}$	Number***
Hahajima	NK		0100	210.7	−10.2	M, S	5.1	12
Hahajima	KK		0150	234.9	22.1	M, S	13.8	6
Hahajima	NS		0250	190.7	8.4	M, S	8.9	13
Hahajima	MY		0100	209.7	−2.2	M, S	5.7	11
Mukojima	MU-1		0100	203.7	−15.0	M, S	9.9	8
Mukojima	MU-2		0100	209.2	14.1	M, S	2.6	5
Yonejima	YM		0100	233.0	−27.0	M, S	37.4	4

*Rock Unit Abbreviations: P = pillow lava, D = dike, T = tuffaceous sandstone B = basalt, OD = outer dike. ID = inner dike

**S = sun compass, M = magnetic compass

***Number of samples

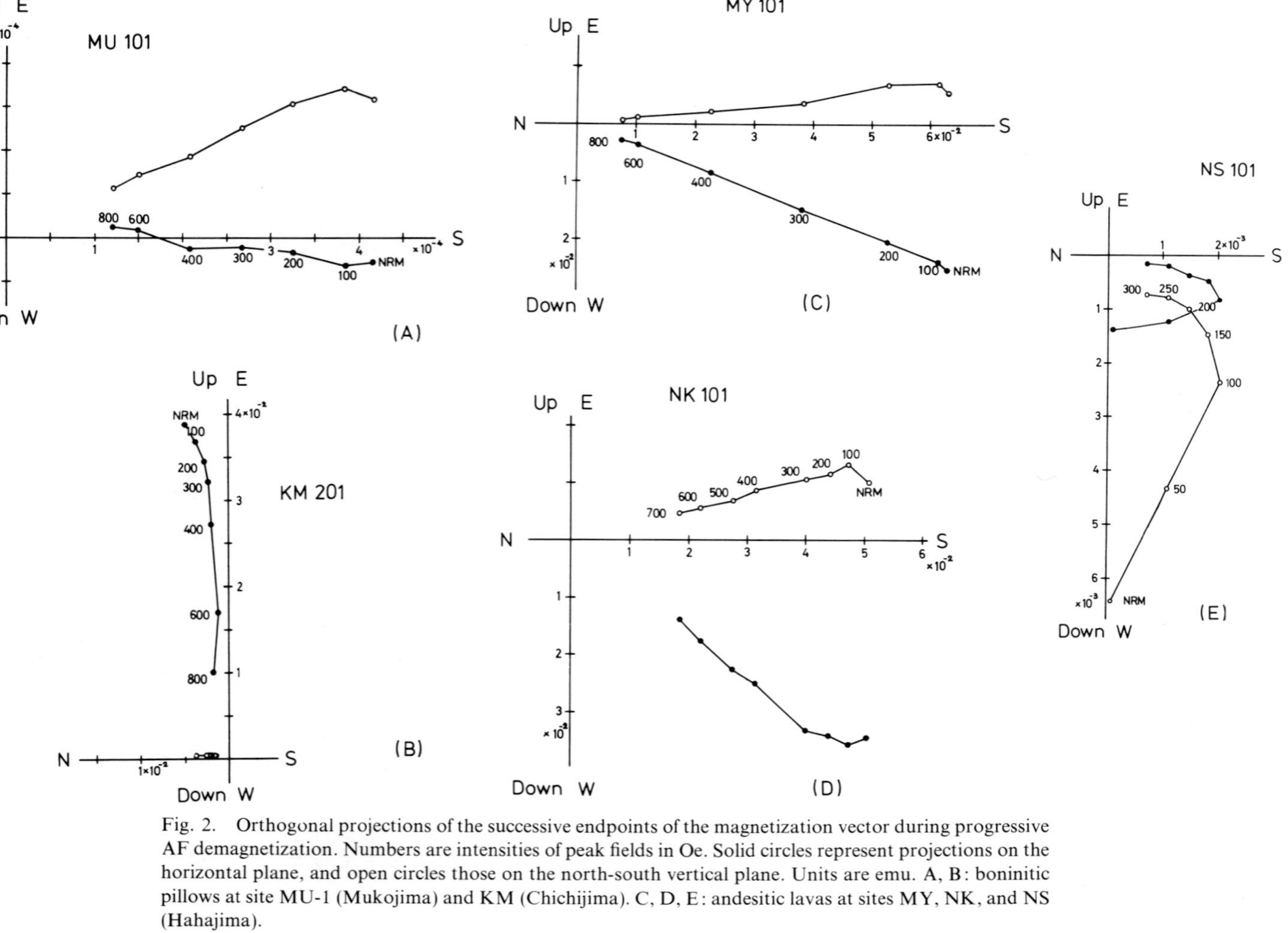

Fig. 2.   Orthogonal projections of the successive endpoints of the magnetization vector during progressive AF demagnetization. Numbers are intensities of peak fields in Oe. Solid circles represent projections on the horizontal plane, and open circles those on the north-south vertical plane. Units are emu. A, B: boninitic pillows at site MU-1 (Mukojima) and KM (Chichijima). C, D, E: andesitic lavas at sites MY, NK, and NS (Hahajima).

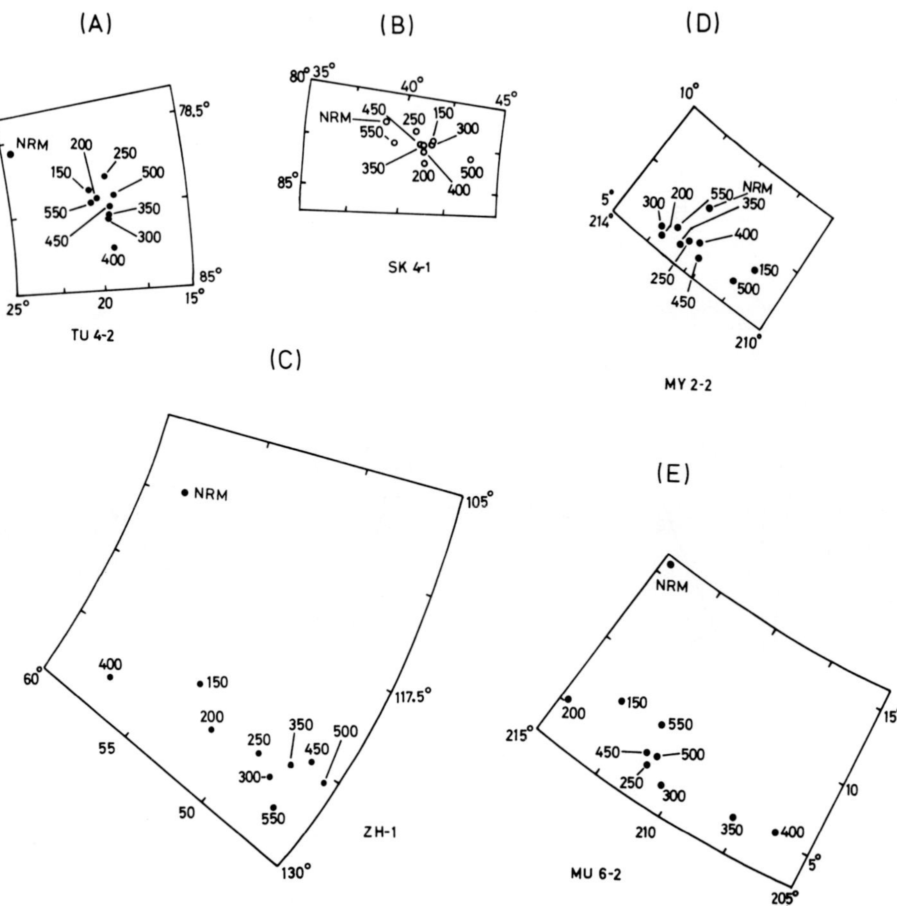

Fig. 3.   Equal-area projections of directions of magnetization through the progressive thermal demagnetization (A-E). Numbers indicate demagnetization temperatures. Solid (open) circles represent
projections on the lower (upper) hemisphere. A, B, E: boninitic pillows at site TU and SK on Chichijima, and
site MU-2 on Mukojima. C: andesitic dike at site ZH on Chichijima. D: andesitic lava at site MY on
Hahajima.

Hc, Jrs/Js, and Hcr/Hc (DUNLOP and HALE, 1977) have diagnostic value in determin-
ing domain structure. The low value of $Jrs/Js = 0.04$ for the boninite indicates
multidomain structure, whereas the moderate value of $Jrs/Js = 0.24$ for the andesite
suggests single domain structure (probably mixed with some multidomain material).
These conclusions probably apply to the carriers of NRM, as Hc correlates reasonably
well with the NRM median demagnetizing field, $H_{1/2} = 200$ Oe.

4.3   Paleomagnetic results

The mean remanence directions of rocks from the Bonin Islands are summarized

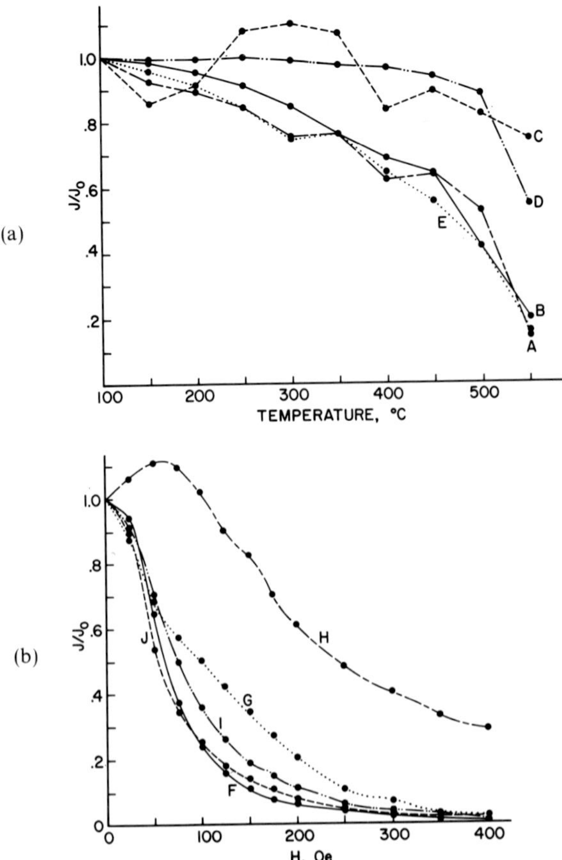

Fig. 4. (a) Thermal demagnetization curves for samples from the Bonin Islands. Samples A, B, and E are boninitic pillows from Site TU and SK on Chichijima and site MU-2 on Mukojima. Sample C is an andesitic dike at Site ZH on Chichijima. Sample D is an andesitic lava at site MYY on Hahajima. (b) Alternating field demagnetization curves for additional Bonin Island samples. Samples F. I, and J are andesitic pillow basalts from sites 12, 26, and 27 on Otootojima. Sample G is an andesite dike from Anijima, Site 16. Sample H is an boninitic pillow lava from Anijima, Site 19.

in Table 1 and Fig. 7. It is obvious that none of the sample directions observed from the Bonin Islands lie near the present geomagnetic field direction. The stable magnetization and the observation of antipodal field directions at several sites lead us to believe that the pole positions we have determined are reliable. Approximately one third of the sites required structural corrections, evidenced by the strike and dip of interflow sedimentary layers (KODAMA *et al.*, 1982). With structural correction, a better between-site mean was obtained. Although there is a scatter in site directions that may be attributed to the geomagnetic secular variation or unresolvable structural disturbances or both, the deflected declinations lead us to suspect that the islands have been affected

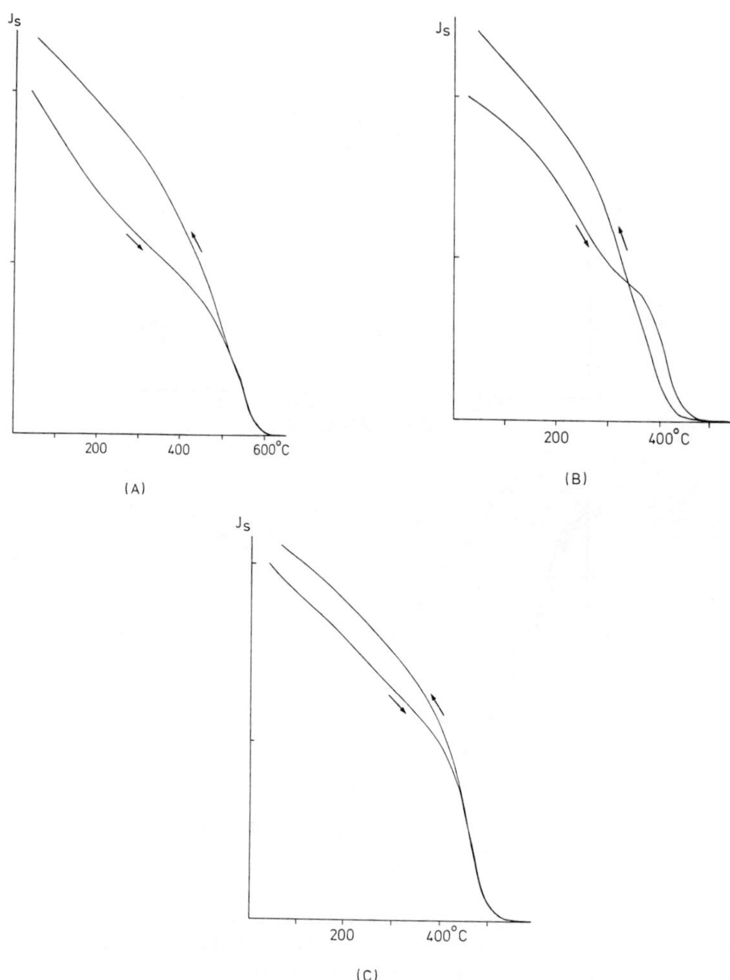

Fig. 5. Curie point analysis curves of boninite pillow (A) at site KM (Chichijima), andesite lava (B) at MY (Hahajima), and (c) andesite dike at ZH (Chichijima); after KODAMA (1981). Vertical axis is relative intensity of saturation magnetization.

by a significant amount of tectonic rotation about the vertical axis. We interpret the results as a clockwise tectonic rotation of the Bonin Islands. We infer a clockwise sense of rotation from the apparent clockwise rotation (obvious in the bathymetry) of the Bonin Ridge away from the Iwo Jima Ridge (see Fig. 8). The observed inclinations are significantly shallower than the present latitude of the Bonin Islands ($I = 46°$ at 27°N). The mean paleo-inclinations observed for each island ranged from 9.40° to $-5.45°$. The mean paleolatitude for all islands was 1.3°, with a standard deviation of 3.2°.

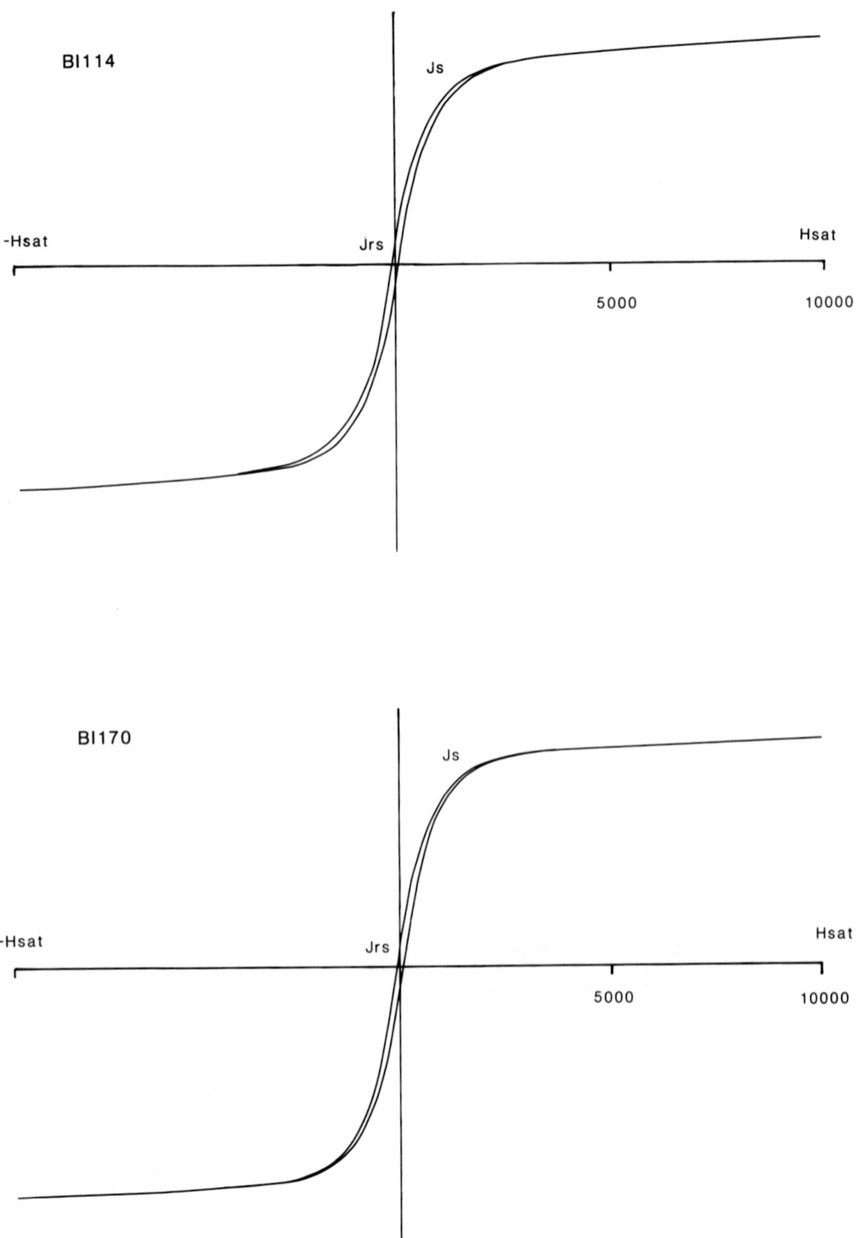

Fig. 6. The magnetic hysteresis loop for Bonin Island boninitic lavas and and andesitic dikes. Sample BI 114 is a boninite from the island of Otootojima and sample BI 170 is an andesite from the same island.

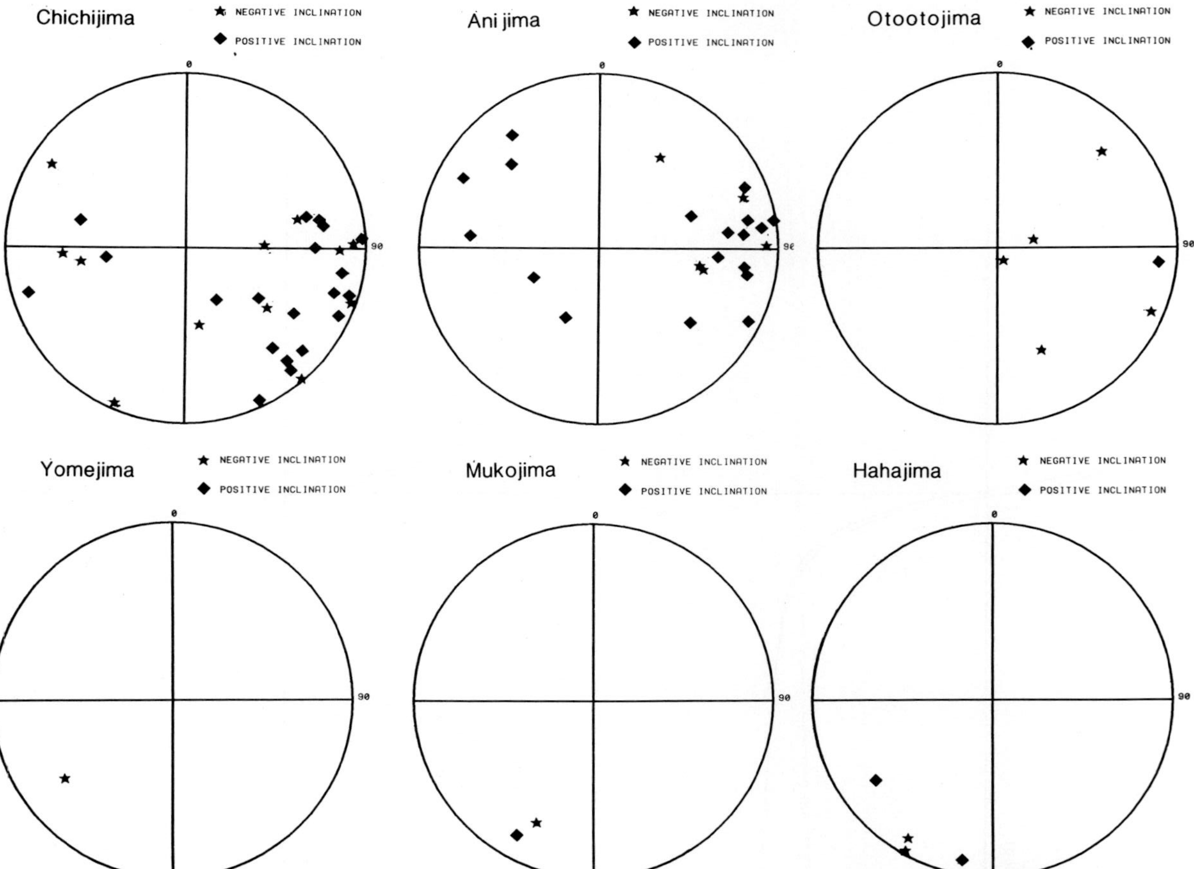

Fig. 7. Mean site directions. Equal-area projections of paleomagnetic mean directions determined for the Bonin Islands. The negative inclinations are marked by a star; positive inclinations by a diamond.

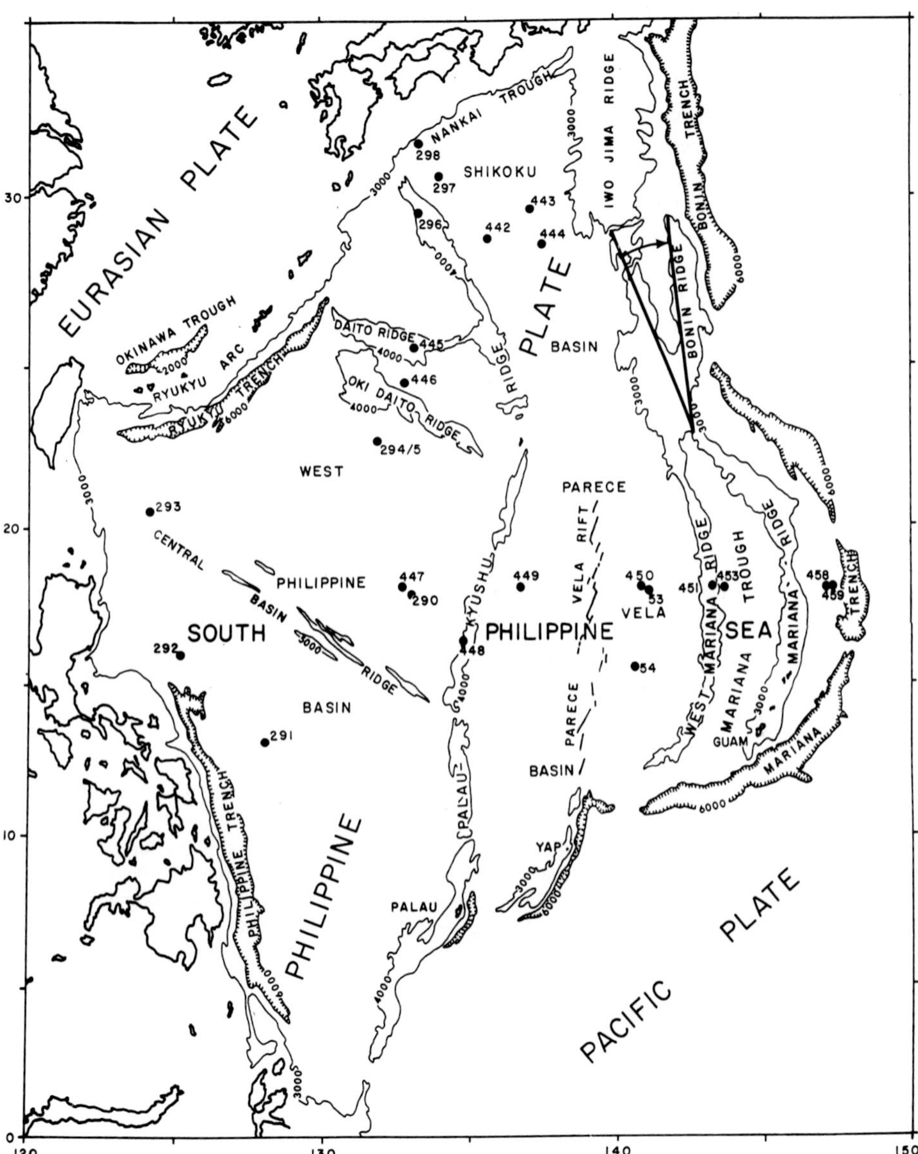

Fig. 8.   Regional setting of the Bonin and Mariana Islands, from SCOTT and KROENKE (1980). Note that the Bonin Ridge appears to be rotated clockwise away from the Iwo Jima Ridge. Land areas are outlined by heavy lines, lighter lines indicate submarine ridges, dashed areas represent trenches, and bold dashed lines show oceanic rift structures. Deep Sea Drilling Project sites are numbered.

5. Conclusion

Results of paleomagnetic studies of the Bonin Islands clearly document large rotations since the Early Tertiary; several tens to over 90 degrees of clockwise rotation has occurred about a vertical axis. In addition, at least $20°$ of northward drift has been documented. We believe that these rotations result from the interaction of the Pacific and West Philippine seafloor plates in the Early Tertiary and are related to the onset of subduction of the Pacific plate under the Philippine seafloor plate. Two tectonic models can explain the observed rotations. The first model would require that the Bonin Island arc act as ball bearings allowing differential movement between the Pacific and West Philippine seafloor plates, similar to the interaction discussed by FITCH (1972). This model will not explain the observed northward motion of the Bonin Arc. The second tectonic model for the Bonin Island arc is that the arc, which was formed roughly 40 m.y. ago, was situated near the equator roughly perpendicular to its present trend. The results of these studies, therefore, are somewhat consistent with the model of KARIG (1975) in that rotations of the Bonin Islands do occur, although the rotations we have observed cannot be accounted for solely by a model such as KARIG's (1975).

It is important to note that these paleomagnetic observations, interpreted as tectonic rotations are not unique to the Bonin Islands. The results of paleomagnetic studies of Guam (part of the southern Marianas) by KOBAYASHI (1972) and a joint Japan-U.S. paleomagnetic group (LARSON *et al.*, 1975) also indicate significant clockwise rotation. Subsequent to the Bonin Island field studies, Keating and others extensively sampled the islands of Guam, Tinian, Rota, and Saipan, and paleomagnetic studies of rocks from these islands are currently underway. Preliminary indications of large rotations were found within the N.R.M. results. Thus, it is likely that large-scale rotations of the Eocene and Miocene volcanics are present in both the Bonin and Mariana islands.

We acknowledge the kind support of the Ogasawara Marine Transportation Co, Ltd. for passage to the Bonin Islands. In addition we thank S. Maruyama for asistance in sample collection and the resolution of structural problems in the field. We are grateful to S. Uyeda for helpful discussions and comments on the manuscript. We also thank Peter Wasilewski for his assistance in the hysteresis studies conducted at the NASA Goddard Space Center. This work was supported by NSF grant EAR-79-20065. Hawaii Institute of Geophysics Contribution No. 1353.

REFERENCES

DUNLOP, D. J. and C. J. HALE, Long-term stability of the magnetic signal of the oceanic crust, *Can. J. Earth Sci.*, 716–744, 1977.

FITCH, T. J., Plate convergence, transcurrent faults and internal deformation adjacent to southeast Asia and the western Pacific, *J. Geophys. Res.*, **77**, 4432–4460, 1972.

HANZAWA, S., On the foraminifera-bearing rocks of Okinawajima and Ogasawarajima, *J. Geol. Soc. Jpn.*, **32**, 461–484, 1925 (in Japanese).

HUSSONG, D., S. UYEDA, and the scientific party of Leg 60, *Initial Reports of the Deep Sea Drilling Project*, **60**, 1982.

INGLE, J. C., Jr., Summary of late Paleogene-Neogene insular stratigraphy, paleobathymetry, and correlations, Philippine Sea and Sea of Japan region, in Ingle, J. C. Jr., Karig, D. E., *et al.*, *Initial Reports of the Deep Sea Drilling Project*, **31**, 837–855, Washington, D.C. U.S. Government Printing Office, 1975.

IWASAKI, Y. and M. AOSHIMA, Report on geology of the Bonin Islands, The nature of Ogasawara, report on scientific and natural monuments of the Ogasawara Islands, the Ministry of Education, 205–220, pl. 1, 1970.

KANEOKA, I., N. ISSHIKI, and S. ZASSHU, K-Ar ages of the Izu-Bonin Islands, *Geochem. J.*, **4**, 53–60, 1970.

KARIG, D. E., Basin genesis in the Philippine Sea, *Initial Reports of the Deep Sea Drilling Project*, **31**, 857–880, 1975.

KARIG, D. E. and G. F. MOORE, Tectonic complexities in the Bonin arc system, *Tectonophysics*, **27**, 97–118, 1975.

KIKUCHI, Y., On pyroxenic components in certain volcanic rocks from Bonin Island, *J. Coll. Sci. Imp. Univ., Japan*, **3**, 67–89, 1890.

KOBAYASHI, K., Reconnaissance paleomagnetic and rock-magnetic study of Guam, Mariana and related sites, in *The Izu Penisula*, edited by H. Hoshino and H. Aoki, pp. 385–390, Tokai University Press, 1972.

KOBAYASHI, K. and N. ISEZAKI, Magnetic anomalies in the Sea of Japan and the Shikoku basin: Possible tectonic implications, in *The Geophysics of the Pacific Ocean Basin and its Margin*, edited by G. H. Sutton, M. H. Manghnani, and R. Moberly, pp. 235–252, American Geophysical Union., Washington, D.C., 1976.

KODAMA, K., A paleomagnetic reconnaissance of the Bonin Islands, *Bull. Earthq. Res. Inst.*, **56**, 347–365, 1981.

KODAMA, K., B. KEATING, and C. E. HELSLEY, Paleomagnetic results from the Bonin Islands, *Tectonophysics*, in prep., 1982.

LARSON, E. E., R. L. REYNOLDS, M. OZIMA, Y. AOKI, H. KINOSHITA, S. ZASSHU, N. KAWAI, T. NAKAJIMA, K. HIROOKA, R. MERRILL, and S. LEVI, Paleomagnetism of Miocene volcanic rocks of Guam and the curvature of the southern Mariana island arc, *Geol. Soc. Am. Bull.*, **86**, 346–350, 1975.

MARUYAMA, S., C. E. HELSLEY, K. KODAMA, B. KEATING, and K. MATSUMOTO, Geology of Anijima and Otootojima in the Bonin Islands, *J. Volcanology*, in prep, 1982.

PETERSON, J., Beiträge zur petrographie von Sulphur Island, Peel Island, Hachijo und Mijakeshima, *Jahrb. Hamburg. Wiss. Anst.*, **8**, 1–59, 1891.

SAITO, T., Eocene planktonic Foraminifera from Hahajima (Hillsborough Island), *Trans. Proc. Paleont. Soc. Jpn., N.S.*, No. 65, 209–225, 1962.

SCOTT, R. and L. KROENKE, Evolution of Back arc spreading and arc volcanism in the Philippine Sea: Interpretation of Leg 59 DSDP results, in *The Tectonics and Geologic Evolution of Southeast Asian Seas and Islands*, edited by D. Hayes, pp. 283–291, 1980.

SHIRAKI, K. and N. KURODA, The boninite revisited, *Chigaku Zasshi*, **86**, 34–50, 1977 (in Japanese).

SYMONS, D. T. A. and M. STUPAVSKY, A rational paleomagnetic stability index, *J. Geophys. Res.*, **79**, 1718–1720, 1974.

TAKAYANAGI, Y. and T. KATAYAMA, Eocene planktonic Foraminifera, Chichijima, Bonin Islands, presented at the 124th meeting of the Paleontological Society of Japan, 1979.

TANAKA, M., T. HOSONO, T. KUBOKI, S. OWADA, and T. TACHIKAWA, Geomagnetic and gravity survey in the Bonin Islands (Chichijima), the Ogasawara Archipelago, *J. Geod. Soc. Jpn.*, **20**, No. 4, 193–208, 1974.

UJIIE, H. and K. MATSUMARU, Stratigraphic outline of Hahajima (Hillsborough Island), Bonin Islands, *Memo. National Science Museum*, No. 10, 5–18, pls. 1–4, 1977.

UYEDA, S. and Z. BEN-AVRAHAM, Origin and development of the Philippine Sea, *Nature*, **240**, 176–178, 1972.

YOSHIWARA, S., Geological age of the Ogasawara Group (Bonin Islands) as indicated by the occurrence of *Nummulites*, *Geol. Mag., N.S.*, **9**, 296–303, 1902.

Index